MW00805770

Cumberland Island National Seashore

Cumberland Island National Seashore

A History of Conservation Conflict

Lary M. Dilsaver

University of Virginia Press

Charlottesville & London

UNIVERSITY OF VIRGINIA PRESS
© 2004 by the Rector and Visitors of the University of Virginia
All rights reserved
Printed in the United States of America on acid-free paper

First published 2004

9   8   7   6   5   4   3   2   1

Library of Congress Cataloging-in-Publication Data
Dilsaver, Lary M.
    Cumberland Island National Seashore : a history of conser-
vation conflict / Lary M. Dilsaver.
        p.   cm.
    Includes bibliographical references (p.   ) and index.
    ISBN 0-8139-2268-2 (alk. paper)
    1. Cumberland Island National Seashore (Ga.)—History.
2. Cumberland Island National Seashore (Ga.)—Management. 3. Conservation of natural resources—Georgia—
Cumberland Island National Seashore. 4. Cumberland
Island National Seashore (Ga.)—Environmental conditions.
5. Environmental protection—Georgia—Cumberland
Island National Seashore.   I. Title.
F292.C94 D55 2004
975.8'746—dc22

                                            2003019677

This book is published in association with the Center for
American Places, Santa Fe, New Mexico, and Staunton,
Virginia (www.americanplaces.org).

CONTENTS

List of Illustrations   *vii*

Acknowledgments   *ix*

Introduction   *1*

1   A Richness of Resources: Cumberland Island to 1880   *8*

2   The Era of Rich Estates, 1881–1965   *36*

3   Creating Cumberland Island National Seashore   *76*

4   Land Acquisition and Retained Rights   *111*

5   Planning and Operating in the 1970s   *137*

6   Resource Management in the 1970s   *165*

7   Contested Paradise: The 1980s and Early 1990s   *195*

8   Hope for the New Century   *239*

Conclusion   *260*

APPENDIX A
National Park Service Officials   *267*

APPENDIX B
Cumberland Island National Seashore Operating Base Budgets   *269*

Notes   *271*

Index   *309*

ILLUSTRATIONS

Figures

Figure credits: Figs. 1.1, 1.3, 2.1-2, 2.8, 3.1-3, 5.3, 6.2, and 8.2 are from the U.S. Department of the Interior, National Park Service Archives, Harpers Ferry, W.Va. Figs. 1.1-2, 1.4, 2.3-6, 2.9-12, 5.1-2, 5.4, 6.1, 6.4, 7.2-3, 8.1, and 8.3 are from the U.S. Department of the Interior, National Park Service, Cumberland Island National Seashore Archives, St. Marys, Ga. Photos by author: Figs. 1.5, 6.3, 7.1, and 8.4.

1.1.   Maritime oak forest   *2*

1.1.   Aerial view of the island   *10*

1.2.   Cumberland Island vegetation   *15*

1.3.   Ruins of Nathanael Greene's mansion   *28*

1.4.   Tabby House at Dungeness   *29*

1.5.   Main Road running the length of the island   *32*

2.1.   Dungeness mansion in its prime   *39*

2.2.   Interior of the Recreation House   *40*

2.3.   Gardens and Recreation House at Dungeness   *41*

2.4.   Lucy Carnegie and her family at Dungeness   *43*

2.5.   The Cottage built for Thomas Carnegie Jr.   *45*

2.6.   Plum Orchard mansion built for George Carnegie   *46*

2.7.   Thomas Carnegie family tree   *47*

2.8.   Ruins of the Dungeness mansion   *58*

2.9.   Picnicking at the beach   *64*

2.10. Fun at the beach for children   *65*

2.11. Cumberland Island Hotel at the north end   *67*

2.12. Primus Mitchell at the Settlement   *70*

3.1.   National Park Service vehicles near High Point in 1957   *82*

3.2.   National Park Service survey team at Lake Whitney   *84*

3.3.   National Park Service survey team at the Stafford Chimneys   *85*

5.1.   Ruins of several generations of automobiles   *141*

5.2.   Dungeness Dock in 1972   *142*

5.3.   National Park Service trams on the south end of the island   *144*

5.4.   Cumberland ferry docking at the island   *144*

6.1.   Dune encroachment on the maritime oak forest   *175*

6.2.   Exterior damage on the Plum Orchard mansion   *182*

6.3.   Ruins of the Recreation House in 2000   *186*

6.4.   First African Baptist Church   *192*

7.1.   Riprap near the Sea Camp Visitor Center   *225*

7.2.   Feral horse in the marsh   *230*

7.3.   Duck House   *233*

8.1.   Trapping of hogs   *241*

8.2.   Plum Orchard mansion   *245*

8.3.   Native American canoe   *246*

8.4.   Tracks from beach driving   *250*

Maps

1.1.   Cumberland Island location map   *4*

1.1.   Land and vegetation on Cumberland Island   *14*

1.2.   Native American and colonial areas   *19*

1.3.   Division of Cumberland by the Greene and Lynch families   *27*

1.4.   Dungeness plantation in 1878   *34*

2.1.   Dungeness estate in 1916   *42*

2.2.   Glidden Company mining proposal   *60*

2.3.   Subdivision plans for the north end   *69*

2.4.   High Point–Half Moon Bluff Historic District   *71*

2.5.   Carnegie land division   *74*

4.1.   National Park Service land acquisition and proposed subdivisions   *114*

4.2.   Retained estates and private land   *133*

5.1.   Cumberland Island in 1972   *138*

5.2.   National Park Service 1971 master plan   *151*

5.3.   Options for a mainland embarkation point   *156*

5.4.   Proposal of 1977 for three wilderness areas   *160*

6.1.   National Register historical and archaeological areas   *166*

6.2.   Shoreline changes since 1857   *168*

7.1.   General management plan proposal of 1981   *197*

7.2.   Revised 1981 general management plan   *211*

7.3.   National Park Service development plan for Point Peter   *215*

7.4.   National Park Service plan for the town of St. Marys   *217*

7.5.   Table Point and South Cut fires   *221*

7.6.   Proposed changes to the Atlantic Intracoastal Waterway Channel   *223*

ACKNOWLEDGMENTS

I began this project seven years ago little realizing how complicated it would become. During that time many people and institutions have helped. First, I wish to thank the three superintendents at Cumberland Island National Seashore who served through those years. Rolland Swain initiated the project and secured early funding. Denis Davis helped through the primary research period with constant information and assistance. Finally, Art Frederick found the funds to enable the completion of the volume. Other personnel at the national seashore also helped me to locate and interpret data and provided the logistical support for my research, including Jennifer Bjork, Dave Casey, Janis Davis, Carol Ferguson, Zack Kirkland, Julie Meeks, John Mitchell, Brian Peters, Joyce Seward, Newton Sikes, and Don Starkey. Andy Ferguson played a particularly important role overseeing my final research and supporting the project through its conclusion.

Other National Park Service employees also provided valuable help, including John Harrington, Tom Piehl, and Richard Sussman at the Southeast Region Office in Atlanta; William Gregg, Janet McDonnell, Barry Mackintosh, and Dwight Pitcaithley in Washington, D.C.; Tom Durant and Dave Nathanson at the Harpers Ferry Center; and the staff of the Technical Information Center in Denver.

Many people outside the Park Service also contributed, led by Thornton Morris, who opened his files and his home to me. He provided valuable insight to the island residents' experience and knowledge. Former congressman Williamson "Bill" Stuckey, former legislative assistant Robert Hurt, and Don Barger of the National Parks and Conservation Association also provided key information. I also wish to thank Ann and James Stacy for their hospitality and encouragement and Patricia Janssen for her editorial assistance.

Support also came from the University of South Alabama in the forms of both funding and release time for research. I particularly wish to thank Deans Lawrence Allen, Margaret Miller, John Friedl, and David Johnson and Professors Glenn Sebastian and Roy Ryder for their assistance. Professor David Allison drafted the original maps, which added immeasurably to the book. I also thank those involved in the publication process, including George Thompson and Randy Jones of the Center for American Places, Penny Kaiserlian of the University of Virginia Press, and the peer reviewers, Professors Craig Colten and William Wyckoff.

Finally, my greatest thanks go to my wife, Marcia Robin Dilsaver, who unceasingly gave encouragement, assistance, and understanding throughout the project. Without her help copying files, discussing the data, making multiple trips to the seashore, typing the final manuscript, and listening to my complaints, I would not have written this volume.

Cumberland Island National Seashore

Introduction

Cumberland Island is an eighteen-mile-long stretch of sand and marsh lying some three miles off the southernmost Georgia coast. It is part of a system of barrier islands stretching from central Florida to Virginia. Individually, the resources of the island do not seem spectacular. Nowhere does it achieve an altitude greater than sixty feet. Its vegetative communities, while delicate, are not especially rare. Its mansions are imposing, even in ruins, but larger and more significant homes are found in many other places. Horses roam free on the island as they also do at Assateague and Cape Lookout National Seashores and many places in the American West. The beach is striking but tainted by a huge jetty and views of an ugly paper mill from the south end. Archaeological sites abound, yet most consist primarily of ancient shells of still common shellfish. Even those have been adulterated by bulldozing for road use and by natural erosion. It seems an odd location for one of the most contentious and bitter resource conflicts in the national park system.

When one looks at the entire assemblage, however, a different picture emerges. The wide, empty beach imparts solitude and wonder to anyone walking it. At night one may witness the millennia-old process of a loggerhead turtle laying her eggs above high-tide line. The maritime oak forest and even the former pine plantations, made impenetrable by a palmetto understory, blanket the island interior and enfold trails and creeks. Freshwater lakes hold rare discoveries including some fairly large alligators. Acres of marsh stretch along the sound side of the island. Raccoons eat crabs along the edges while myriad sea birds come and go. A long, bumpy sand and shell road runs the length of the island, an alternative to the beach for those seeking an easy path. It connects several historic complexes made as fascinating by their years of seclusion and limited use as by their architecture.

More than anything else, the diversity of Cumberland Island, the almost overwhelming sense of natural beauty mingled with mystery, and its enveloping calm make rabid loyalists of nearly all who live on, visit, or manage it (fig. 1.1). During 1998 a meeting was held for adversaries representing

Fig. 1.1. One of the most important features of Cumberland Island is its extensive maritime oak forest. (National Park Service photograph by Richard Frear)

historic preservation, wilderness advocacy, and recreation development organizations. Near the meeting's conclusion, the combatants gathered to sit in an old African-American church at the north end of the island. The participants had come several miles by either the main road under the towering forest or by boat through twisting lanes in the marsh, according to their conservation beliefs. All sat in the few pews of the tiny church as a silence descended. Finally, someone acknowledged that he understood the passionate beliefs of the other side even though he could not agree. The opposition offered a similar view. The talk turned to the values that they shared. Every person in that abandoned church loved Cumberland Island and sought the absolute best for it. Their understandings of best would continue to clash, but for a moment they saw themselves as a united group, united by passion for a special place. That passion for this paradise makes Cumberland Island National Seashore an unfortunate arena for conflict. It is also what makes it one of the most rewarding of the nearly 390 national park units.

This book is a historical geography, which is to say it is the biography of a place.[1] Cumberland Island is one segment of the complex, interrelated earth, subject to a unique combination of environmental forces and processes. For at least 6,000 years the island has absorbed the modifying effects of human habitation as well. Until 1972 these societies manipulated the land for utilitarian purposes. Beginning in 1972, however, the pattern changed. The arrival of an organization with a mandate to preserve natural processes signaled the most significant change in human activity since the arrival of Paleo-Indians.

This book is also a history. Specifically, it is a history of the National Park Service and its practice of conservation. The federal bureau came to Cumberland armed with a complex set of policies and a philosophy based on voluminous legislation, procedural tradition, and an agency culture formed from evolving management. The 1916 act that established the Park Service charged the agency to preserve resources, both natural and historical, but also to encourage and support public visitation. If either of these mandates is carried to its ultimate extreme, it becomes contradictory. Pure preservation of natural ecological processes precludes nearly all nonresearch visitors. Development to meet all public recreation demands unquestionably would destroy the ecological resources. Throughout its history the National Park Service has struggled to achieve a balance between these extremes.[2]

On Cumberland Island the issues are especially complicated. The island's rich ecology and long, colorful history have left a spatial pattern of intermingled resources that resembles two decks of cards shuffled together (map 1.1). Management for historic preservation inevitably hampers the purity of natural resource protection. Alternatively, management only for the natural environment and unchecked natural processes demands that man-made artifacts be ignored or even eliminated. Establishment of wilderness status on the north end seriously complicates any proactive procedures aimed at protecting either the human or the natural resources. Added to these conflicting management directives is the continuing presence of private landowners as well as long-term leased estates on lands sold to the Park Service. Having rights to modify historic structures on their estates, drive over much of the island, and otherwise use its many resources, these residents cannot help but impact all of Cumberland's many assets.

Cumberland Island, then, presents a remarkable collision of historic preservation, environmental protection, and numerous legal restrictions and caveats. All other units in the national park system face these issues to some degree. However, this idyllic, semitropical island presents one of the most complex and controversial collisions of value, law, and emotion in the

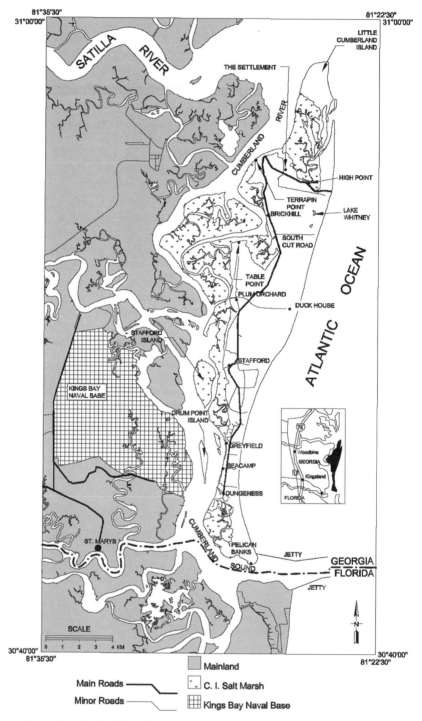

Map 1.1. Cumberland Island location map

system—in fact, among all the preserved lands of the United States. Thus, the histories of this fascinating place and of these different philosophies come together to shape, but also to embroil in conflict, one magical place.

Three related questions address the story of Cumberland Island. The first concerns the level of human occupation and the repeated escapes from massive development that the island has experienced. Native Americans on the island seldom numbered more than a few hundred. Their use of fire and their disposal of garbage left only faint evidence. When the Spanish arrived, their mission efforts threatened Cumberland with a high population of acolytes imported from the mainland. However, this project collapsed because of English harassment. The latter constructed two forts on the island but never fully developed them. In the antebellum plantation period, Cumberland Island reached its population apogee, yet nearly all were slaves whose movements were proscribed. The Carnegies added a chain of estates but rejected suggestions to transform the island by strip-mining, clear-cut logging, and residential subdivision. Finally, upon acquisition of the island, the National Park Service developed a plan for massive recreation development. The public soundly rejected it. That same environmentalist-led public pushed the Park Service to seek and receive wilderness designation for the northern half of the island.

Today, Cumberland Island has less human presence and pressure on it than at any time since the American Revolution. The designation "national seashore," in the Park Service lexicon, means a recreation area. Yet only 300 tourists per day are allowed to visit more than 16,000 acres of land. The island has no motels, no restaurants, no stores, and no true visitor center. Is this or is this not a recreation unit?

A second question addresses the legacy of human activities, the infrastructure of more than 6,000 years of use. Mansions, gardens, simple shacks, airfields, roads, trails, burial mounds, foundations, walls, and cemeteries dot the island. Most have achieved at least one criterion for addition to the National Register of Historic Places: they are more than 50 years old. From 1972, the National Park Service has faced decisions on how many of these features to preserve. Many factors influence the choices made for each of the more than 200 structures. Which era of human activity is most important historically and should be emphasized in visitor interpretation? The answer will determine which buildings and relics to save. If all of them are critical, how can the agency fund their maintenance? Should any be removed from an area zoned for reversion to a natural ecosystem?

Perhaps the most troubling concern is how far should the government go to preserve a building that has suffered decades of neglect and damage? Cumberland Island is a platform for widely dispersed and radically different types of cultural resources developed over many centuries. How much of this human legacy should be maintained and at what cost? Is this national seashore a historic unit or not?

The third question addresses the natural ecosystem that has existed at Cumberland Island for thousands of years. Vegetation and animals in complex communities carried out their life cycles together. Geological and climatic forces shaped the island's coasts and landforms. As soon as humans arrived, changes began. Native Americans increased the frequency of fire and created artificial habitats with their shell and bone deposits. Europeans caused much more change, bringing horses, cattle, pigs, old-world crops, and devastating diseases. During the plantation era some species disappeared from the island while a host of exotic plants and animals were deliberately introduced. What natural vegetation remained was spatially manipulated. Even during the relatively benign Carnegie years, the acreage of salable pine increased dramatically at the expense of oak woodland and other biotic communities.

When the National Park Service arrived, it sought to return the island to an approximation of the biotic system prevalent before Europeans arrived. This meant unfettered natural processes as well as an assemblage of plants and animals quite different from the one that met it in 1972. However, many elements of the human presence on the island directly or indirectly stunted this effort. Not only did historic structures disrupt the natural ecological processes, but their maintenance demanded transport across the island. Retention of estates further diluted ecosystem preservation. And then there are the matters of the financial and political prices of returning to nature. How much will it cost to permanently remove feral pigs? How about tamarack? Will the public ever accept the removal of horses? What efforts, if any, should be taken to halt erosion? Can natural processes ever approximate their prehuman form while historic buildings, residents, and visitors are on the island? What visitor activities can be allowed and by how many before unacceptable environmental change occurs? Is this island a natural resource unit of the park system, or is it not?

The following eight chapters reveal the actions taken and decisions made by several centuries of people on Cumberland Island. But they focus particularly on the stresses and strains of trying to preserve the island for a tripartite of worthy goals. The first two chapters deal with the ways the island reached its transformed stage by 1972. The first describes the earliest inhabitants up to the end of the plantation period while the second chronicles the years when the island served as a vacation retreat for the wealthy Carnegie and Candler families. Chapters 3 and 4 explain the creation of Cumberland Island National Seashore and land acquisition by the Park Service. Chapters 5, 6, and 7 relate the myriad issues and conflicts the agency encountered in its management from 1972 through the mid-1990s. Finally, the last chapter brings the reader through a tumultuous few years when the Park Service aggressively tried to solve many of the most pernicious problems. The conclusion reflects on the state of the island today and on the future of its evolving government management.

A Richness
of Resources:
Cumberland
Island to 1880

Congress and President Richard Nixon established Cumberland Island National Seashore on October 23, 1972, after an unexpectedly controversial campaign.[1] Most of the landowners favored the seashore and willingly sold their property at reduced prices in order to save the land. However, they exacted a price in the form of retained estates and driving rights for up to three generations. Others elected to keep their land private. Once the seashore was established, environmental organizations sought wilderness status to transform the island into a nature preserve while historic preservation groups insisted that the entire island merited nomination to the National Register of Historic Places. Another group, dominated by local residents on the mainland, demanded that the island be developed as a national recreation resort. One of the most compelling characteristics of this largest of Georgia's "golden isles" is that it contains the resources to satisfy all these competing visions.

This chapter explains the genesis and formation of those resources up to 1880. The first section describes the natural resources and the environmental processes responsible for them. Thereafter, a chronological survey of the human history on the island illustrates the modifications effected by Native Americans, Spanish missionaries, English adventurers, and antebellum plantation owners and slaves. Together with the Carnegies and Candlers, whose era is the subject of chapter 2, these groups of people altered the island's landforms and ecosystems while adding their relics to create a landscape markedly different from the prehuman one. The National Park Service must protect and preserve that accumulation of human designs, as well as the dynamic natural system upon which they repose, "unimpaired for future generations."[2]

Cumberland Island's Genesis and Formation

Cumberland Island is part of a system of barrier islands stretching along the Atlantic coast from southern Virginia to central Florida (fig. 1.1). It lies immediately north of the Florida border and is the largest of the Sea Islands of Georgia. Including Little Cumberland Island, which is separated from the larger isle by salt marsh and a narrow creek, it is approximately 18.5 miles long and from half a mile to more than 3 miles wide. The upland terrain totals roughly 16,400 acres (including Little Cumberland). In addition, some 9,400 acres of salt marsh stretch from much of the western coastline out into Cumberland Sound. The highest elevations occur in the dunes east of the Plum Orchard mansion, reaching over fifty-five feet at several points. More than 1,600 acres were classified as freshwater lakes in the early 1970s, but the depletion of groundwater flowing from the mainland has diminished their size. The remaining freshwater habitats lie behind the dunes from Lake Whitney southward along the center of Cumberland Island.[3]

The geologic story of the barrier islands of Georgia begins with the Appalachians. The ancient range reaches its highest elevations along the most southeastern ridge known as the Blue Ridge. To its east lies a hilly plateau of similar resistant rock called the piedmont. These geologic features supply eroded material that is carried by scores of rivers toward the southeast and the ocean. At the edge of the piedmont, the slopes briefly steepen, signaling the beginning of the Atlantic coastal plain, which consists of the eroded materials of the past 25 million years. There the rivers quicken into a series of small waterfalls and rapids. This piedmont–coastal plain boundary is known as the fall line.[4]

The coastal plain is a shallow, sloping depositional surface that continues under the sea to the edge of the continental shelf eighty miles east of the present Georgia shoreline and 160 feet below the present sea level. Over the millennia sea level has periodically changed, causing the shoreline to reach as high as the fall line during the Miocene (from 25 million to 7 million years ago) and as low as the edge of the continental shelf 20,000 years ago. During all this period the rivers continued to carry eroded material to the sea and deposit it on ancient beaches.

When rivers reach the ocean, they drop their sediment load within a few miles of the shore. The accumulated silt and sand then is washed along the coast by wave action and coastal currents. Along the Georgia coast the sea generally moves material southward. Geologists have offered several

Fig. 1.1. An aerial view of part of the eighteen-mile-long ocean beach with maritime oak forest and the Dungeness estate in the background

scenarios for the formation of barrier islands off the Georgia coast. Some propose that sediment deposited at each river mouth is washed against the adjacent headlands, forming spits that reach across the bays. A slight rise in sea level cuts off the spits, forming barrier islands. Subsequently, the sound between the islands and the mainland continues to collect sediment, and salt marsh vegetation colonizes the margins of the land. If sea level remains relatively constant, the marshy sound may eventually fill. Whether or not this occurs, a drop in sea level will expose the area as dry land with discernible eastern ridges composed of the former barrier islands.[5]

Another scenario holds that a sandbar in the shallow, near-shore sea serves as a collection point for sediment that builds to near the water surface. A drop in sea level exposes the bar as a barrier island. Thereafter the sound is colonized by salt marshes, and the entire landward margin is exposed once sea level drops far enough. In either of these scenarios, the barrier islands themselves manifest irregularities in their surfaces that become the starting points for sand dunes. In this way the elevations of the barrier islands increase.[6]

The Georgia coast experienced many sea-level changes over the period from 25 million to 18,000 years ago. The latter date corresponds with the maximum point of the most recent glacial advance and the lowest sea level. Periodic stabilization has created seven identifiable lines of former barrier islands on the coastal plain. These descend from the fall line in a series of terraces or steps composed of former marshlands and islands. In the last 2 million years, two of these barrier island formations have created the present Georgia Sea Islands. During the Pleistocene (1.8 million to 10,000 years ago), a barrier island chain known as the Silver Bluff formation appeared. The most recent period, the Holocene (10,000 years ago to present), has witnessed an accelerated recession of the world's glaciers and concomitant sea-level rise that developed another barrier formation. The Holocene barriers then migrated shoreward and became welded onto the Silver Bluff islands. Cumberland Island is composed primarily of Silver Bluff material, but the eastern edge, especially in the north (including virtually all of Little Cumberland Island) and the extreme south, are Holocene additions.[7]

One side effect of this constant deposition of eroded material at a fluctuating coastline is the layering effect of the sediments. During the millions of years, the rivers carried different types of material to the sea, depending on their source. These were deposited along the sea floor. Subsequent periods saw ancestral rivers deposit different types of material on top of the previous layers. Under Cumberland Island a number of distinct strata reflect different geologic periods and different source materials. All these layers, like the surface of the coastal plain, slope downward toward the edge of the continental shelf. Two of these strata primarily contain porous material that allows groundwater to flow easily through them. Others are composed of more impermeable material like clay that does not contain groundwater and blocks penetration by water upward or downward from the more porous layers. The principal groundwater layer on Cumberland Island and around the coastal region is composed of Miocene era limestone. The earliest deep well to tap this eastward-flowing source on Cumberland Island reached the water table more than 500 feet below the surface in 1887. The water, freed from the confined permeable layers, gushed out of the well at 800,000 gallons per day. A second aquifer flows through Pleistocene sands some 90 feet below the surface. The National Park Service accesses this water source for the campgrounds. Recently, fears have arisen that extensive water removal on the mainland will deplete the eastward-flowing groundwater before it can reach the wells on Cumberland Island.[8]

The topography of Cumberland Island is affected by the dual barrier island systems of the past 20,000 years. More than 90 percent of the island is composed of fine-grained sands. As one approaches the ocean beach from the sea, the slope gently rises at a rate of four feet per mile. Once ashore, a traveler crosses the wide beach and encounters the foredune, a low sand ridge that forms the first line of defense against the sea. Beyond the foredune lies an interdune meadow followed by the much higher rear dune. The highest elevations on the island can be found on the latter. Typically these dunes are anchored by vegetation and remain relatively stable except when a severe storm strikes. However, on Cumberland Island destruction of dune vegetation has allowed the rear dune to move westward and encroach upon the next zone, an area of flat ridges and depressions that contains freshwater lakes and marshes. The foredunes and some of the rear dunes consist of Holocene sediment.[9]

Beyond the zone of depressions, one encounters ridges and intervening swales of the Pleistocene Silver Bluff era. The highest of the old dune ridges are found in the center of the island. A portion of the ancient dune formation reaches the sea at the north end of the bigger island, accounting for the bluffs that face Little Cumberland. Continuing to move westward, the island slopes downward, occasionally interrupted by minor ridges until it meets the tidal mudflats and marshes of Cumberland Sound. Both the Pleistocene and Holocene dune ridges are lowest in height at the narrow southern end of the island.[10]

The two most important mainland rivers for Cumberland Island are the Satilla on the north, which separates the Cumberland group from Jekyll Island, and the St. Marys on the south. The St. Marys inlet separates Cumberland from Florida's Amelia Island. The Satilla historically brought sediment to the beaches of Cumberland, but the amount has been vastly reduced by dam construction and water diversion upriver. On Cumberland Island itself, several creeks cross from the rear dunes to the sound. The three most significant geographically are Christmas Creek, which separates Little Cumberland from the larger island; Old House Creek, which empties into the sound near Stafford Island; and Beach Creek, which flows just south of Dungeness and through the marshes on the southwestern part of the island.[11]

Cumberland's Ecology

The biota of Cumberland Island is rich and diverse despite the inherent stresses of island life. At the height of the last ice age, the Pleistocene Cumberland Island was a low ridge many miles inland from the coast. A cooler and much drier climate supported species of plants that today inhabit areas in the Appalachians or several hundred miles to the north. During the centuries of warming, but before the island was cut off from the mainland, numerous terrestrial species of plants and animals colonized the future island. By 5,500 years ago, a rising sea level returned the Silver Bluff ridge to island status. Subsequently, additional plant species with adequate seed dispersal techniques, as well as animals that could fly or swim, also reached the island. Finally, throughout the period of the island's emergence, Native Americans lived in the region and visited Georgia's primary golden isle. They likely brought valuable plants and animals with them to increase the range of these species in order to ease their subsistence hunting and gathering. By 3,000 years ago these favorites included agricultural crops.[12]

At the time of European contact, Cumberland Island boasted an assortment of biotic communities adapted to the climate and various microhabitats of the island. Southeastern Georgia, including Cumberland Island, has a classic humid subtropical climate with hot, muggy summers, marked by frequent convection thunderstorms, and mild winters with periodic frontal storms. It is a climate conducive to the growth of species from both the semitropical and temperate latitudes. Five major groups of vegetation communities that occur along the island's geomorphic profile were identified by Hillestad et al. (map 1.1). On the eastern side the dunes contain three grass and scrub communities dominated respectively by sea oats and other grasses; shrubs including saw palmetto, Spanish bayonet, and bayberry; and a gnarled buckthorn–live oak–pine assemblage. The interdune flats contain grass-sedge, shrub, and pine-mixed hardwood communities, depending on the degree of protection from salt spray and the time since the last disturbance. The third group of communities clusters in and around the freshwater lakes in the zone of depressions. Among these communities are various aquatic plants, emergent sedges and grasses, and a riparian forest dominated by water-loving species such as willow, sweet bay, and bayberry.[13]

The category of vegetation communities most prized by early explorers and modern campers includes several upland forest assemblages. Hillestad et al. described five different communities that reflect soil type, drainage, successional stage, fire history, and various human activities. The three predominant ones are oak-palmetto, oak-pine, and oak–mixed hardwood (fig. 1.2). The human manipulation of these forest ecosystems has caused the most significant changes in the visual landscape throughout the last 5,000

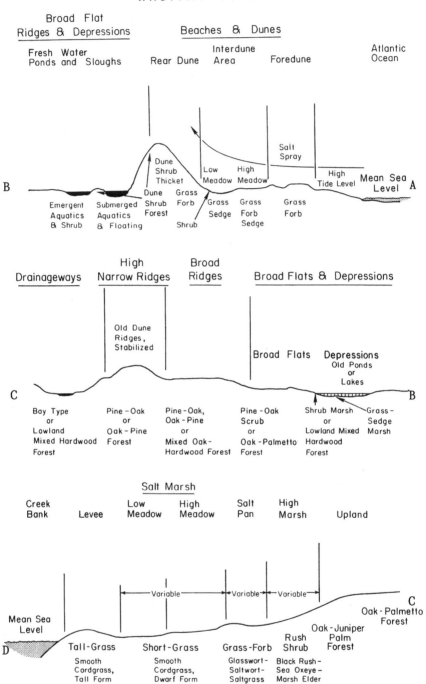

Map 1.1. Generalized profile of land and vegetation on Cumberland Island. (Original map in National Park Service, 1975, *Draft Environmental Statement, General Management Plan, Cumberland Island National Seashore*, CINS Library)

Fig. 1.2. A dense understory of palmetto makes large portions of the forest virtually impenetrable.

years. A final group of communities occurs on the western edge of Cumberland Island. These include several types of salt marsh as well as shrub- or tree-dominated transition zones between the marshes and the upland forest. Important salt-marsh species include salt grass, cordgrass, and various bulrushes and reeds. Saw palmetto, bayberry, and live oaks occur in the transition zone. Recently, National Park Service resource management specialists identified more than 500 species of plants on the island, some 95 percent of which are native to the region.[14]

Elements of these vegetation communities have existed on Cumberland since it regained island status. However, their spatial distribution and the percentages of the island that they cover have changed repeatedly. Two significant natural phenomena, lightning fires and severe storms, have destroyed vegetation from time to time, initiating succession on the disturbed areas. Early colonizers like grasses are replaced by other communities until a stable one, known as a climax community, comes into equilibrium with external environmental factors. Nevertheless, natural fires are so prevalent during the summer season that some of the most extensive and seemingly permanent communities on the island are successional stages. The oak-palmetto and some of the pine forest communities are cases in point.

Another environmental factor is the influence of salt spray on dune communities, which works to limit the shoreward extent of sensitive species. Thus, when a hurricane causes coastal erosion or deposition along the island's shoreline, it sends a ripple of change through the dune communities. These natural processes have manipulated the vegetation geography of the island for centuries. However, when humans arrived, they brought exponentially greater changes.[15]

The animals of Cumberland Island have faced the limits of a barrier island ecosystem for thousands of years. One result is less diversity than the mainland. Another is the smaller size of island species like white-tailed deer compared to their mainland counterparts. Nevertheless, the Park Service lists approximately 450 species of animals that can be seen on the island, more than two-thirds of which are birds. As the largest of the Georgia's Sea Islands and a protected environment, Cumberland serves as a major stopping point along the Atlantic flyway. Native terrestrial mammals present today include white-tailed deer, raccoon, Virginia opossum, and various bats, squirrels, gophers, and mice, a few of them endemic to the island. Marine mammals include manatees and dolphins. The largest reptile on the island is the alligator, but various species of snakes, lizards, frogs, and turtles also occur. Hillestad et al. identified 52 species of reptiles and amphibians on the island, half the number found on the adjacent mainland. Loggerhead and, occasionally, other species of sea turtles lay their eggs on the beach as well. Resident bird species number more than 100, with wild turkeys being the visitors' favorite. In addition, Hillestad et al. identified more than 220 species that seasonally or occasionally visit, including bald eagles and peregrine falcons.[16]

During the centuries of human occupation, species appeared and disappeared. Some, such as black bear, Florida panther, and bobcat, vanished within the last two centuries. The principal introduction by Paleo-Indians was the dog. Since Europeans arrived, however, the entire Cumberland ecosystem has been drastically altered. Europeans and Americans have turned loose horses, pigs, and many species of birds. They have also facilitated the arrival of armadillos and other species. Floral exotics arrived in the forms of crops, unusual garden plants, and a variety of pernicious weeds that accompanied livestock and grain seed.

The Cumberland Island of 13,000 years ago was a heavily forested upland. The forest composition and distribution on the island was likely different than that found by the earliest European visitors. When the Europeans arrived, they still found a more impressive forest than today's. All reports by the Spanish, English, and early American settlers describe towering, thick pine trees, especially on the north end of the island, and magnificent oaks with huge limbs, perfect for ship timbers. Palmetto and other understory species were less prevalent than at present due to shading and other types of competition from the extensive forest canopy.[17]

130 Centuries of Native Americans

The arrival of hunting and gathering peoples in North America is a topic that excites considerable scholarly debate. Recent archaeological finds as far south as Chile have overturned a cherished theory that humans arrived within the confines of the present United States only 12,000 years ago. A date of 20,000 B.P. (years before present), coinciding with the onset of glacial retreat, is now widely accepted.[18] Although no data prove that Paleo-Indians arrived that early on the Georgia coast, evidence from Florida suggests that 13,000 years ago is not unreasonable. Between 18,000 and 12,000 B.P., the shoreline remained close to the edge of the continental shelf, some seventy miles east of its present location. Nomadic hunters and gatherers were few in number and possessed a simple technology. Their most dependable sources of food would have been the estuaries and shallow waters of the sea. A rapid rise in sea level between 12,000 and 3,500 B.P. repeatedly inundated their littoral sites as it pushed the shoreline upward and westward. Recently, archaeological remains from southwest Florida have been dated at 12,030 B.P. plus or minus 200 years. A considerably earlier date is suggested by the nearby discovery, at a depth of more than eighty-five feet below current sea level, of the remains of a long extinct giant tortoise that appears to have been killed by a wooden stake driven through its shell.[19]

After Cumberland Island reappeared and sea level stabilized, use of the island by Native Americans was probably sporadic and seasonal. National Park Service archaeologist John Ehrenhard and others propose that intensive use of the current coastal region began around 4,000 years ago during what is known as the Bilbo archaeological phase. This culture eventually gave way to other groups about which little is known. One persistent cultural trait, however, was a focus on the sea and marshes for subsistence.

The next stage began about 1,450 years ago and is known as the Deptford phase. These people showed distinct advances in ceramics manufacture that may indicate a diffusion of ideas from elsewhere. This diffusion might also have brought maize horticulture to the region, allowing people more leisure to experiment with their handicrafts. A further sign of the diffusion of ideas is provided by the appearance in Georgia of burial mound construction, a cultural practice that came to dominate much of the central and southeastern United States. The Deptford people, like those who preceded them, show cultural affinity with peoples north and northwest of Cumberland Island.[20]

At some time between the appearance of the Deptford phase and the arrival of Europeans (1,450 to 450 years ago), an infusion of people and their culture traits from southern Florida took place. European explorers found that people on the island spoke a Timucuan dialect similar to those found along the Atlantic coast of northern Florida. On the mainland north of Cumberland Island lived a people known as the Guale, bitter enemies of the Timucuans.

The people on Cumberland referred to their particular group as the Tacatacuru. Like their predecessors who inhabited Cumberland Island for centuries before, the Tacatacuru lived in villages on the sound side of the island and exploited shellfish, sea animals, and various terrestrial plants and animals. Their primary settlement was located at the area now known as Dungeness, but they had other significant sites at Table Point and Brickhill (map 1.2).[21]

The Dungeness site is a large one with shell deposits stretching from the dunes on the east along the edge of the marsh to the shore of the Cumberland Sound and then north beyond Sea Camp. Furthermore, coastal erosion has removed the bulk of the village site, leaving only a fraction of the aboriginal relics. Early groups amassed huge communal shell mounds. Ehrenhard has suggested that much of the Dungeness area might have been five or six feet higher than at present. Interestingly, the later Tacatacuru shifted to a rectangular village form of dispersed houses, each accompanied by a small, individual mound. This sprawling village spanned an area from the modern Dungeness Dock to north of Sea Camp. General Nathanael Greene, his heirs, and the Carnegies mined the mounds for shells to build tabby structures and roads. Greene himself leveled one of the largest mounds to build his huge tabby mansion called Dungeness. The ruins of the late nineteenth-century Carnegie mansion still occupy the site.[22]

The impact of the Native Americans on the natural ecology of Cumber-

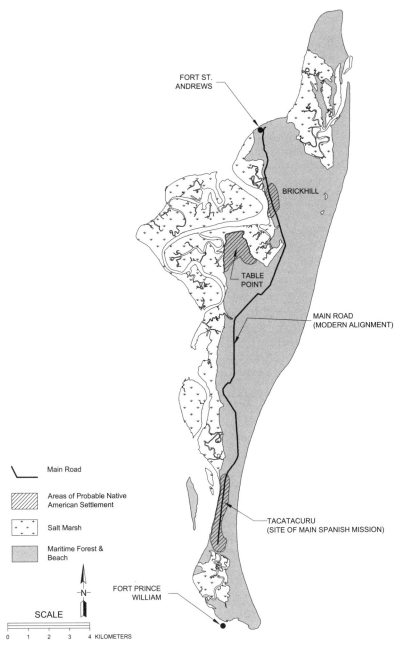

FORT ST.
ANDREWS

BRICKHILL

TABLE
POINT

MAIN ROAD
(MODERN ALIGNMENT)

Main Road

Areas of Probable Native
American Settlement

Salt Marsh

Maritime Forest &
Beach

TACATACURU
(SITE OF MAIN SPANISH MISSION)

N

SCALE

0   1   2   3   4 KILOMETERS

FORT PRINCE
WILLIAM

Map 1.2. Native American and colonial sites on Cumberland Island

land Island is difficult to determine definitively. Many centuries of island use and habitation have destroyed or corrupted much of the archaeological record. Nevertheless, some Native American environmental impacts can be confidently proposed. First, Paleo-Indians caused, or at least contributed to, the extinction of the Pleistocene megafauna. Animals such as mammoths, giant sloths, saber-toothed cats, and giant tortoises disappeared in a few thousand years due to the skills of the hunters as well as the animals' inexperience with human predators. The actions of animals partly determine the structure and species content of ecosystems. Elimination of these megafauna sent a ripple effect through the natural world that resulted in new geographies of both plant and surviving animal species.[23]

Native Americans also deliberately manipulated the ecology of Cumberland Island by introducing desirable animals and plants and by frequent burning. Timucuan peoples, like other cultures, burned to clear or maintain open fields for agriculture, improve browse for desirable animals like the white-tailed deer, drive animals during a communal hunt, and create an open area around a village for defense. In addition, Native Americans established a latticework of trails and village sites readily adopted by later European and American settlers. Hence when French explorer Jean Ribault visited Cumberland Island in 1562, he found an island already deeply humanized.[24]

A Colonial Prize

The arrival of the Spanish in North America initiated two processes that forever changed the region. First, the Native American population underwent extensive decline and redistribution. European diseases decimated the tribes of coastal Georgia as they did Native Americans throughout the continent. This catastrophic population decline radically altered their interaction with and impact on the natural environment.

Second, direct conflict between the Spanish and the English over the Georgia coast, an area some historians call "the debatable land," incited raids that led to construction of forts and, ultimately, permanent settlement in the region. These processes took place on Cumberland Island as well as on the other Sea Islands and the adjacent mainland. Although no historic structures survive on the island from this era of imperial design, these European actions began the large-scale modification of Cumberland's landscape.[25]

The first Spanish incursion in the southeastern portion of North America came in 1513 with a visit to the Florida coast by Juan Ponce de Leon. The Spanish followed with more probes of the Atlantic coast by Francisco Gordillo and Lucas Vasquez de Ayllon. The latter established a short-lived settlement near present Savannah, Georgia. All of these forays quickly turned hostile. Hence, Spanish-Indian contact was brief.

It is uncertain whether European diseases took hold in the region at those times. It is certain that the arrival of Hernando de Soto in 1539 brought deadly pathogens to the native peoples. Although his massive 600-man expedition came no closer to the Georgia coast than the western side of Okefenokee Swamp, the four-year sustained contact with Native Americans across the South unquestionably introduced European diseases. The Indians themselves then diffused these maladies through trade and other contacts. By 1565 the population of Native Americans in the Southeast had dropped by at least half. The effects of this holocaust included elimination of some tribes and villages as independent entities and the drastic reduction of such environmental modifiers as hunting, farming, and burning the timberlands.[26]

Ironically, the first attempt to settle the Atlantic coast of Florida came not from Spain but from France. French Huguenots established a colony on St. Johns River, presently called Fort Caroline. The Spanish answered by massacring the Protestant French and establishing their own fortified settlement at St. Augustine in 1565. From there they attempted to build missions, presidios, and towns up the coast to modern South Carolina. Among the sites chosen was Cumberland Island, which the Spanish named San Pedro. The missionary effort started badly. When the first three Jesuit missionaries from Europe inadvertently landed on San Pedro, they were slaughtered by the Timucuans, who favored the French.[27]

Ultimately, Franciscan missionaries succeeded where the Jesuits did not. Beginning in 1587, they established six missions along the Georgia coast, including two on San Pedro. The main mission, called San Pedro de Mocama, stood somewhere between Dungeness and Sea Camp on the western side of the island. Park Service archaeologists posit that it lay close to the modern Dungeness Dock. Nearby, the Spanish had a small fort. One historian, citing the records of a missionary priest, claims that the mission included seven pueblos (towns) with 384 converts. In addition, mainland Timucuans also attended services on the island. The Franciscans established a second and subsidiary mission on the northern portion of the island called San Pedro y San Pablo de Puritiba. In 1602 Father Baltasar Lopez reported 792

Christian Indians on the island. They depended more heavily on agriculture than their mainland counterparts and hence were more sedentary. The missions on the Georgia coast continued to succeed until pressure from the English and their Creek allies caused the Spanish to abandon the indefensible coast in 1686. Subsequently, St. Marys River became the unofficial border between Spanish and English territory. The missions on San Pedro at that time quickly fell into disrepair.[28]

The Spanish missionary period on Cumberland Island lasted ninety-nine years but left virtually no lasting imprint. The churches and other buildings were of simple and temporary construction. According to historian Michael Gannon, "Pine tree trunks held up the roof and walls, and between these rough-hewn pillars, small posts were interwoven with horizontal wattles, tied with leather thongs. Clay was then daubed on the latticework and, when dry, it was whitewashed on the interior. Palmetto thatching served as roofing."[29]

This type of structure would have rapidly disintegrated, particularly if the Christian Indians left with the missionaries. Archaeological investigations so far have not found any evidence of the structures. Small sherds of Spanish pottery are the primary artifacts of their presence. Nevertheless, change in the natural ecosystem unquestionably spread through the island. After the initial population collapse from new diseases, the Spanish caused an increase in native settlement, heightened exploitation of timber and other resources, and expanded agriculture to support its growing missionary base. In addition, various old-world plants and animals including cattle, horses, and hogs, accompanied these settlements. Some of those animals have never left the island, and they have negatively affected Cumberland's natural vegetation, native fauna, and even dune morphology.[30]

The Spanish retreat left a vacuum that the English filled from their base in Charleston, South Carolina. The two nations and their Native American allies continued to skirmish over Georgia's coast until 1763, although Spain never reoccupied the region. Instead, the conflict centered on raiding across St. Marys River in search of loot, slaves, livestock, and political hegemony. Major attacks into each other's territory consistently failed.

The initial English thrust into Georgia came in 1733 with the arrival of James Oglethorpe and his utopian colonists. He established Fort Frederica on St. Simons Island by 1736 and then elected to build forts on San Pedro, which he renamed Cumberland Island after William Augustus, the duke of Cumberland. Scottish Highlanders under the command of Hugh Mackay built the first fort on a bluff at the north end of the island. After six months of labor in summer and fall of 1737, Oglethorpe named it after St. Andrew, the patron saint of Scotland.[31]

Various reports describe the fort as a star-shaped structure with an underground powder magazine and nearby buildings capable of housing 200 men. Later, a small settlement called Barrimacke was constructed for married soldiers and their families. The proximity of this town to the fort is uncertain. Park Service historian Louis Torres cites two sources, one of which places it very near to the fort and another that claims it was seven or eight miles away. There is some archival evidence that the village survived Spain's destruction of the fort in 1742 by at least fourteen years. Detailed Spanish reports of the abandoned fort's destruction omit any mention of the settlement, which seems to lend credence to the theory of its location miles to the south.[32]

Oglethorpe constructed a second fort at the south end of Cumberland Island, which he named Fort Prince William. Ultimately this became the more important of the two forts as Oglethorpe ordered the Fort St. Andrews detachment to abandon it and take up station in its southern counterpart. In addition, Oglethorpe apparently built a hunting lodge of unknown size north of Beach Creek, which he called Dungeness. Traditional lore suggests it was in the area of the old Timucuan village of Tacatacuru, specifically on the site of the two subsequent Dungeness mansions.[33]

In 1742 the Spanish, responding to a raid by Oglethorpe on St. Augustine, arrived at Cumberland Island to attack its two forts. They found Fort St. Andrew abandoned but Fort Prince William stoutly defended. The Spanish lost two ships unsuccessfully trying to take the fort and retreated to St. Augustine when a fleet of English ships appeared. Despite their victory, the English considered the southern coast of Georgia to be indefensible. They chose not to rebuild Fort St. Andrew, while the detachment at the southern fort dwindled to little more than an advance reconnaissance team. Some of the English soldiers had started small plantations on the island, but these quickly deteriorated because of constant danger from Spanish, French, and Indian raids. By 1748 the island became part of a "no man's land" between the Altamaha and St. Marys Rivers. For a time, a settlement of bandits, pirates, and other lawbreakers existed on the island. These ne'er-do-wells came from both the English and Spanish colonies. They occupied Cumberland through the 1750s, and a few may have stayed well beyond that time.[34]

In 1763 the Treaty of Paris delivered Florida to the English and ended colonial conflict in Georgia. Almost immediately, Georgians poured south-ward into the former "no man's land" seeking claims for plantations and other economic pursuits. On Cumberland Island the British government granted nearly 11,500 acres of property but withheld the sites of the two di-lapidated forts. At least twelve individuals received plots ranging from 50 to 3,270 acres. One landholder, Jonathan Bryan, subsequently bought the lands of most of the other owners. Eventually he was able to offer 10,870 acres for sale at once. His advertisement noted the tremendous opportuni-ties for growing "corn, rice, indigo, and cotton." He also noted that the is-land contained "extraordinary range for cattle, hogs, and horse."[35]

Despite Bryan's optimistic claim, the only truly successful enterprise on the island was "live-oaking." Cumberland's enormous oaks provided some of the best ship's timber in the American colonies as well as lumber, shingles, and other products for both the navy and the burgeoning mainland settle-ments. Even live-oaking was in its early stages, however, while other eco-nomic activities remained insignificant. As a result, two men, Thomas Lynch and Alexander Rose, managed to acquire undivided ownership of nearly the entire island by 1770.[36]

The English legacy in the landscape is even fainter than those of their predecessors. No evidence exists for the sites of either English fort, the settle-ment of Barrimacke, or the camps of the lawless gangs of the 1750s. Archival and geological research and fruitless archaeological searches have led some scholars to conclude the sites may no longer be part of the island. Fort Prince William on the southern end of Cumberland was almost certainly several dozen yards south of the present beach. Modification of the coastline by an extensive jetty on the island's southeast corner as well as erosion by the St. Marys River outwash has erased all traces. The hunt for Fort St. Andrews is more intriguing and frustrating because both English and Spanish records place it on a prominent upland. Most archaeologists assume that it lay near the site of the Cumberland Wharf, but no visible signs or significant relics pinpoint it. Furthermore, while the English did accelerate the cutting of timber on the island, their agricultural activities did not approach the level attained by Native Americans or the Spanish mission system.[37]

The Plantation Period

The centuries of human activity on Cumberland Island before the Ameri-can Revolution dramatically impacted its ecology and biogeography. Yet few physical relics remain to show the island's rich history up to that time (see map 1.2). That situation changed as newly American settlers transformed Cumberland into a base for plantations producing sea island cotton, citrus fruit, and other horticultural products. The eighty years between the end of the Revolution and the beginning of the Civil War saw more than half the island's land converted to agricultural fields. Fence lines, dikes, and ditches remain. Homesteads, some quite rich and elaborate, have left buildings, foundations, roadways, cemeteries, walls, and chimneys across the landscape. The plantation period was the apogee of human modification of the island. At no time in the succeeding decades has Cumberland Island been so distant from its natural character.

Virtually all settlement and commercial activity on Cumberland Island ceased with the onset of the American Revolution. For nearly eight years the island served as a foraging ground for both American and British militias. The island's primary landowners, Thomas Lynch and Alexander Rose, both faced economic strains and viewed the island as an investment that could readily be sold. By the end of the war, Rose's half interest passed through several hands before being bought by General Nathanael Greene, a Revolutionary War hero.[38]

The war had created severe financial problems for Greene. When the American government could not feed and supply his troops, the general used personal funds and signed promissory notes for provisions and equipment. After the war Greene moved his family from Rhode Island to a plantation north of Savannah known as Mulberry Grove given to him by the state of Georgia, and he bought the Cumberland land as a speculative investment to recoup his fortune. He planned to sell the island's rich timber as well as the land itself.[39]

Unfortunately, neither Mulberry Grove nor Cumberland Island could provide enough income to pay off Greene's debts. Creditors hounded the general to his premature death in 1786 and then pursued his widow, Catherine. She subsequently married Phineas Miller, the general's personal secretary and tutor for their children. The Yale graduate was a promising businessman who later formed a company with Eli Whitney to sell the latter's new invention, the cotton gin. However, an economic recession in the young country and an inability to protect Whitney's patent exacerbated the Greene-Miller family's financial woes. Between 1798 and 1800 Phineas and Catherine sold their primary residence and lands at Mulberry Grove to pay creditors and moved to Cumberland Island. Four years later Phineas died, leaving Catherine alone for a second time. This continued a process that would become characteristic of the island: the premature deaths of men leaving strong and intelligent women to shape Cumberland's future.[40]

In the meantime, Thomas Lynch, co-owner of most of Cumberland's property, died in 1776, and his son and heir died only three years later. The younger Lynch left his share of the island's land to his sisters' children. They also faced economic problems, so that by the late 1790s a proper division of the lands held by the heirs of Greene and Lynch was imperative. Two surveys were conducted in 1798 and 1802. The latter survey produced a map showing land use at that time as well as acreage figures. The court and the two families used this information to divide the land equitably into twelve sound-to-ocean strips on both Great Cumberland and Little Cumberland Islands (map 1.3). This division anticipated a similar one carried out by the heirs of Thomas Carnegie 160 years later. Because of the variance in island width, soil quality, and timber resources, the two families selected parcels in a noncontiguous pattern. The Greene heirs received parcels 1, 3, 5, 8, 9, and 12. Parcel 1 included the Dungeness area, and parcel 12 was Little Cumberland Island. The Lynch group acquired parcels 2, 4, 6, 7, 10, and 11. Small landholders owned some plots along the sound, and the state government attempted unsuccessfully to hold the areas on either end of Cumberland Island. The latter were the sites of the decaying British forts, and Georgia had hoped to reserve them for possible future use.[41]

Over the next several decades, property continued to change hands as the descendants of Greene and Lynch parried financial blows by selling parcels of their land. Others who established plantations on the island included Robert and Thomas Stafford, Peter (Pierre) Bernardey, George McIntosh, and Henry Osborne. In 1831 Phineas Miller Nightingale, an heir of Catherine Greene Miller, was able to buy five remaining Lynch properties, ending that family's presence on the island. He promptly sold many of those parcels to small landowners anxious to participate in the island's lucrative sea island cotton production.[42]

In addition to developing a successful plantation, Catherine Greene Miller and her heirs built a large home and estate on parcel 1. The house, called Dungeness after the earlier hunting lodge, was a four-story, symmetrical structure composed of blocks of tabby, a mixture of coquina shells, lime, and sand that hardens like concrete (fig. 1.3). Catherine and Phineas Miller began the construction shortly after their 1799 move to the island, but various monetary troubles delayed its completion for more than twelve

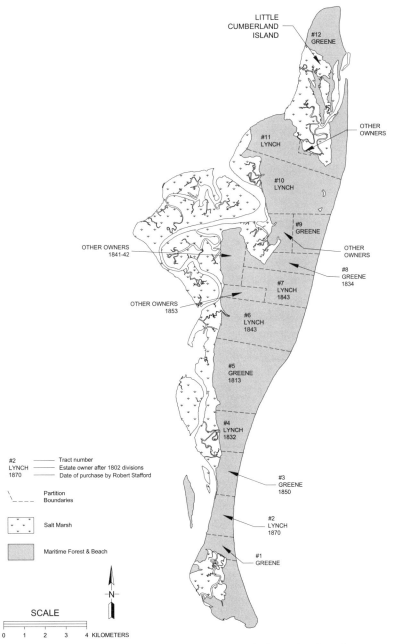

LITTLE
CUMBERLAND
ISLAND

#12
GREENE

OTHER
OWNERS

#11
LYNCH

#10
LYNCH

#9
GREENE

OTHER OWNERS
1841-42

OTHER
OWNERS

#8
GREENE
1834

#7
LYNCH
1843

OTHER OWNERS
1853

#6
LYNCH
1843

#5
GREENE
1813

#4
LYNCH
1832

#2   Tract number
LYNCH   Estate owner after 1802 divisions
1870   Date of purchase by Robert Stafford

Partition
Boundaries

Salt Marsh

Maritime Forest & Beach

#3
GREENE
1850

#2
LYNCH
1870

#1
GREENE

-N-

SCALE

0   1   2   3   4 KILOMETERS

Map 1.3. The division of Cumberland Island property by the heirs of Nathanael Greene and Thomas Lynch in 1802 and subsequent acquisition by Robert Stafford. (Data from Mary R. Bullard, 1993, "Uneasy Legacy: The Lynch-Greene Partition on Cumberland Island, 1798–1802," *Georgia Historical Quarterly* 77, 4, 757–88, used with permission of the author, who based it on the 1802 McKinnon map housed at CINS Archives, and John E. Ehrenhard and Mary R. Bullard, 1981, *Stafford Plantation, Cumberland Island National Seashore: Archaeological Investigation of a Slave Cabin,* National Park Service, Southeastern Archaeological Center, Tallahassee, Fla.)

Fig. 1.3. The ruins of General Nathanael Greene's Dungeness mansion in the late 1870s

years. Not long after construction ended, the family suffered another financial setback when an 1813 hurricane blew off the roof, necessitating an expensive new one made of copper. Catherine also built many supporting
structures in the Dungeness area including a small tabby house that may
have served as the Millers' temporary residence while construction of the
mansion was under way. This small cottage later served as an office for the
Carnegie estate. It lies a few dozen yards from the ruins of the Carnegie
mansion and is the oldest standing building on the island. Today it is simply
called the Tabby House (fig. 1.4).[43]

Catherine and her youngest daughter and heir to Dungeness, Louisa
Greene Shaw, also developed the grounds around the new mansion. They
created a formal garden devoted to both vegetables and ornamental plants
between the mansion and the Beach Creek marshes. Later Thomas and
Lucy Carnegie maintained the basic layout of this garden. The National
Park Service is currently restoring it to the Carnegie period. Louisa Shaw
was an accomplished botanist and horticulturist, widely known along the
Georgia and South Carolina coasts. She introduced a variety of semitropical and temperate-latitude crops to the estate. By the time of her death in
1831, the Dungeness plantation was producing significant commercial crops
of oranges and olives as well as lemons, figs, dates, and pomegranates.[44]

Fig. 1.4. The Tabby House at the Dungeness Historic District was built in the early 1800s.

Both Catherine Greene Miller and Louisa Shaw maintained busy social lives in spite of their isolation. Visitors included prominent businessmen like Eli Whitney, politicians, English gentry, and many notables from southern society. Revolutionary War hero General Henry "Light-Horse Harry" Lee (the father of Robert E. Lee) landed on the island in 1818 and promptly died. Louisa buried him in the Greene-Miller cemetery near the Dungeness mansion. Nearly a century later the state of Virginia removed his body and reinterred it by his son's grave on the campus of Washington and Lee University in Lexington. Even during the War of 1812, Catherine maintained a formal social schedule with officers of the occupying British forces.[45]

Upon Louisa Shaw's death in 1831, her favorite nephew, Phineas Miller Nightingale, inherited the Dungeness plantation as well as one called Oakland near the modern Duck House Road. He continued the mixed economy of timber harvesting and farming cotton, oranges, and olives. A planter at this time had many sources of income. For example, Nightingale also owned productive land on the mainland, leased his slaves for various projects, and invested in stocks and other financial pursuits. Despite this diversification, economic troubles continued to follow the descendants of Nathanael Greene. After his purchase of the Lynch properties, a severe frost in 1835 badly damaged crops on the island and seriously strained his financial means. Nightingale tried to sell land to relatives in order to keep it in the family. However, as his fortunes continued to decline, both neighbors and strangers snapped up the property.[46]

The primary buyer was Robert Stafford Jr., son of Thomas and nephew of Robert Stafford, who first appeared on the island in the early 1780s. The younger Stafford's productive tenure on the island began with his purchase of Catherine Greene Miller's 600-acre Littlefield tract in 1813. This became the site of his home, primary agricultural buildings, and a substantial group of slave quarters marked today by their surviving chimneys. Stafford was an extremely successful planter, and he continually added to his lands, ultimately owning 8,125 acres from the area around Sea Camp to north of Table Point (see map 1.3). During the 1850s Stafford reportedly owned up to 348 slaves. His aggregation of formerly scattered and tiny holdings enabled him to consolidate cotton production, adapt transportation, and achieve an economy of scale that matched the high quality of his product. It also erased some of the landscape features that marked the boundaries of former plantations.[47]

Sea Island Cotton Production

From the end of the Revolution to the Civil War, the backbone of the island economy was sea island cotton. This type came originally from South America, where wild relatives still grow in southern Ecuador. Its use dates back to textiles found in the Chilean desert from more than 5,500 years ago. Paleobotanists propose that the plant, *Gossypium barbadense,* spread to the West Indies at least 1,000 years ago. By 1650 the British cultivated it on plantations in Barbados, hence its species name. From there planter-settlers brought it to the South Carolina coast in the 1670s. Because of its ability to flower outside the tropics, it was an immediate success in the mild climate of the Carolina and Georgia islands. Sea island cotton produces fine and silky lint much desired for luxurious fabrics. Its severe nutrient demands forced Sea Island planters to rotate the cotton production, leaving worn fields fallow or in restorative crops. On a typical plantation, therefore, only a portion of the farmland would be planted in cotton at any one time. This sometimes led to forest cutting for new fields. Another adaptation was the application of fertilizer. On the Sea Islands the most common compound consisted of marsh mud, crushed shells, and manure. Cultivation of sea island cotton continued in the United States until the advent of the destructive boll weevil in the early twentieth century. During its peak production on Cumberland Island, it supported a population of 65 whites and 455 slaves.[48]

Robert Stafford Jr.'s Cumberland Island cotton commanded extraordinary prices of up to 75 cents per pound. Island historian Mary Bullard has estimated that he had more than 260,000 pounds of unginned cotton on hand and cultivated at least 75 acres of the work-intensive staple. The proceeds allowed him to live a wealthy and comfortable life. He built a substantial planter's house that would later serve as a mansion for one of the Carnegie sons until it mysteriously burned in 1900. Stafford never married but received from a neighbor a mulatto slave by whom he fathered six children. When the Civil War and emancipation came, Stafford was an old man in his seventies. Unlike his neighbors, he refused to abandon his home. Although he repeatedly appealed to Union forces for protection from his former slaves, the Northern army gave him only sporadic attention. However, he did stay on the land, which saved him from having it declared as "abandoned" and divided among the newly freed slaves. Nevertheless, the war destroyed the plantation economy on Cumberland Island. The former slaves migrated to the mainland or the northern part of the island, which left Stafford with no workforce. Most other owners sold out for a pittance and moved on.[49]

Sea island cotton production required many island developments, both technological and residential, that influence the modern Cumberland landscape. Planters and their workers installed ditches to drain fields, mark boundaries, and serve as moats to keep out livestock and feral animals. The spoils from these ditches became the bases for fences to further exclude animals and unwelcome visitors. Occasionally these spoil-walls served as dikes to prevent saltwater intrusion. Various functional structures were scattered through the island, including equipment sheds, storage buildings, and foundations for cotton gins and cotton rams. In addition, many buildings housed or provided services for the slaves, who typically outnumbered the white residents at least ten to one. Slave quarters consisted of small individual cabins, each containing a fireplace and chimney. Families or segregated groups of single men and women occupied each. A typical unit included space for a dooryard garden and a yard for pigs and chickens. Other buildings might be reserved for medical purposes or church services.[50]

On Cumberland, several collections of chimneys mark the sites of former slave cabins. The most notable lies due east of the present Stafford house perpendicular to a road traversing the island from east to west. Three parallel

Fig. 1.5. The Main Road runs the length of Cumberland Island. Portions of it date from the earliest English settlement of the island.

rows of five cabins stretched south of Stafford Road while a few other buildings paralleled it. That road was part of a latticework of wagon roads and horse trails that connected the fields and plantations. Several of these roads became part of the modern Main Road that runs the length of Cumberland Island (fig. 1.5). Connected to it were numerous spur roads, a half dozen of which crossed the island from the sound to the dunes. In addition, pathways served each field, allowing access to the fences, ditches, and crops.[51]

The varied uses of cleared fields have left a landscape legacy as well. Some, like Stafford Field, were never reforested. The Carnegies adapted them for various uses, including recreation and new experiments in agriculture. Landscape historian Peggy S. Froeschauer has shown that the crop history and soil types of each field influenced the type and density of natural vegetation that reclaimed it.[52]

Even the extensive adaptation of the island for plantation agriculture, however, did not remove all of the original forest. A reconstructed map of the Stafford slave settlement by archaeologist John Ehrenhard and historian Mary Bullard shows the site nearly surrounded by woods. After the war statistics compiled by Freedmen's Bureau agent William F. Eaton showed that 36 percent of the land on seven "abandoned" plantations was woodland. These forests served as an auxiliary source of income, a reserve for fuel and building material, and a haven for feral livestock and deer.[53]

After the Civil War

The war years laid waste to the fields, utility buildings, and homes. The emancipation of slaves created such a financial loss for the planters that most could not afford laborers or repairs to their land and property. After 1865 Cumberland Island never recovered its antebellum agricultural wealth. Some small landowners like William R. Bunkley and Rebecca Clubb on the north end easily reclaimed their land. Larger landowners had more trouble. In some cases their lands were redistributed to freedmen. However, the former owners refused to employ them, and they lacked the resources to establish their own farms. Some planters chased off the freedmen and burned the slave cabins while others waited until the former slaves left for jobs on the mainland. Robert Stafford Jr. lived on until 1877, an impoverished and bitter old man. After his death and subsequent litigation among his heirs, his land passed to two nephews, John Tomkins and Thomas D. Hawkins. Stafford's six children by his slave Elizabeth did not receive any land on Cumberland.[54]

Even before the war the constant grinding financial decline of Phineas Miller Nightingale took its toll on Dungeness. The great house fell into disrepair and became uninhabitable. Nightingale sold pieces of land and finally surrendered the famous plantation to creditors in 1870. He died one year later. In 1880 W. G. M. Davis, a former Confederate general, purchased the Dungeness plantation (map 1.4). Davis toyed with several ideas to revive the estate's economy, including olive production and horse raising. How-

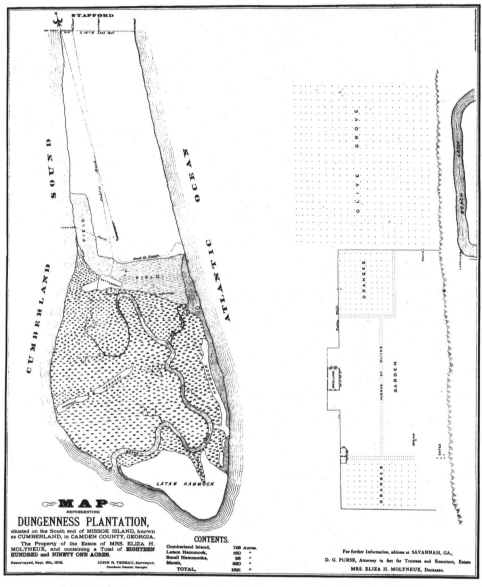

Map 1.4. The Dungeness plantation in 1878. The map legend explains that this is "The Property of the Estate of Mrs. Eliza H. Molyneux." (Georgia Archives, acc. no. 69-501, box 105, folder 11-1-003)

ever, the only business that seemed to work was tourism. The decaying mansion of the famous Revolutionary War hero Nathanael Greene became a popular sight for travelers through the region. A Savannah newspaper summed up the state of Cumberland Island in 1876:

> Cumberland Island, which before the war had ten or twelve large plantations devoted to the production of the valuable staple, has now not one acre in cultivation. . . . The houses have been burned, the fences have rotted, and the fields have grown up in weeds. Reconstruction and radicalism freed and made citizens of the laborers who formerly made the island fields fertile, and enriched the world with the fruits of their toil. The freemen and the citizens abandoned the cotton plantations and retired to the coast of the mainland, and the sea-island cotton has become almost a thing of the past.[55]

The end of the 1870s saw the last gasps of an agricultural era on Cumberland Island. At its height the cotton business cleared two-thirds of the forest, ditched and fenced fields from one end of the island to the other, and imprinted roads and structures on the island. It also created a settlement geography that would influence later development. The year 1881 saw the beginning of a new and radically different era, one in which the forest reclaimed many fields and the island turned from productive farmland to idyllic retreat.

The Era of
Rich Estates,
1881–1965

Three significant events during the period 1881 to 1965 dramatically shaped Cumberland Island. The first was the purchase of land on the island by Thomas Carnegie and his wife Lucy. The second was Lucy Carnegie's death in 1916 and the implementation of a complex trust arrangement designed to keep the island available for her heirs' enjoyment. The trust made it very difficult to sell or subdivide before the death of all nine of her children. The third was the death of that last child, Florence Carnegie Perkins, in 1962, which ended the trust's restrictions and led to the division of the island in 1965. Thereafter, Lucy's grandchildren and their heirs were free to dispose of their portions as they saw fit. During the periods between these three events, the Carnegie family developed the island, built a substantial infrastructure that included five mansions, and then suffered the erosion of both the infrastructure and their lifestyle with the decline of the trust's assets. Despite this long and worrisome economic slide, each generation of heirs lived at least part of their lives on Cumberland Island and developed traditions, relationships, and a deep love of place that has carried through to the present.

At the same time the Candler family of Atlanta, whose fortune came from Coca-Cola, acquired most of the northern tenth of the island, which had suffered its own economic roller-coaster ride with concomitant ownership changes. Ultimately, they transformed it into a substantial vacation estate. In the process the Candlers too fell under the spell of Cumberland Island's mystical beauty and addictive lifestyle. They too became protective of its resources and proprietary over its future. The combined ownership of the Carnegie and Candler families held the island in a relatively undisturbed state and made it possible for the National Park Service to consider it for a national park.

The Carnegies Come to Cumberland Island

At the conclusion of the Civil War, the South became a source of fascination to people from the North, especially the wealthy who could afford a leisurely tour of its sights. The flow of tourists grew from small parties of scientists such as William Bartram and social critics like Frederick Law Olmsted to include well-to-do businessmen, industrialists, and political figures. Many of these visitors found the winter environment of the South attractive and its land readily available for purchase. Popular second-home and tourist destinations included Jekyll Island to the north and Amelia Island to the south of Cumberland Island.

Among the visitors to the Georgia coast in 1880 were the future owners of Dungeness, Thomas Carnegie, the younger brother of the famed steel magnate Andrew Carnegie, and his wife, Lucy Coleman Carnegie. Thomas had been born in Dunfermline, Scotland, in 1843, eight years after his brother. Economic woes beset the family, and they migrated to Pennsylvania in 1848. William Carnegie, the boys' father, died within a few years, leaving his wife Margaret to be supported by the teenaged Andrew. The older son proved to have uncommon ability and relentless drive in the business world, eventually accumulating one of the largest fortunes in American history. With brother Thomas he founded Carnegie Brothers & Company, a conglomerate of iron and steel plants plus associated manufacturing and financial operations. Thomas was an excellent businessman himself, and Andrew relied on him heavily, especially during his increasingly frequent trips abroad. The younger Carnegie knew the business well and was better liked by both their partners and their employees. Yet, Andrew was a harsh critic of his brother and sent frequent letters full of overly detailed and demeaning orders.[1]

In 1866 Thomas married Lucy Coleman, the twenty-year-old daughter of a businessman and neighbor who supplied coal and coke to the Carnegie factories. She was a popular young woman and, according to one heir, included Andrew Carnegie among her potential suitors. Thomas and Lucy had nine children between 1867 and 1881. The six boys and three girls lived with their parents in Pittsburgh at Andrew's former home, which he had given to the newlyweds as a wedding present.[2]

When Thomas and Lucy visited Cumberland Island, they were probably familiar with its reputation. Many northern tourists visiting Florida or cruising the channels between the Sea Islands and the mainland stopped to view the ruins of Catherine Greene's Dungeness mansion. In addition, Lucy had spent part of her childhood in a boarding school in nearby Fernandina, Florida. She may even have visited the island during that time. Furthermore, tradition claims that Lucy became fascinated with the island after she read a well-known article on its history by Frederick A. Ober in *Lippincott's Monthly Magazine*.[3] After their visit to the island, Lucy resolved to buy the Dungeness estate for a winter home and as much island land as possible with it. She convinced Thomas to pay whatever it took to secure the property.

Negotiations with estate owner General W. G. M. Davis initially proved difficult. The retired Confederate officer did not want to sell to a "Yankee." However, the accidental death of his grandson soured the general on the island while intermediaries sent by Carnegie convinced him that the "Yankee" would care for the island. Davis sold his 1,891-acre holding to Thomas Carnegie on November 17, 1881, for $35,000. A year later Carnegie and business associate Leander Morris bought the adjacent 8,240-acre Stafford plantation, including its mansion, from the cotton baron's heirs for $40,000. In 1886 Lucy acquired Morris's interest in the Stafford land for a little over $38,000. She would continue to add island lands to her estate through the rest of her life, ultimately controlling the southern 90 percent of the island.[4]

Shortly after his purchases on Cumberland Island, Thomas Carnegie retired from active business, although he remained on the board of Carnegie Brothers & Company. He planned to enjoy life and develop his new estate on Cumberland Island. Milton Meltzer, in his recent book on Andrew Carnegie, claims that the stress of working for his zealous and censorious brother had taken its toll on Thomas. He began drinking just before retirement and died in 1886 a few weeks before his mother. He was forty-three. His death left Lucy with nine children, an unfinished island retreat, and a large fortune with which to support the family and continue developing the estate. She sold the family home in Pittsburgh and established her main residence at Cumberland Island while continuing summer trips to the North in order to escape Georgia's summer heat.[5]

Before he died, Thomas spent lavishly to construct a suitable mansion on the ruins of Catherine Greene's Dungeness house. It was a relatively modest structure for a very wealthy man, subsequently described as "about 120 by 56 feet, two stories high with an attic, and built in the Queen Anne and Stick styles. A tower at the east end was 90 feet high. The outer walls consisted of a light-colored granite and the roof was covered with Vermont

Fig. 2.1. The Dungeness mansion during the life of Lucy C. Carnegie

slate." Later, Lucy Carnegie employed the Boston architectural firm of Peabody and Stearns to enlarge the house. What resulted was a massive structure of 250 by 150 feet in an elegant Italianate style. It contained more than fifty rooms (fig. 2.1).[6]

Thomas also began shaping the larger landscape of the new estate before his death. He maintained the existing north-south road but developed elaborate driveways through adjacent forested lands. He kept the Tabby House, the Greene-Miller cemetery, and the Stafford house and farm but cleared away most of the other ramshackle structures and debris. Thomas Carnegie also tried to revitalize the various orchards on the island while widely distributing ornamental exotics to beautify his land.

Lucy Carnegie proved no less ambitious in her plans for Cumberland Island. From 1890 through 1905, while she added to the Dungeness mansion, Lucy Carnegie also paid for a variety of other buildings to provide for the family's every want. Ultimately, the area adjacent to the main house, which came to be known as the Dungeness complex, included more than twenty buildings plus assorted walls, decorations, and a pergola (a colonnaded walkway). The most architecturally significant was a recreation and guest

Fig. 2.2. The Recreation House at Dungeness was one of the architecturally significant structures on the island as well as a favorite of the Carnegie family.

house in the Queen Anne style erected around 1900. Lying just east of the mansion, it held a heated pool, steam room, recreation room, squash court, and guest bedrooms (fig. 2.2). It became the traditional abode of several Carnegie bachelor sons when they stayed on the island. The Recreation House was wood frame with cedar shingles and sported a variety of innovative architectural features.[7]

In keeping with her desire to have a self-sustaining estate, Lucy Carnegie spared no expense in building and staffing an elaborate residential plantation. The upper terrace of the original garden maintained its basic form and purpose, that of a scenic ornamental display (fig. 2.3). However, she had the lower garden expanded by diking and filling more marshland and converted it to a subsistence garden of vegetables and fruit. It also boasted a complex of greenhouses, primarily for cut flowers, and a waste dump at the edge of the marsh, carefully screened from the house. To the northeast of the mansion, Lucy continued to cultivate the olive orchard that came with

the property until a frost killed the trees in 1895. Still farther east, near the
dunes, was a fenced dairy pasture. The estate also raised poultry, beef cattle,
and pigs plus assorted crops on the developed fields of the old Dungeness
and Stafford plantations. Some of the latter were cash crops including sea
island cotton that won a state fair award in 1895. Still, the estate required
daily runs of its boats to the mainland in order to supply the elaborate
lifestyle of the Carnegies and their guests.[8]

Lucy Carnegie employed a staff of more than 200 who lived and worked
on the island. Thus the estate ultimately included a variety of other buildings. There were separate dormitories for white and black laborers, a dairy
manager's house, a poultry manager's house, a boat captain's house, and a
modest residence named The Grange that served as home for estate manager William Page and his wife. Functional buildings included a large carriage house and stable, a kitchen-dining building, an icehouse, a water
tower and cisterns, a pump house, a woodworking shop, a smokehouse, a
laundry building, greenhouses, a large garden shed, and a boathouse near
the Dungeness Dock (map 2.1). All together the work buildings and gardens
of the "village" consumed some 250 acres.[9]

The beauty of Cumberland Island and the lavish lifestyle of the Carne-

Fig. 2.3. This photo shows the Recreation House and a portion of the estate gardens in the
early twentieth century.

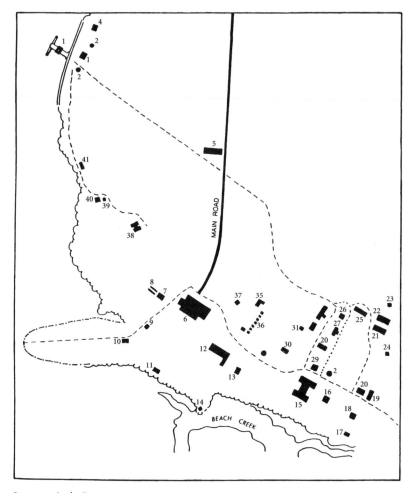

Structures in the Dungeness Area

1. Dungeness Dock
2. Cistern
3. Ice house (boat house)
4. Captain's house
5. Carnegie cemetery
6. Carnegie house ruins (big house)
7. Tabby House
8. Pergola
9. Garden house
10. Greenhouse
11. Water wheel
12. Pool house/ Bachelor apartments
13. Manager's house
14. Little dock
15. Carriage house
16. Gas pump
17. Miller-Shaw cemetery
18. Kennel
19. Dairy barn
20. Feed barn
21. Lumber shed
22. Carpenter shop
23. Chimney for carpenter shop
24. Silo
25. Quarters for black male help
26. Commissary
27. Dairy manager's house
28. Quarters for white male help
29. Ice plant
30. Steam laundry (restrooms)
31. Recreation hall
32. Dining room for white help
33. Dining room for black help
34. Chimney for bakery
35. Poultry manager's house
36. Chicken houses
37. Nancy Carnegie's playhouse
38. House of T. & C. Carnegie (NPS ranger residence)
39. Garden shed
40. Laundry for cottage
41. Unknown

Map 2.1. The Dungeness estate in 1916. (National Park Service, Southeast Region Office, 2000, *Draft Introduction to Planning Effort and Environmental Impact Statement,* vol. 1, CINS Superintendent's Office)

Fig. 2.4. Lucy C. Carnegie (in black) and her nine children in front of the Dungeness mansion

gies became a source of constant interest and envy to the mainland neighbors in poverty-stricken Camden County, Georgia. A who's who of notable businessmen, led by Andrew Carnegie, various politicians, rich friends and associates, and notable figures in the arts and sciences visited the estate. There they would relax at the mansion or recreation house, go picnicking at the ocean beach, hunt, fish, and enjoy lavish parties. Various house servants, both white and black, ensured that every want was satisfied.[10]

As her children matured, seven of them married (fig. 2.4). For each of these, Lucy provided a substantial wedding gift which most used to build a home on the island. Oldest son William received the Stafford house and a cash gift to help renovate it upon his marriage to Margaret Gertrude Ely. On January 5, 1900, a fire of unknown origin destroyed the antebellum mansion. A year later William built another house on the site and continued to call it Stafford. The new place resembled the old one in form, a two-story structure with a gable roof and an open porch across the front. However, it was built to better resist fire and sported a white-painted stucco exterior unlike its predecessor. Various outbuildings stretched across the estate to the standing chimneys at the site of the former slaves' cabins. Subsequently, heirs of William's brother Andrew II built a more modest home adjacent to the ruins that is itself called The Chimneys. Next to The Chimneys lay one of the larger fields of the old Stafford plantation, which William turned into a nine-hole golf course. Later it became an airfield.[11]

Daughter Margaret (called Retta) married wealthy scion Oliver G. Ricketson. Because she already had spent her wedding gift money, she used her husband's fortune to build a residence two miles north of Dungeness. This large two-story frame building, called Greyfield, passed to Margaret's daughter Lucy Ferguson. Her heirs now operate the only privately owned mansion on the island as a popular and upscale inn. It too lies amid a variety of outbuildings, most of them family residences and employee quarters rather than functional structures. Subsequently, Lucy Ferguson had a second area of structures built on land she received after the division of the island. This complex, called Serendipity, also remains in private hands.[12]

Thomas Carnegie II received a twenty-nine-room mansion called The Cottage upon his marriage. Unlike Stafford and Greyfield, this house lay only a few hundred feet west of the Dungeness mansion. This elaborate two-story home featured porches on three sides of the bottom story and two sides of the top one (fig. 2.5). Because of its location on the Dungeness estate, it required no outbuildings of its own, although young Thomas III maintained a zoo near the house for a period of time.[13]

In the late 1940s the same Thomas III started a fire that completely destroyed the house. According to Mrs. Carter Carnegie, The Cottage was being closed anyway owing to the cost of its upkeep. During the winter of 1951–52, after the estate bulldozed or buried the remains, heirs built a modest home in its place. In later years that home has served as the residence of the national seashore's superintendent or other Park Service personnel.[14]

The remaining mansion, Plum Orchard, was second only to Dungeness in both size and magnificence. Lucy Carnegie ordered it built for her son George on the site of a former orchard. Peabody and Stearns, the same firm that renovated Dungeness, designed the mansion. The occasion of this enterprise was George's marriage to Margaret Thaw in 1898. Eight years later the Thaw name became a household word in America as a result of a sensational murder by Margaret's brother Harry. Jealous that his young wife, Evelyn Nesbit Thaw, had been "kept" by the much older Stanford White a few years earlier, he accosted and shot dead the famous architect in front of

Fig. 2.5. The Cottage, built for Thomas Morrison Carnegie Jr. and his family ca. 1901

dozens of witnesses at Madison Square Garden. In the ensuing circuslike trial, Thaw's mother spent much of her vast fortune securing him a verdict of insanity.[15]

Despite the huge fortune she spent on son Harry, Mrs. Thaw had enough left to enable her daughter to enlarge Plum Orchard substantially in 1906. Adding a wing on either side of the house, Margaret Thaw Carnegie expanded it to a sprawling, 240-foot mansion with thirty principal rooms, twelve bathrooms and lavatories, and numerous smaller rooms (fig. 2.6). As befitted the second largest mansion, Plum Orchard had more than a dozen outbuildings and a dock. George died childless in 1921, and his wife wandered off to Europe to marry a French count. George's sister Nancy Carnegie Johnston then moved into the house, and her heirs maintained it until 1970.[16]

In addition to these four major centers, the Dungeness-Cottage village, Greyfield, Stafford, and Plum Orchard, the Carnegies also erected other facilities to serve their needs and wants. Chief among them were several "duck houses" located close to the dunes and the ocean beach. The earliest of these coastal houses was built in the 1880s, but it was abandoned by the 1920s be-

Fig. 2.6. Plum Orchard mansion, built for George L. and Margaret Thaw Carnegie, date unknown

cause of poor location and general deterioration. A subsequent one built across the island from Plum Orchard around 1900 was used as a hunting camp through the 1920s, but it was destroyed in the 1980s by a careless camper's fire.[17]

The Carnegie Family

The youngest of Thomas and Lucy Carnegie's nine children was born the same year they bought land on Cumberland Island (fig. 2.7). In chronological order they were William, Frank, Andrew, Margaret, Thomas, George, Florence, Coleman, and Nancy. Frank and Coleman never married while William and George married but did not have children. The remaining five had families who would become the five branches to divide the island and secure retained rights when it became a national seashore. Andrew Carnegie II had two daughters, Nancy who married James Stillman Rockefeller and Lucy who married Phineas S. Sprague, divorced him, and married Jack Rice. Hence she became Lucy Carnegie Sprague Rice. One of her daughters, Lucy Carnegie Sprague, married a man named Foster and later became the matriarch of Stafford.

Thomas M. Carnegie II (known as Morris) had two sons, Thomas III

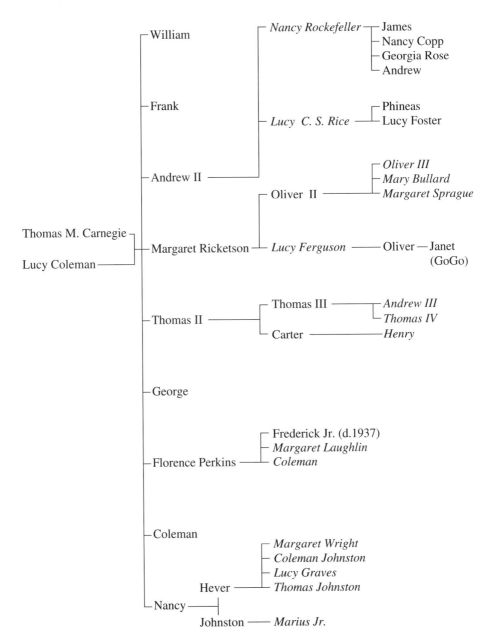

Fig. 2.7. A family tree of the Carnegies. This chart is not complete but shows all the children of Thomas and Lucy, all the grandchildren, and selected others who appear in the text. The names of those who inherited portions of the island are in italic type.

and Carter. This branch of the family, ironically the only one to continue the Carnegie name, suffered the worst financial decline, making retention of their portion of the island after the trust ended an unaffordable luxury. Margaret married Oliver Ricketson and had two children, Lucy, who married Robert Ferguson and became the dominating figure on the island during her later life, and Oliver II. Florence (called Floss) married Frederick Perkins and gave birth to three children, Frederick Curtis Jr., Coleman, and Margaret (called Peggy). Finally, Nancy eloped with Dungeness stableman James Hever and had four children, Margaret, Coleman, Thomas, and Lucy. The oldest later married and became Margaret Wright, the most vocal proponent of national park status for Cumberland Island. After the death of Hever, Nancy married Marius Johnston, the island doctor, who sired Marius II. These five divisions are called the Andrew II (subdivided into Rockefeller and Rice groups), Ricketson, Thomas II, Perkins, and Johnston branches of the Carnegie family. The fourteen grandchildren named above became the principals who would await the death of the last of their parents to decide the ultimate disposition of Cumberland Island.[18]

The Design of the Trust

During her long life as a widowed mother and dominating head-of-family, Lucy Carnegie shaped the future of both her children and her estate. None of her six sons ever held a job. Instead, they engaged in the recreational life of the very wealthy. Oldest son William became an avid golfer playing around the world as well as on his own course at Stafford. The others traveled, sailed, maintained rich social lives, and always returned to the enfolding embrace of their mother and her idyllic island retreat. Later James S. "Pebble" Rockefeller Jr., her great-grandson, reflected on the effect this had on the Carnegie sons: "I also came to appreciate the underlying sadness inherent in some of the Carnegie clan, exiled as they were to a remote Island with a powerful, matriarchal figure in charge, leaving them nothing to do but hunt and fish, drink too much, chase after what women could provide, and think what excitement was planned for the coming day. I can remember my grandfather [Andrew II] telling me, . . . 'Pebble, I wish I had had a 9:00 to 5:00 job like your father. It is not healthy to not have a job.'" Pebble's mother, Nancy Carnegie Rockefeller, later castigated her great-uncle, the senior Andrew Carnegie, for not furnishing jobs for her father and uncles in his business as he had for her grandfather.[19]

Lucy Carnegie took various steps to ensure that she would control the island estate. When Thomas Carnegie died, he left his fortune to Lucy. However, his children, as heirs, might have challenged that eventually. Thus in February 1899 the nine children legally ceded all interest in the lands and property of the island, as well as other holdings in Pittsburgh, to her. Thirteen years later she had a complex and binding will designed that sought to forestall any irresponsibility by her children with the island and the rest of her fortune. She set up a complex trust consisting of two major parts. The first ordered that the Carnegie office building in Pittsburgh be rented and the income from this used to maintain the island estate for her heirs. Any residual annual income was to be distributed among the surviving children. The second part prevented those heirs from dividing or selling any island land until the death of her last child unless they unanimously agreed to do so. The trustees were to be the five oldest children or a bonded, professional trust officer. Shortly after this, oldest son William married his nurse and was banned from the island. An angry Lucy Carnegie created two codicils eliminating him from trusteeship.[20]

When Lucy died in 1916, seven of the other children survived her, Coleman Carnegie having died five years earlier. Within another five years Frank and George also died childless. Meanwhile Florence and Nancy elected to refuse trustee duties. That left a family trusteeship consisting only of Andrew II, Thomas II, and Margaret. By 1946 only Andrew remained as trustee. At seventy-six years of age, he sought relief from the onerous business of caring for the island and its financial upkeep. He, Florence and Nancy, his only living siblings, and the grandchildren secured the services of the Peoples First National Bank in Pittsburgh. Later dissatisfaction with the trust officer, Robert D. Ferguson (not to be confused with the unrelated Robert W. Ferguson, husband of Lucy), caused the family to shift the trust arrangement to the First National Bank of Brunswick, Georgia, in 1954.[21]

A New Era for the Carnegies: The Trust Years

For the first nine years of the trust, Lucy Carnegie's huge fortune comfortably maintained the surviving children and their families. In 1926 only six children remained to draw upon the trust income and enjoy the Cumberland Island estate. A report compiled by the Pittsburgh firm of Patterson, Crawford, Miller, and Arensberg that year showed that the Carnegie office building had brought in a total of nearly $600,000 in rental fees from 1917 to 1925. Added to this were payments from the remaining stocks and a meager $24,830 produced on the island. Over the same span of years, as the island's economically productive activities waned and the land returned to a wilder appearance, maintenance had demanded $316,135. The highest costs were nearly $67,000 for the Stafford farm, some $57,000 for general expenses, and $38,665 in taxes. The first two were predominantly for wages.[22]

The trustees divided the surplus income among the heirs, and through 1924 it averaged more than $42,000. However, in 1925 the residual income plummeted by nearly a third to $31,500. In response to this sudden downturn, the family decided to close the Dungeness mansion and forgo its heavy maintenance costs. Four years later the collapse of the stock market further strained the resources available to support both Cumberland Island and an idle lifestyle for the heirs.[23]

In the mid-1920s, between the deaths of George Carnegie and Margaret Carnegie Ricketson in 1921 and 1927 respectively, Andrew Carnegie II wrote a general letter to his siblings on the costs and potential earning power of Cumberland Island. He suggested that the surviving heirs form a corporation to run the island. He further proposed that the corporation be chartered to "buy and sell or lease or rent property—to conduct cattle and hog raising—lumbering—planting—sawmilling—fishing (shrimping), lay out 'developments'—build hotels—roads—operate freight and passenger boats—aeroplanes—amusements, etc. Endeavoring in every way to make money by developing The Island's resources which now go to waste."[24] He concluded by urging his brothers and sisters to exclude husbands, wives, and widows of deceased siblings from the business. Although Andrew II's recommendation for a corporation to run Cumberland Island would not be acted upon for another three decades, many of his ideas for developing the island's resources were adopted. Unfortunately, no combination of economic activities came close to offsetting the cost of maintaining the idyllic retreat.

The heirs considered a variety of economic resources and functions on Cumberland Island. Invariably they sought expert advice from foresters, agronomists, soil scientists, and appropriate government specialists. As a well-educated group of men and women, they knew the value of science in determining the most profitable economic functions while also recognizing the environmental limitations of an island. Over the nearly five decades of the trust, they studied numerous plans that, if acted upon, would have made Cumberland Island unsuitable for a national park unit. The Carnegie heirs were not averse to altering the landscape in their pursuits, but they also remained sensitive to unsightly change. In that half century Cumberland Island ran a gauntlet of threatening suggestions. The family's inability to agree on any of these landscape-altering plans saved the island.

Among the most obvious and available resources of Cumberland Island was its forest. Lumbering and naval stores had been the first resource to draw non-Indian settlers, and the island held thousands of acres of woodland. In 1927 Thomas Carnegie II invited the James D. Lacey Company of Jacksonville, Florida, to survey the family land and design a plan to manage the island as a "permanent forest estate." Company representative S. J. Hall reported five classes of forested land with different values. Of greatest value, he claimed, were 2,300 acres of old-growth longleaf and slash pine. These trees could be worked for turpentine before being selectively cut in such a way as to "bring the greatest return consistent with the perpetuation of a continuous timber crop."[25]

A second category of timber consisted of "pond pine" or "pocosin." Hall noted that these trees were stunted and scattered and might contain "red heart rot." He blamed this on a hardpan in the soil fairly close to the surface in these areas. He suggested digging exploratory holes to determine the extent of the hardpan and then blasting it in order to "shatter the impervious strata which is responsible for the acid condition of the soil." Hall reported that the island held 700 acres of this type of forest.

Old and abandoned agricultural fields accounted for another 2,000 acres. Hall noted that these were being reclaimed by loblolly pine and live oak. Thomas Carnegie II wanted to clear the oak in order to improve pasturage, increase the parklike vistas, and favor the more marketable pine. Hall suggested trying to break even on the cost of cutting the oak by selling it for cordwood in Jacksonville. He then recommended the establishment of seedbeds for longleaf and slash pines for periodic transplantation to these areas in the future. These seedbeds, he suggested, should be in low places to "enhance the natural beauty of the fields."

A fourth timber resource consisted of 5,000 acres of "hardwood hammock land" (oak-palmetto community). The predominant species in these areas had little value themselves, but Hall thought that commercial hardwoods might be scattered through these groves. He suggested a careful study to locate these trees. Then they could be selectively cut to pay for clearing the rest of the vegetation. Once cleared, the area could be planted in valuable pines. A final category of land included 1,000 acres of beach, sand dunes, old rice fields, and areas developed for residential and recreational use.

Hall concluded his forestry plan with various ancillary suggestions. His plan for fire prevention included wide firebreaks around all the commercially valuable timber stands as well as a fire tower at least eighty feet tall. He suggested leasing the forestlands for hunting and trapping in order to derive extra income while emphasizing the lumbering operation. Finally, Hall assured the Carnegies that all activities would be managed carefully to preserve the island's natural beauty.

For most of the next two decades, intermittent logging on Cumberland Island roughly corresponded to Hall's plan. Old-growth pine proved particularly valuable for telephone poles. Loggers floated up to 2,000 at a time to the mainland. Through most of that period, Hall and partner F. W. MacLaren managed the operation through their new company, Forest Managers. Cutting proceeded under strict rules to leave 100 feet of forest along both sides of any road in order to preserve the image of wildness. Still, from time to time problems occurred. In 1941 Carnegie granddaughter Margaret Wright complained to a forestry manager: "We all hope lumbering is not leaving such dreadful places as it did before and that they are really sticking to dead and infected trees. We are so glad you are there to check on them, as we were very sad over the way it was done last time."[26]

By 1954 reforestation had increased the land dedicated to commercial pine to more than 5,000 acres. Slash and loblolly were the most common species, with a scattering of longleaf pines in the northern portion of the island. Major cuts had taken place in 1939 and 1946, and another was under way. These timber harvests left numerous stumps throughout the logged land. Trust manager Leo Larkin of the Peoples First National Bank in Pittsburgh sought to sell the stumpage, but it had deteriorated too much, according to a sorrowful note from the Hercules Powder Company.[27]

Meanwhile, as the Darcy Lumber Company conducted a 1953–54 cut for both lumber and pulp, the Georgia Forestry Commission issued an ominous report. Its research on Cumberland Island showed poor growth of new timber in cutover areas over the previous two decades. Any future for a sustainable timber operation on the island would require burning nearly all of the land and replanting pine that would mature in twenty-five to thirty years. The financial situation on Cumberland Island had become precarious, and the Carnegie heirs faced hard choices about balancing their need for income with their desire to preserve the land's character and beauty.

Wildlife presented another group of resources for exploitation on Cumberland Island. To the heirs the most consistent and seemingly inexhaustible of these were the feral pigs. Throughout their years of owning the island, the family hunted pigs for their own use as well as for sport. In addition, they allowed commercial hunters to take pigs for a percentage of the sales. As early as 1928 Frank E. Dennis of Jacksonville sold seventy-four island hogs for $8.60 each. Later Carnegie grandchild Lucy Ferguson and others penned pigs and fed them for sale while at the same time releasing better breeds into the forest to improve the feral stock.

By 1954, however, the efficiency of Ferguson foreman J. B. Peeples reduced the number of pigs to a point where hunting them was no longer cost-effective. Ironically, only a few years earlier the Carnegie estate manager had complained that too many pigs were destroying the forage for deer and other more desirable animals. Apparently, even such a prolific and pernicious pest as the wild pig could be reduced to economic uselessness in the limited island ecosystem.[28]

Occasionally other animals were sought. Pig hunters took deer also, although this was sporadic and not very remunerative. Estate employees rounded up horses for sale on rare occasions while introducing stock from the mainland to improve the quality of the island herds. Commercial shellfishing was a brief possibility, but a sudden depletion of marine life around the island in the late 1930s ended that plan. In 1937 Thomas II tried unsuccessfully to get an injunction against a Fernandina, Florida, pulp mill, claiming that its pollution had caused the decrease of marine life. One useful, if not directly profitable, enterprise was the practice of allowing the sheriff and other Camden County officials to seine and fish the island waters.[29]

Raising cattle was the most successful animal operation tried on Cumberland Island. Like pigs, cattle had grazed the island for nearly 200 years. They furnished a ready source of extra income for Lucy Carnegie throughout her life. In 1951, as the cost of maintaining the estate spiraled upward, Lucy Ferguson and her husband, Robert W. Ferguson, submitted a plan to the other heirs to turn the entire island into a cattle ranch. The Fergusons pointed out that they had purchased 10 Hereford cows and a bull for $2,500 in 1945. After six years they had sold 76 cattle for $10,000 and had a herd of 100 cows and calves plus 2 bulls worth a total of $30,000. They suggested that Cumberland Island could support 1,000 cattle if they cleared the island fields of vegetation and planted legumes or new grasses such as the centipede grass they had introduced at Stafford Field. The Ferguson proposal required each branch of the Carnegie family to pay a share of the costs for additional cattle and the labor necessary to modify the island and work on the ranch.[30]

Initially the other heirs responded positively. They unanimously agreed to further investigate the "cattle plan." Lucy and Robert W. Ferguson suggested that each of the five family groups select a portion of the upland acreage for their home and recreational use while the remaining 10,000 acres would be open range. After citing the various financial and tax benefits of the plan, the Fergusons warned "this is probably the last chance the family will have to put a feasible plan into operation that will enable them to hold the property together."[31]

Despite these exhortations, unanimity was impossible to preserve. The heirs were now a widely dispersed group of people, and many only knew their relatives from occasional meetings or short visits to the island. Some family members wanted to sell the island outright, which the Fergusons called illegal under the trust arrangement. Technically it was not, but it too would have required unanimity. Other heirs favored different forms of land use, including the rising possibility of mining. Still others, such as the wealthy Nancy Carnegie Rockefeller, did not want to relinquish the island's traditional role as family playground.

In their frustration, Lucy and Robert W. Ferguson questioned bank trustee Robert D. Ferguson about inviting outside investment in the project. He advised against this step as long as the trust remained in effect. Eventually the grand "cattle plan" faded away, although cattle operations continued as a mainstay for Lucy Ferguson until the advent of the national seashore. A 1958 real estate appraisal, sponsored by the National Park Service, estimated between 300 and 400 cattle roamed Cumberland Island.[32]

The impact of the cattle on the island's landscape is still a matter of study and debate. The most obvious features are the infrastructure of the cattle business. Fences and cattle dips are found at several sites north of the Ferguson estate at Greyfield. Of more significance but less certainty are the lasting biological effects. Research sponsored by the National Park Service suggests that overgrazing by cattle contributed to open forests, destabilization of sand dunes near the ocean beach, and intense competition for forage that impacted the populations of deer and feral horses. Not only does this mean that cattle changed the biotic communities of Cumberland Island, but that they have made assessment of horse and deer carrying capacities more difficult.[33]

Crop Experimentation

With so much land at their disposal, it was only natural that the family members considered a variety of crops for their profit potential. They also showed an unusual willingness to experiment. Before 1916 the estate grew oranges and olives for sale as well as vegetables and other fruit for home consumption. However, both commercial crops disappeared by the early twentieth century.

After the death of Lucy Carnegie, Thomas II and various estate managers sought information on or experimentally planted several other citrus fruits, tobacco, indigo, sweet potatoes, taro, figs, and chayote, a tropical American vine that produces a pear-shaped fruit. During the late 1940s and early 1950s, the issue of sea island cotton reappeared, but a flurry of hopeful correspondence failed to return the old mainstay to Cumberland Island.[34]

Of all the experimental crops, tung nuts seemed to hold the most promise. Oil from these nuts was widely imported from China for use in paint, caulking, and wood preservatives. Carter Carnegie owned several orchards and processing plants on the Florida mainland and encouraged Thomas II to try the crop on Cumberland Island. From the 1930s through the 1950s, estate managers on the island experimented with tung trees planted on the edges of Stafford Field and the field immediately northeast of the dock at Dungeness.[35] The trust realized small profits from the enterprise, usually in the hundreds of dollars per year. Eventually petroleum-based products eliminated the market for tung oil as well as these minimal profits. Despite their poor performance as a cash crop, tung trees did succeed as invasive exotics on Cumberland Island. Today they must be periodically removed by hand from their old areas of commercial production.[36]

Island Maintenance

As the Carnegie heirs struggled to find a profitable business for Cumberland Island, the cost of maintaining its elaborate and aging infrastructure soared. Furthermore, as the trust income declined, the estate had fewer and fewer employees to carry out that maintenance, operate the various agri-

cultural projects, oversee the logging and hunting activities, and cater to
the occasionally demanding whims of different Carnegie family members.
Roads, yards, gardens, cemeteries, and an airfield had to be cleared of brush
and fallen limbs. The Main Road periodically required dozens of truckloads
of shell from middens at Table Point. Wind and water damage to more than
100 homes, outbuildings, fences, wells, cisterns, and recreation facilities had
to be repaired. All the buildings required painting as often as every five
years. Four beach crossings through the dunes opposite the Dungeness,
Greyfield, Stafford, and Plum Orchard houses had to be "kept open at all
times." Several dozen vehicles and two substantial yachts had to be in working order. Electric generators, water lines, and various docks needed periodic inspection and repairs. Keeping residents, visitors, and employees in
supplies demanded daily trips to the mainland for groceries, ice, kerosene,
gasoline, and spare parts. Garbage had to be collected daily from all the
homes. Yet in 1960 a letter from bank trustee Edward Gray Jr. to new estate
manager John H. Stanley listed only thirteen full-time employees of the estate plus two employed by Lucy Rice at Stafford.[37]

In addition to the wages for full-time and occasional temporary or part-
time workers, the estate also faced more serious repairs requiring contract
specialists. Each building required periodic roof and foundation repairs.
Termite damage posed the most insidious and serious problem. Between
1948 and 1963 every structure on the island required treatment, sometimes
repeatedly. The unoccupied ones, chiefly the Dungeness mansion and the
Recreation House, received almost no maintenance and suffered accordingly. An evaluation of the estate in 1951 or 1952 reported that the mansion's
roof "is in very bad shape, most of the windows are out and in general
everything is falling apart." As for the Recreation House, "the roof in general leaks badly, it is in [a] very bad state . . . the hurricane [in 1949] broke
down the wood columns on the front porch, they were badly eaten away by
termites." The Chimneys house near Stafford Field needed a new floor, and
the Stafford house required major repairs to its basement, windowsills, and
pool house.[38] The fire that destroyed The Cottage in 1949 was almost a bless-
ing because termites had badly damaged it as well. Even the modest house
built on the site of that burned mansion in the early 1950s required major
termite treatment within ten years.

During his short and unpleasant tenure as estate manager, J. Pat Kelly
was exceptionally blunt about the condition of the island and the difficulty
of managing it under the unusual trust arrangement. He began his tenure

in 1949 by summing up his first impression: "I have never seen a place so completely run-down and dilapidated as this one is. This applies to buildings, grounds, furnishings and equipment. The labor is of a mediocre or inferior type, slow, lackadaisical and ignorant. Constant supervision is necessary at all times."[39]

His remaining correspondence during the year he managed the estate did not become any more positive or cheerful. When questioned about a sudden rise in payroll cost, he responded that "most of the increase was caused by the various members of the family coming to the Island and each and every one wanting this and that done prior to their arrival, and all wanting something by a given time and all at the same time." Still later he referred to his difficulties in pleasing the family members: "I of course realize that it is most difficult to please everybody, particularly when you have twelve of a different idea to please, therefore I have done the things that I thought would be of most value to the most people involved. For instance, Mrs. Rockefeller says, 'I am against a big garden.' Mrs. Perkins says, 'I think it is wonderful.' At any rate, or by any sense of evaluation, I have saved both of these heirs some money by having a large garden."[40]

Eventually the costs of satisfying their individual wants on the island as well as the maintenance of their particular homes necessarily shifted to the heirs themselves. Some of them were ill prepared to absorb the financial burden that had hitherto been borne by the trust. Despite this unpopular adjustment, the strain on estate resources continued.

The elaborate Dungeness estate was a particularly difficult burden. All attempts to maintain the gardens were abandoned by the early 1950s. Facing forbidding repair costs or the disintegration of the Dungeness mansion, either the family or trust officer Robert D. Ferguson requested a bid to demolish the house and remove the rubble from the island, hopefully for sale on the mainland. In 1952 Fernandina Beach contractors N. S. Hernandez and George T. Davis proposed several different plans for how to accomplish the work and estimated a price of $12,500 to do it. This evidently proved too much for no further action was taken.[41]

Eventually the matter of the Dungeness mansion was settled in an unexpected way. For years the estate, and especially Lucy Ferguson, had suffered increasing poaching and malicious vandalism. They responded by "hunting" the poachers to chase them off the island. After employee J. B. Peeples shot at one aggressive poacher, things came to a head. On the night of June 24–25, 1959, a fire erupted in the big house and completely destroyed it. Sev-

Fig. 2.8. The ruins of the Dungeness mansion after fire destroyed it in 1959. (National Park Service photograph by Richard Frear)

eral Florida men with a history of poaching were suspected, but no arrests were ever made. Thus the most magnificent and significant of the island's structures escaped a humiliating fate and went out in a blaze that could be seen for miles up and down the mainland coast (fig. 2.8).[42]

Strip-Mining Cumberland Island?

Finally, the efforts to make Cumberland Island self-sufficient drove the bank trustees and some family members toward an apparent solution that posed the greatest threat to the island's beauty and character in its long history. Pittsburgh trustee Robert D. Ferguson invited the National Lead Company to survey the island for useful minerals. In particular, it was to look for an ore called ilmenite, which contained titanium that could be used in the manufacture of paint. The ore was being mined successfully on nearby mainland areas of both Georgia and Florida, and the geologic prospects seemed promising for Cumberland Island too. National Lead Company officials visited the island in spring 1954. Over the next year the trusteeship passed to the First National Bank of Brunswick in Georgia. Thus in March 1955 National Lead approached the bank's president, A. M. Harris, to negotiate a mining lease.[43]

Harris immediately contacted six other mining companies, and two, Union Carbide and American Cyanimid, expressed interest in bidding. The three companies agreed to share the $46,000 cost of a series of comprehensive tests by the Humphrey Gold Company of Jacksonville. At an April 6, 1956, meeting, the heirs unanimously approved further testing but insisted on competitive bidding before they would approve a lease to strip-mine the island.[44]

Later that month Nancy Rockefeller asked that the American Smelting and Refining Company, for which her husband served as director, also be allowed to survey the island and bid. At the same time Coleman Johnston requested a similar opportunity for the Glidden Company of Cleveland, Ohio. Eventually, the original three companies agreed that all five should split the cost of the tests, which had ballooned to $85,000. When the test results came in, they showed that a 7,000-acre segment along the central spine of the island contained economically worthwhile ore with a little over 3,000 acres of exceptional quality (map 2.2). Each of the five mining companies received a copy of the report, and the Carnegie descendants also secured one.[45]

On December 1, 1956, National Lead, American Smelting and Refining, and Glidden submitted bids to strip-mine Cumberland Island. National Lead's bid was not competitive. American Smelting and Refining offered $2 million for the outright purchase of all Carnegie holdings on the island, a plan that probably would have been impossible to achieve under the stipulations of the trust and the inability of the heirs to agree.

Glidden, however, made an extraordinary offer. For a twenty-year lease it would pay $1.20 per ton of titanium-bearing ilmenite plus 10 percent of the value of any other minerals it recovered. Furthermore, it would guarantee the heirs $2,225,000 even if it did not carry out the mining operation. In a later court hearing, mining experts testified that the family should easily realize twice that much over the span of the contract. In addition to the financial rewards, the heirs would benefit from free electricity brought to the island for the mining operation and free use of Glidden's boat service to and from the mainland. Glidden president Dwight Joyce called the offer "the highest I have ever heard of." Most mining experts agreed.[46]

This rich offer would have solved the various heirs' problems of main-

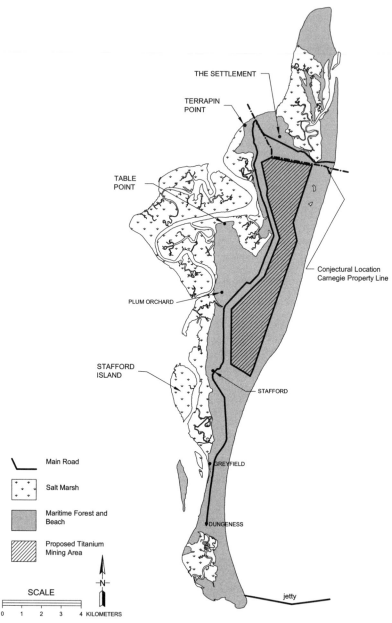

THE SETTLEMENT

TERRAPIN
POINT

TABLE
POINT

Conjectural Location
Carnegie Property Line

PLUM ORCHARD

STAFFORD
ISLAND

STAFFORD

GREYFIELD

DUNGENESS

jetty

Main Road

Salt Marsh

Maritime Forest and
Beach

Proposed Titanium
Mining Area

SCALE

—N—

0   1   2   3   4 KILOMETERS

Map 2.2. The titanium mining proposal offered by the Glidden Company. (Data from Robert M. McKey, 1958, "Appraisal: Cumberland Island Properties and Little Cumberland Island, Camden County, Georgia," National Records Center, Denver, Colo., acc. no. 079-97-0009, box 3)

taining the island. However, their reaction was not universally favorable. Indeed, there were two different negative responses. Carter Carnegie led one contingent of heirs who belatedly tried to introduce another bidder. Apparently he had taken the family's copy of the geology test results and approached a San Francisco firm, the American Exploration and Mining Company. This group suddenly offered to buy 10,764 acres of Carnegie land for a maximum price of $9.7 million, to be paid as $1 million at the outset and the rest in eighteen annual installments. This plan would have left the heirs approximately 1,500 upland acres in the four primary housing areas on which they could continue their island residency. However, the company reserved the right to cancel the contract after paying a minimum of $2.5 million.[47] The San Francisco company may have thought it could strip-mine the valuable area in two or three years and then cancel the contract. This would allow it to secure the resources for less than Glidden offered.

Trustee Harris of the Brunswick bank and the Glidden Company reacted with outrage to this belated offer and to the lawsuit brought by the Carter Carnegie group to reopen the bidding. Harris testified that the San Francisco proposal and the lawsuit broke faith with the companies that had shared the cost of the tests, bid competitively among themselves, and provided a copy of the geology report to the family in an effort to be fair. He urged other mining companies to "blacklist" American Exploration and Mining Company.[48]

A second group of Carnegie heirs was appalled by the potential impact of the proposed Glidden project on the island's landscape. The company planned a $9 million installation on the island. The *Brunswick News* reported, "They would employ 100 persons, pay $37,500 rental on a 150-acre village site, build 70 residences, fence the area to keep employees from trespassing on the rest of the estate, pay up to $50 [presumably per acre] to reforest the mined section and shape the land into lakes or terraces as desired." Three of Lucy Carnegie's grandchildren, Nancy Rockefeller, Lucy Rice, and Margaret Wright, filed suit to block the destructive project. Nancy Rockefeller, the wealthiest of the heirs, bankrolled the legal fight.[49]

Camden County Superior Court judge Douglas F. Thomas heard the cases during two intense days in April 1957. According to the *Brunswick News*, heirs who supported the Glidden plan included the last survivor of the original children, Florence Carnegie Perkins, and the family's only full-time island resident, Lucy Ferguson. After acrimonious legal presentations, Judge Thomas ruled in favor of Glidden and trustee Harris. Immediately the other factions filed appeals that the Georgia Supreme Court rejected the following September. Disappointed heirs again appealed and filed a new lawsuit in federal court to end the trust arrangement despite the fact that one of Lucy Carnegie's children was still living. The Glidden Company agreed to delay mining operations through the rest of 1957 when it hoped the latest round of suits would be concluded.[50]

Two events finally derailed Glidden's plans to strip-mine Cumberland Island. First, one of Nancy Rockefeller's appeals succeeded. Glidden and trustee Harris had worked out the lease for twenty years. However, the last child, Florence, was seventy-eight years old and in poor health. She could not be expected to live another twenty years. The trust, a party to the contract, would cease to exist at her death. Hence, the contract was ruled invalid.

The second factor was a sudden, precipitous drop in the world price of titanium. Not only did it make the Cumberland Island project uneconomical, but it forced Glidden and other companies to halt their mainland operations as well. Thus ended the most serious threat to the island's resources, serenity, and potential as a park. However, the experience created tension between the Carnegie heirs that would influence later decisions about the island, especially its future as a unit of the national park system.[51]

Property on the Island

Other opportunities to alter the island's character and future also arose from time to time. Various individuals and agencies approached the Carnegie family seeking property on Cumberland Island on a regular basis. In 1917 the U.S. Navy inquired about the south end of the island as a site for a "naval aircraft station." Estate manager William Page responded with a detailed letter in which he described open areas, potential airstrip sites, and the positive aspects of the island for such a use. However, the navy did not pursue this option any further. Between 1923 and 1924 Thomas II corresponded with land agent A. A. Ainsworth of New York about the possible purchase of land by the latter's client. Subsequently, Ainsworth sought to purchase the entire island, hoping to ride the speculative wave of the "Florida boom." He hoped to build a major development and cited Coral Gables, Florida, as an example of what he had in mind. In this early and unsuccessful bid for a resort, he presaged later Carnegie nemesis Charles Fraser. During the depression years Cumberland Island was briefly considered as a location for a Civilian Conservation Corps camp. However, the lack of a causeway to the mainland sank this idea.[52]

In 1954 real estate agent Esther Angwin wrote to estate manager H. H. Sloss about a client interested in land on the island and intimated that many more would follow. Other private individuals and agents continued to approach the family throughout the life of the trust. To each of these propositions, the heirs cited the wording of Lucy Carnegie's will and the conditions of the trust as justification to decline. However, neither the Carnegie family nor the trust could stop the U.S. Army Corps of Engineers from taking 518 acres at the southwestern end of the island in the mid-1950s. The corps claimed it needed the land to dump spoils from dredging the Intracoastal Waterway. After condemning the land, the agency deposited the spoils that then blocked the Beach Creek entrance to the Dungeness area by all but the smallest boats. This forced other craft to enter by a much lengthier and more difficult route closer to the ocean side of the island. Subsequently, the Brunswick trustee unsuccessfully sued to have the government dredge the opening to Beach Creek because it was the only safe all-weather anchorage for the family yacht.[53]

Later the U.S. Air Force and the National Aeronautics and Space Administration (NASA) considered Cumberland Island as a launch site for the early space program. In 1961 the two agencies studied seven sites and quickly eliminated all but Cape Canaveral, Florida, and Cumberland Island. Although NASA favored the Georgia island, the air force lobbied successfully for the Florida site adjacent to its air base. Had NASA succeeded in locating on Cumberland Island, the National Park Service could not have gained, nor the Carnegies maintained, any presence there.[54]

Ironically, at one point during the long, grinding decline of the family fortune and the frequent tantalizing offers for all or parts of the island, the Carnegie heirs actually sought to increase their holdings. In 1930, when the hunting club at the north end of the island offered to sell its property, the family trust offered $25,000 for it. The offer was politely declined, and a counterproposal was made for $45,000. Nothing more came of this last effort to bring the entire island under Carnegie ownership.[55]

Life on Cumberland Island

The lifestyle during the early years of the Carnegie era was an idyllic one filled with a mix of exciting, sometimes dangerous outdoor activities and all

Fig. 2.9. Picnics at the beach were a favorite activity of the Carnegies. (Photograph by Joe Graves ca. 1934)

the comforts that wealth and servants could provide. The former included hunting, fishing and crabbing, horseback riding, and swimming in the lakes, creeks, ocean, and pools. Roping sea turtles to small wagons for "races," dragging alligators out of the water, and baiting rattlesnakes were less approved excitements.

Investigating the daily workings of the estate was an enjoyable activity for children. They watched the carpenters, dairymen, gardeners, and stablemen, often interfering with the work. Picnics at the beach, attended by liveried servants, were always popular (figs. 2.9 and 2.10). Exploring the vast estate was a never-ending source of fascination. Children enjoyed its many environments, abundant wildlife, and, for most of the years of the trust, the exciting possibility of exploring the closed and vaguely forbidding Dungeness mansion. Thomas Carnegie IV later recalled: "The island was a young boy's paradise. It welcomed everybody and rejected nobody. It was a private world, running on a timetable all its own. Though I was often alone on Cumberland Island, I was never lonely."[56]

More formal outdoor recreation included golf at the course William Carnegie built at Stafford Field, skeet shooting at Old House Field, tennis and squash at the Recreation House, and croquet. Sedentary activities included dances, masquerades, card games, and singing. Nancy Rockefeller

Fig. 2.10. The fun, family, and excitement of Cumberland Island created a deep love of place for the Carnegie children generation after generation. (Photograph by Joe Graves ca. 1934)

recalled a "playhouse" built for the children's amusement and the fun she had with favorite cousin and playmate Thomas Carnegie III climbing trees and buildings and playing impish tricks. Meals were often elaborate affairs, and dinner was always formal.

Guests were frequent and sometimes stayed for lengthy periods. Later Cumberland Island became a place to socialize between siblings and cousins who had dispersed all over the United States and even to Europe. Throughout the Carnegie years a pleasant pastime was visiting the other island mansions when their residents were present. Parties of every sort took place. Friendships, romances, and marriages began on the island.

Life on Cumberland wasn't all rosy. Children attended school or tutorials on the island. There were duties to perform, chief among them organizing all the social and recreational events, overseeing the care of the households, and dealing with the nearly 200 employees the estate required during the golden age of "Mama" Lucy Carnegie. During the difficult trust period, the contraction of employment and services meant visiting heirs had to perform more and more of the utilitarian tasks themselves.[57]

Bad things happened too. Broken bones, gunshot wounds, and a fair sprinkling of deaths occurred on the island. Illnesses, occasionally serious, meant some Carnegie family members spent their days on the island in pain and weakness. For many, Cumberland Island became their final resting place, primarily at the family cemetery near Dungeness. Each burial there further cemented the generational tie to the island. Those ties are echoed in the words of Lucy C. S. Foster, great-granddaughter of Thomas and Lucy Carnegie, who recalled:

> Cumberland Island is far from the maddening crowd . . . it's where my roots are. . . . It was . . . smelling the land way before the "Dungeness" [the yacht] docked . . . galloping ponies . . . dashing down to Greyfield to sit on the red settee to listen to the guests as they finished dinner. It was the music of the different dinner gongs . . . riding sea turtles through the surf . . . swinging on the rings at Dungeness Pool. It was sneaking up on ducks with my son at Lake Retta; throwing the net with Nate [a servant] for mullet house parties. It was the smell of magnolias; rattlesnakes being skinned; pool-hopping sprees; buggy rides at Plum; dashing for Easter eggs at the beach . . . a treasure of memories, lives of children, burying grandparents, family and friends, then the thrill of holding grandchildren to begin again.[58]

James "Pebble" Rockefeller agreed: "Cumberland was, is, and will always be like that—placing her mark on so many of us who have spent time exploring her many worlds, or meeting kindred spirits who have been attracted to her shores. Her symbiosis of beauty, nature, and people make her an island to herself."[59]

The North End

While the Carnegie era unfolded on the southern part of Cumberland Island, the northern tenth evolved in a very different way. The tourism business, begun during the antebellum period, continued to enjoy a modest success after the Civil War. The Clubb family sold its hotel, the Oriental House, to M. T. Burbank in 1881.[60] He enlarged the operation and changed the name to the High Point Hotel. Meanwhile, on adjacent properties, both George W. Benson and William R. Bunkley began similar businesses. The Bunkley House at the site of the present High Point Compound was such a success that in 1883 its owner began constructing a horse-drawn tramway from the Cumberland Island Wharf to his hotel. Later it would be extended to the ocean beach. During the rest of the 1880s, Bunkley built cottages, adding more rooms to those in the two-story hotel.[61]

Fig. 2.11. The Cumberland Island Hotel at the north end of the island ca. 1898

Despite this successful run, in 1890 William R. Bunkley decided to devote more attention to his properties and businesses on the mainland. He first set aside 30 acres for his children and for a Methodist church for whites. Then he sold some 1,000 acres of highland, 600 acres of marsh, and the ho-tel to the Macon Company. In 1891 the new owners reopened the resort as the Cumberland Island Hotel to banner newspaper notices. Over the next decade the hotel enjoyed its busiest time (fig. 2.11). Many visitors came with conventions, the largest of which was the Georgia Teachers Association. Some groups brought nearly 150 guests at a time and stayed for up to two weeks.[62]

This apparent success, however, did not forestall financial problems for the Macon Company. In 1901 the company was forced to turn over the ho-tel to creditors in Macon, Georgia. That group in turn lost the complex to the estate of William R. Bunkley, holders of the mortgage. Robert Bunkley, like his father, preferred to live on the mainland, so he leased the hotel to L. A. Miller. Miller chose to operate the hotel more as a year-around hunt-ing and fishing lodge than a site for teachers' conventions and genteel vaca-tions. Eventually Robert Bunkley reassumed management, sold the hotel, bought it back again, and in 1921 sold it once more as a hunting lodge to a group known as the Cumberland Island Club. This was not an "arm's length" sale because both Bunkley and L. A. Miller were members of the club. A few years later Atlanta businessman Howard Candler Sr. and his son joined the club. When it faltered financially in 1930, Candler bought the High Point property and buildings for a family retreat.[63]

Another planned development at the north end might have significantly altered the island's future. In 1891 a group called the High Point Cumberland Island Company filed a plan to subdivide property it had acquired a decade earlier near the site of Fort St. Andrew. This ambitious scheme would have divided the property into 125 blocks, most of them in a tight grid pattern, and sold 50-by-150-foot lots in "the coming Saratoga of the New South" (map 2.3). The plan also included a large hotel and a 2,000-acre "game and hunting park." The owners auctioned lots from July 9 through July 11, but apparently there were too few takers to enable the scheme to go forward.[64]

One sequence of events that did result in more settlement began in the 1890s. Island lore holds that the dispossessed slaves of Stafford and other plantations wandered to the north end of Cumberland Island where they eked out a miserable living. In 1890 M. T. Burbank acquired 5 acres along "Old Clubb Road," subdivided them into 50-by-100-foot lots, and sold them to former slaves in the area. Among the buyers were Charlie Trimmings, William Alberty, Quash Merrou, Morgan Hogendof, and Primus Mitchell (fig. 2.12). Some were able to afford more than one lot. Three years later the new owners set aside part of this land to be the site of the First African Baptist Church. In 1937 the residents of the tiny community tore down the original log structure and built the present church with wood from a house razed on the High Point property. It is widely believed that Burbank and others established this subdivision, colloquially called the "Settlement," in order to secure a pool of laborers for their hotel businesses. The tiny hamlet would later become the focus of private land acquisition and historic preservation controversy.[65]

During the half century after their purchase of High Point, the Candlers further developed their estate into an idyllic retreat, purchasing many small properties from their former neighbors. They carefully preserved the hotel and some other buildings, erected new ones, and turned the old tramway into High Point Road. They too became entranced by the island and developed a deep attachment that spans generations. The rest of the north end suffered depopulation as its residents died or moved to the mainland for more economic opportunity. The 1930s depression drove off most of the white settlers. A few African Americans remained as employees of the

Map 2.3. The proposed subdivision in 1891 on the north end of the island by the High Point Cumberland Island Company. Note that the Cumberland River and north are at the bottom of the map. The map is part of a brochure advertising a land auction for the lots. (Georgia Archives, acc. no. 69-501, ser. 11, folder 11-1-010)

Fig. 2.12. Slave descendant Primus Mitchell at his home in the Settlement at the north end of the island

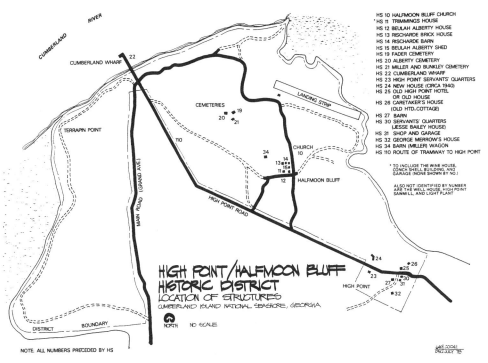

Map 2.4. The High Point–Half Moon Bluff Historic District. (National Park Service, Denver Service Center, July 1978, Map 640/20041)

Candlers. However, the Settlement too decayed under the onslaught of abandonment, storms, and termites. By the time the National Park Service arrived in 1972, only four unoccupied houses still stood.[66]

The history of the north end contrasts with that of the rest of the island. The Carnegie lands consolidated into a unified and palatial estate and then suffered a declining fortune and a legal inability to divide or develop the land. Meanwhile, the north end began as a variety of parcels, many with commercial tourism operations, that the Candlers consolidated into another elaborate estate. One northern parcel became a poor settlement for former slaves and their dependents, while others lay dormant, their owners having vacated the island for better, if not richer, lives on the mainland (map 2.4).

The End of the Trust and the Division of Cumberland Island

Eventually the trust would end, and the grandchildren of Thomas and Lucy Carnegie pondered that event. When Andrew Carnegie II relinquished administration of the trust in 1946, it spurred a lot of soul-searching about the long-term future of Cumberland Island. Pittsburgh trustee Robert D. Ferguson wanted to pursue profit-making activities aggressively. Some family members worried about the effect of such activities on their island retreat. The ultimate issue was whether the family could keep the island estate together when the trust ended by forming a family corporation. The alternative would be division of the property, which would open the door to piecemeal sales.

In 1944 Oliver Ricketson II addressed the possibilities in an exhaustive twenty-one-page letter to the rest of the heirs. He began by summarizing the five branches of the family, their attachments to the various remaining homes, and his estimate of what they would want to do with the island if given the choice. Then he listed his reasons for believing that retention of the island would not work. First, Ricketson predicted that unanimity among the fourteen grandchildren was impossible. In fact, he doubted whether a simple majority would opt to keep the island.[67]

A second reason was the possibility that putting all the island holdings on the market at one time might coincide with an economic downturn, resulting in a low price. He suggested that dividing the island would allow those who could afford it to wait for the best time to sell. Furthermore, he suspected that such a "great diversity of interests, finances, and wishes among so many related heirs might result in bitter litigation" if they still owned the island in common. Finally, Ricketson pointed out that those who wanted to continue to live on the island could be "forced to move against their will and without adequate previous preparation for a change in residence." Ricketson's desire to act was reinforced that same year by the deaths of Carnegie sons William, who was not part of the trust, and Thomas II. He urged that a division of the island take place immediately so that each heir would know which parcel he or she would have and could plan for its eventual disposition.[68]

Despite Ricketson's lengthy evaluation and strenuous urgings, too many heirs still wanted to keep the estate together. They feared that a division might reduce each individual's access to other parts of the island, especially those who only visited occasionally. The only action the heirs took toward eventual division was to commission a careful survey by the Watts Engineering Company. After the survey Watts produced a large (48 by 96 inches) map of the island. It showed land use, natural vegetation categories, and buildings and other infrastructure as well as base points that could be used in a later division. The Watts map became a pivotal tool when the trust did finally end.[69]

After the divisive mining controversy, only one original Carnegie child remained alive. In 1959 Florence Carnegie Perkins was nearly eighty years old, in failing health, and clearly near the end of her life. The end of the trust had to be faced. Finally the family formed a corporation, the Cumberland Island Company. Florence died on April 15, 1962. Within forty-eight hours company president Coleman Perkins notified banks, businesses, and the press that the trust had ended and the Cumberland Island Company would assume control of the island and other assets. In reality, the Brunswick bank did not officially end its trusteeship until February 9, 1963.[70]

During the three years before Florence's death, the family worked out a plan to divide the island among its five branches. Each branch could then further divide its land among its members. The island was to be divided into ten segments, five in the narrow southern portion of the island where most of the structures were located and five in the wider, less developed northern portion. Each family branch received one southern segment and one northern one. The acreages and land values were theoretically equal. The surviving homes, Greyfield, Stafford, Plum Orchard, the new Cottage, and The Grange (a small house near the Dungeness mansion ruins) would go with their segments to the branches that occupied them at that time. All heirs were entitled to use the entire Main Road, the Dungeness Dock and its surrounding area, and the family cemetery.[71]

A few problems arose in the divisions within each branch. Nancy Rockefeller of the Andrew II branch gave up her claim to Stafford to preserve amity with her sister, Lucy Rice. In some cases the grandchildren had died, and subdivision among their heirs was necessary. In the Ricketson branch granddaughter Lucy Ferguson received one-half of the holdings in the northern and southern segments. Her deceased brother, Oliver Ricketson II, left two daughters and a son. Oliver III, Mary Bullard, and Margaret Sprague inherited their father's half in common but later divided both their northern and southern segments into thirds. In the Thomas II branch, both of his sons, Thomas III and Carter Carnegie, had died, the former leaving two sons and the latter one adopted son. Some members of the corporation sought to exclude the adopted son, Henry Carter Carnegie. He sued to be included and won the right to his father's tenth in Camden County Superior Court in July 1962. Judge Douglas Thomas, the same one who had presided over the titanium case, ruled that the trust was established under

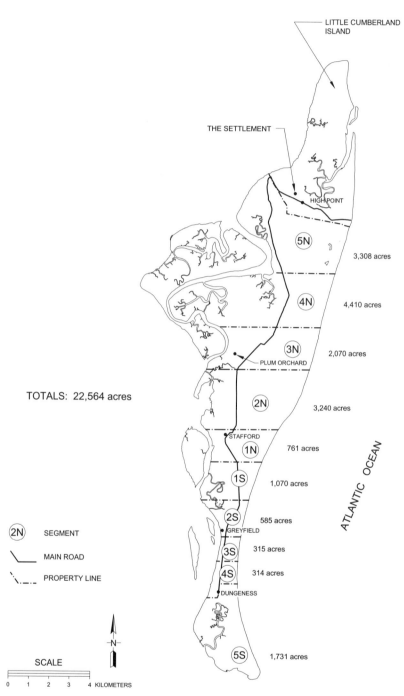

LITTLE CUMBERLAND
ISLAND

THE SETTLEMENT

HIGH POINT

(5N)   3,308 acres

(4N)   4,410 acres

(3N)   2,070 acres
PLUM ORCHARD

TOTALS: 22,564 acres

(2N)   3,240 acres

STAFFORD

(1N)   761 acres

(1S)   1,070 acres

(2S)   585 acres
GREYFIELD

(3S)   315 acres

(4S)   314 acres
DUNGENESS

ATLANTIC OCEAN

(2N)   SEGMENT

\   MAIN ROAD

`\._   PROPERTY LINE

(5S)   1,731 acres

SCALE

-N-

0   1   2   3   4 KILOMETERS

Map 2.5. The land division by the heirs of Lucy C. Carnegie. (Data from a map of the
same title, Georgia Archives, acc. no. 69-501, ser. 11, folder 11-1-009)

Pennsylvania law, which stipulated that adopted children have the same rights as natural ones. A final conflict arose when Thomas M. C. Johnston, who lived in France, refused to pay his share of the court costs and was sued by his relatives.[72]

Finally, in July 1964 the case for dividing the island went to Camden County Superior Court. The petition appeared to be an adversarial one with Mary Bullard and fourteen other heirs against Lucy Ferguson and two others. In fact this was merely a vehicle to get the case before the court. Judge W. D. Flexer readily accepted the family's division plan, complimenting them for their foresight. However, the judge did notify the family that the court would retain jurisdiction until it received a "detailed and appropriate [survey] description and easements" on the partitioned land. These were submitted in May 1965, and on June 1 of that year the final supplemental decree ended the era of a single 15,000-acre Carnegie estate (map 2.5).[73]

Once the division was final, the full diversity of opinion over the island's future quickly surfaced. Lucy Ferguson continued as always to live at Greyfield and ranch on the island. She, her family and guests, and her pigs and cattle roamed over the entire island as before. The Nancy Carnegie Johnston branch, led by Margaret Wright, entertained the National Park Service and sought its protection for the island. Coleman Perkins devised a family corporation, the Table Point Company, to carry out further logging operations in the northern part of the island. The Rockefellers built new homes on their segment and continued to use the island as a playground. The descendants of Lucy Rice did the same at Stafford. However, the less wealthy grandsons of Thomas II looked for ways to sell their holdings. Oliver Ricketson III, living in New Mexico, also cared little for the island and looked to sell. The glorious estate of Thomas and Lucy Carnegie would never again function as a unified and idyllic family retreat. Cumberland Island suddenly and jarringly faced opportunities and threats from every direction.

Creating
Cumberland
Island National
Seashore

During the later years of the Lucy Carnegie trust, the heirs considered a variety of options for the island's future: cattle ranching, hotel and recreation development, titanium mining, and outright sale to developers. Each of these decisions conflicted with a core belief that the futures of the Carnegie family and Cumberland Island were irrevocably linked. This attachment was by no means universal among the heirs, but it was a strong bond that crossed generations in the five family branches.

In the important meetings of the late 1940s and early 1950s, the heirs sought a way to maintain the island's character and their presence on it. These desires crystallized into an invitation to the National Park Service to consider Cumberland Island as a possible new park. The agency responded quickly and enthusiastically, initiating a nearly two-decade process that led to the creation of Cumberland Island National Seashore in 1972. As is typical with efforts to establish new units of the national park system, especially those where the federal government does not already own the land, the road to legislative establishment was littered with obstacles, unwelcome competition, and considerable division of opinion among all the people with a stake in Cumberland Island's future.

The National Park Service and Coastal Recreation

Congress established the National Park Service on August 25, 1916, to manage an aggregate of thirty-five national parks and monuments located primarily in the West.[1] The secretary of the interior chose two men to lead the young agency, the first director, Stephen Mather, and his lieutenant and successor, Horace Albright. They were ambitious and capable men who initiated policies and practices in the young agency that would last for decades. One of those policies was an aggressive attempt to broaden the system of park units throughout the United States. This effort led to a doubling of the number of park units and the establishment of a process to fill out the system with units to satisfy all American needs by the time Albright left the service in 1933.[2]

The first coastal units actually came in the weeks leading up to the National Park Service Act of 1916. First, President Woodrow Wilson created Sieur de Monts National Monument in Maine in early July 1916. Three years later Congress added more land and changed the name to Lafayette National Park. In 1929 the rugged coastal unit was again renamed, this time as Acadia National Park. On August 1, 1916, Wilson signed a bill establishing Hawaii National Park on the islands of Maui and Hawaii. In 1960 these were split into Haleakala and Hawaii National Parks, the latter renamed Hawaii Volcanoes National Park a year later. During the tenures of Mather and Albright as directors, two more units with coastal frontage entered the system: Katmai and Glacier Bay National Parks, both in distant Alaska.[3]

Each of these five coastal parks shared two characteristics. First, each was established to preserve and display extraordinary scenery and natural features, specifically coastal geology. Second, the satisfaction of the public's need for active coastal recreation was an insignificant factor in their establishment or management. This followed an early and persistent agency belief that inspiration and education, not active recreation, were the purposes of the national parks. The public, however, often had other ideas. Soon park managers met strong resistance in their evangelical efforts to inspire visitors and deny them access to common entertainments and amusements.[4]

Faced with an incessant demand and a persistent need for active recreation among the public, the Park Service took steps to encourage its provision elsewhere. Stephen Mather took the lead in the first conference on state parks held in Des Moines, Iowa, in 1921. The director and his national park superintendents believed the state parks offered an excellent opportunity for active recreation and allowed the national parks to be reserved for the deeper, more sophisticated purposes of inspiration and education. Throughout the 1920s the Park Service continued to promote state parks vigorously, providing training and expertise to their staffs and occasionally planning actual development. Hence the agency became the de facto leader in studying and coordinating the response to all the recreation needs of the country.[5]

The agency's leadership in recreation planning and its efforts to expand the system into the heavily settled East placed the Park Service squarely in the spotlight after the Great Depression began. A part of the wave of social thought and legislation that came with Franklin D. Roosevelt's election to the presidency was concern about the physical health and morale of the common man and woman. The new administration ordered the Park Service to conduct studies of the recreation needs of the nation. Later, Congress passed the Park, Parkway, and Recreational Area Study Act in 1936, which ordered the agency to compile a report that identified and ranked various areas according to their recreational value. Armed with funds from the Civilian Conservation Corps and other federal programs, the Park Service launched an array of studies across the United States. In each state it surveyed resources in all types of environments and ranked them according to national, state, or local significance. Those of national significance were supposed to contain significant scenic values worthy of visits by people from all over the country. Ultimately, the agency reported to Congress in 1941 with a summary entitled *A Study of the Park and Recreation Problem of the United States.* In laudable detail the Park Service provided data on a state-by-state basis. These activities led directly to the origins of many state park systems, especially in the South.[6]

As part of its charge to identify recreation areas of national significance, the Park Service focused on coasts. During the preceding half century, a majority of the land on the Atlantic, Gulf, Pacific, and Great Lakes coasts had been purchased by private owners, and public access had become worryingly scarce. At the same time, a day at the beach was clearly one of the most popular forms of recreation. Hence, early in its survey of recreation needs, the National Park Service conducted studies on the Atlantic, Gulf, and Pacific coasts. The latter never appeared as an organized report, but in early 1935 the agency sent a summary of its recommendations for the Atlantic and Gulf coasts to Secretary of the Interior Harold Ickes. The report identified ten sites on the Atlantic and two on the Gulf of Mexico. Among them were Hatteras Island in North Carolina and Sapelo Island in Georgia. In the case of Sapelo, park planners recommended an area of 44,100 acres with twelve miles of beachfront. They projected a total cost to the federal government of $1.3 million and estimated that 3,283,000 people lived within 200 miles, a number sufficient to justify its protection as a recreation site.[7]

Although the Park Service continued to monitor these places, little action resulted from the study. Only one of the areas, Cape Hatteras, received serious attention from Congress, which authorized it as a national seashore in 1937.[8] Even then, problems in land acquisition froze the new coastal park as an idea rather than a reality for fifteen years. Great Smoky Mountains, Shenandoah, and Mammoth Cave National Parks had all been authorized in 1926. However, Great Smoky Mountains took eight years, Shenandoah nine years, and Mammoth Cave fifteen years to establish fully. Land acquisition at each was expensive and controversial. Their legislative acts stipulated that the respective states had to acquire and donate most of the land. However, local farmers facing eviction quickly focused on the federal government as the source of their troubles.

The act authorizing Cape Hatteras National Seashore also contained this provision. However, the government of North Carolina had been burned by the Great Smoky Mountains fracas and was reluctant to start another land campaign. With the advent of World War II, all attention to new parks ended. After the war Congress had other concerns, and funding for the existing parks remained at the 1941 level. Ultimately, the Eisenhower administration and Congress supported a massive, decade-long federal program called Mission 66. It focused on improved infrastructure for visitors and employees.[9] Although the Park Service revitalized its interest in new parks, including coastal areas, there was little money to study or acquire them. Cape Hatteras continued to languish as an unrealized congressional idea.

Into this quagmire stepped the Old Dominion and Avalon Foundations of Andrew Mellon. A confidant of Andrew Carnegie, he supplied the money to purchase lands at Cape Hatteras. Meanwhile, the Park Service negotiated an agreement to exclude the local towns on Hatteras, Bodie, and Okracoke Islands from the seashore park and promised them full control of all tourist lodging and dining facilities. Cape Hatteras National Seashore was fully established in January 1953.[10]

With this agreeably wealthy ally, the National Park Service began to reconsider its search for coastal recreation areas. However, nearly two decades had passed since the original report on the Atlantic and Gulf coasts. Once again the Mellon foundations stepped forward to finance a resurvey of the eastern coasts and new surveys on the Pacific and Great Lakes coasts. Over the years since the first study, most of the original twelve proposed sites on the Atlantic and Gulf coasts had been developed or otherwise removed from the list of available lands. The new initiative charged Park Service officers to evaluate the remaining opportunities and identify those of national significance. One survey team member, Bill Everhart, had been part of the group that wrote the first master plan for Cape Hatteras. When he asked survey leader Al Edmunds how to recognize a site of national significance, the latter "smiled and said I'd have no trouble recognizing one when I saw it."[11]

Actually, two criteria were used: suitability and feasibility. The former meant the unit had to have sufficient size, grandeur, pristine quality, and content to be of significance to the entire nation. Feasibility referred to the ability of the Park Service to acquire the land, manage it without severe external or internal threats to the resources, and secure its establishment through local and national support and congressional action.

Carrying these sometimes vague and subjective values, the Park Service began the resurvey of the Atlantic and Gulf coasts in 1954. It culminated with a report entitled *Our Vanishing Shoreline*, issued in June 1955. The results of the study were stark. Of the 3,700 miles of shoreline along the Atlantic and Gulf coasts, federal and state governments preserved only 240 miles or 6.5 percent for recreation. More than half of the 240 miles were in Cape Hatteras National Seashore. Furthermore, of the twelve areas recommended in 1935, only one had been preserved while most of the remainder were "ghosts of departed opportunities."[12]

Of more immediate importance, the survey team recommended fifteen new sites and repeated its proposal for Padre Island, Texas. Two in particular stood out, Cape Cod, Massachusetts, and Cumberland Island, Georgia. Of the latter, the report stated: "Cumberland Island in southeast Georgia is considered by the survey to be the best of its type—the low-lying lands separated from the mainland by stretches of marsh and rivers or estuaries. . . . This 'sea island' is thought to contain practically all the desirable features for public enjoyment. . . . The possibilities of developing Cumberland Island for public recreation and cultural enjoyment are considered to be exceptional."[13]

Our Vanishing Shoreline was widely read, and its release accelerated the surveys of the remaining two coastal zones. The National Park Service issued reports on the Pacific and Great Lakes beaches in 1959. The public was disturbed by the loss of nearly all the Atlantic and Gulf areas proposed in 1935. These reports galvanized the campaign for recreation areas, particularly coastal ones. State agencies as well as the federal government sought public beach areas under a cloud of desperation. If they did not move quickly, various experts testified, there would be no coastal lands left for the public. Ironically, the report also served to identify the best areas for tourism development, as team member Howard Chapman had predicted while carrying out the Great Lakes survey.[14]

The National Park Service Comes to Cumberland Island

Before *Our Vanishing Shoreline* appeared, members of the Carnegie family approached the National Park Service. Of the five branches of the family, the Johnston group took the lead in encouraging the National Park Service to consider the island for park status and in exhorting the other heirs to pursue this conservation option. On June 9, 10, and 11, 1954, a Park Service survey team led by Al Edmunds conducted a "reconnaissance" of Cumberland Island by air and on the ground. Estate manager H. H. Sloss drove the members around the island. During the tour they met with a "Mr. Maury Johnson" (possibly T. M. C. Johnston or Marius Johnston Jr.)[15]

Four months later a follow-up team including Bill Everhart spent a day inspecting the island. Everhart later wrote, "Cumberland Island was an obvious choice [for national seashore status] from the time I jumped into the surf from a Coast Guard boat and waded ashore (I found a dead deer in a decaying mansion)."[16] On June 18, 1955, just as *Our Vanishing Shoreline* was issued, a high-level Park Service team including Director Conrad Wirth; senior officials Ronald F. Lee, Al Edmunds, and Ben Thompson; and Paul Mellon, trustee of the Andrew Mellon Foundation,[17] toured the island and met with T. M. C. Johnston (fig. 3.1). On the basis of glowing reports by all three visiting teams, the Advisory Board on National Parks, Historic Sites, Buildings, and Monuments resolved on September 9, 1955, to endorse the acquisition of Cumberland Island as a national park, citing it as one of two areas left on the Atlantic coast suitable for that purpose.[18]

Meanwhile, the state of Georgia also became interested in Cumberland Island as a recreation site, albeit a much more developed one. In 1947 the state had acquired Jekyll Island and turned the former rich man's retreat into a state park. However, the cost of leasing one of the 550 lots on that island was significantly higher than the average Georgian could afford. Cumberland Island came to the attention of the state legislature in January 1955 when it resolved to name the island's strand Griffen Beach to honor the incumbent governor. State representative John Odom of Camden County followed a week later with a proposal to establish a Cumberland Island Authority to acquire the larger island as a resort for the "average man."[19]

Various details of Odom's plan were subsequently explained in the

Fig. 3.1. A 1957 National Park Service survey team's vehicles sit among the live oaks near High Point.

Camden County Tribune.[20] A bridge terminating at Harriet's Bluff would connect the island to the mainland. There it would meet a road that Odom expected to be part of a transportation bill that awaited the governor's signature. In a lengthy quote Governor Samuel M. Griffen supported the idea of the road, the bridge, and a state park for Cumberland Island, but only if acquisition of the latter could be done by some sort of Cumberland Island Authority "without expenditure of state funds at this time." On February 17, 1955, the Georgia House of Representatives established the Cumberland Island Study Committee to determine the island's qualifications and feasibility for a state park or beach under these constraints.[21] The committee consisted of Odom and five other state legislators.

The next year the Cumberland Island Study Committee issued a *Report to the 1956 Session of the Georgia Assembly* on the island. The committee had cruised the coasts of the island, interviewed county officials and an advisory group appointed by Odom, and spoken to Robert W. Ferguson, "representing one of the heirs," and the Brunswick Bank and Trustee Company, trustees for the estate. According to Odom, Ferguson was "greatly in accord and in agreement that a road and bridge to Cumberland Island would be highly beneficial to the people of Georgia, as well as to the heirs of the Estate." There is no evidence that the committee consulted any contingent of the family other than the Fergusons, whose cattle plan had recently been dropped. Odom and his committee recommended that the legislature create a Cumberland Island Authority with six members, including one from Camden County. This authority would then work to obtain the land on the island, create a state park, develop the real estate, build roads "to and on the island," and "propagate it as a vacationland for all the people of Georgia and the U.S.A."[22]

These legislative activities occurred during the time when the family debated titanium mining on the island. That possibility dampened the fervor of the National Park Service for a national park on the island. Director Wirth wrote to Assistant Secretary of the Interior Wesley D'Ewart, "This kind of mining will result in the complete destruction of the surface vegetation and other surface features of the island since it is done by floating dredge that leaves nothing but the rejected sand in its wake."[23]

Nevertheless, the agency continued to investigate the island for a national seashore (figs. 3.2 and 3.3). A ready rapport quickly developed between Margaret Wright, the oldest of Nancy C. Johnston's children, and Park Service officials. In late March 1956 Director Wirth wrote to her and re-

Fig. 3.2. Part of the 1957 survey team at Lake Whitney. Future Park Service associate director Ben Thompson is the hatless member. Note the dune encroachment on the lake.

iterated the agency's interest in the island but warned that titanium mining was a severe threat. He then explained how such an area would be established as a national park unit, that land acquisition could occur through purchase, donation, or both, and that it might be possible for heirs to retain their residences for a specified period of years or for their lifetimes. He further suggested that these retained residences would be "best accomplished by mutually agreeing on a suitable area that could be used as a family retreat without interference from the public."[24]

Over the next several years, the Park Service maintained contact, primarily with the Johnston branch of the family, and bided its time. Park officials met with Margaret Wright and Nancy Rockefeller in Washington, D.C., as well as on Cumberland Island. Various park teams also discussed the potential national seashore with members of the Perkins and Ferguson branches, as well as with Howard Candler on the north end.

Surveyor Robert McKey conducted a preliminary appraisal of the island. He valued island property at $2,148,000. This was over 30 percent lower than the estimate by the state's Cumberland Island Study Committee.[25]

In the meantime, Odom's idea for a Cumberland Island Authority died, perhaps because of the titanium mining threat. In 1960 and 1961 Little Cumberland Island was sold twice and wound up in the hands of a group of conservation-minded investors who functioned as the Little Cumberland Island Association. This group planned to develop 200 to 300 residences and to maintain the rest of the island in a natural state.[26]

During 1959, as the Carnegie heirs established the Cumberland Island Company in preparation for the death of Florence Carnegie Perkins, the Park Service moved ahead with its plans to acquire seashore areas around the four coasts of the nation. On May 20, after receiving approval from the Bureau of the Budget and the Eisenhower administration, the agency prevailed upon Senator Richard Neuberger of Oregon to introduce S. 2010, a bill aimed at preserving "not more than three national seashore recreation

Fig. 3.3. The survey party among the ruins of the slave cabins of the Stafford plantation. The Stafford Chimneys archaeological site is now listed on the National Register of Historic Places.

areas." The bill left it to the Park Service to determine which three areas but restricted their acreage to a total of 100,000 acres and their cost to $15 million. Senator Neuberger had already introduced a bill to create Oregon Dunes National Seashore and clearly expected it to be one of the three. Section 3a of the bill noted that the land could be secured by donation or purchase and that the authority to acquire the lands "shall include authority to condemn." Subsequently, eighteen senators introduced a similar bill, S. 2460, that specifically mentioned Cumberland Island as one of the possible seashore areas.[27] Four similar bills also were introduced in the House of Representatives, most mentioning Cumberland Island.[28]

The response of the Carnegie heirs to these bills was one of concern. The chairman of the family's Cumberland Island Company wrote in their first newsletter: "While it is perhaps too early to determine what this will mean, the importance of the bills to the family is obvious. . . . It may be desirable for the family to act as a body rather than as widely scattered individuals. There has probably never been a time when a calm and united front has been more critically needed by the family than is the present case."[29]

On January 8, 1960, senior Park Service officials, including Chief Counsel Jackson E. Price and Assistant Director Ben Thompson, met in Washington, D.C., with officers of the Cumberland Island Company, including Coleman Perkins, Ferguson son-in-law Putnam B. McDowell, and Joseph C. Graves Jr., who represented the Johnston branch. The park officials explained what they planned for the island as a national park unit and that a national seashore was less restrictive and more flexible than a national park and could allow mining and hunting. The Cumberland Island group brought up other options including a wildlife sanctuary. Its members did not want the island to become a site of extensive hunting.

The size of the unit came under discussion as well. Ben Thompson noted that the National Park Service had suggested to bank trustee Harris that it would seek a national seashore, exclusive of mineral rights, north of a line through the mouth of Old House Creek (between the Greyfield and Stafford mansions). In response to questions from the Cumberland group, Thompson explained that the agency did not discount the value of the southern part of the island but thought that members of the family might want to keep homes there indefinitely, and this afforded them that opportunity. Joseph Graves responded that the Park Service was underestimating the willingness of the families to sell.

As for development, Park Service officials suggested that overnight accommodations, campgrounds, and access to the beach and other points of interest would be advisable. Places where food was provided also would be necessary. They reiterated their long-held desire to prevent the construction of a bridge and suggested that this would control the number of visitors, which island residents feared could become unmanageable.[30]

Throughout the discussion the delicate issue of land acquisition repeatedly arose. Coleman Perkins reported that most of the island people were against S. 2010 and any bill that gave the government the right to condemn land. Joseph Graves Jr. suggested that if the power to condemn existed, it would seriously reduce the bargaining position of the owners of Cumberland Island. Ben Thompson responded that the agency did not want to use the power as a club. Graves then said that these bills and some of the Park Service publicity had been inadvisable. He explained that it was hard for the family to view such bills in a friendly manner and that he and his relatives wanted to have extensive input in a bill for Cumberland Island.[31]

Thompson then referred to a bill to establish Cape Cod wherein the power of eminent domain was suspended in the towns where satisfactory zoning ordinances were adopted. He said that while Cumberland Island had no town government, this bill indicated a willingness of the Park Service to be flexible in the matters of negotiated purchases and retained life estates. After a few more details, the meeting concluded amicably. The Park Service resolved to meet with the heirs again and to work out the Cumberland Island project to satisfy everybody concerned.[32]

A New Dawn for Cumberland Island Negotiations

When the trust ended, members of the Johnston branch renewed their efforts to have the National Park Service acquire the island. In September 1963 Joseph Graves Jr. notified estate manager John Stanley that the Johnston family would sponsor a trip to Cumberland Island by Secretary of the Interior Stewart Udall and his aide Max Edwards. Soon local congressman J. Russell Tuten began to back the national seashore proposal and work to secure a visit by Udall. Finally, the secretary scheduled a visit to several of Georgia's Sea Islands for November 1965. The local media in Camden and adjacent Glynn Counties anticipated the secretary's visit with various stories and opinions about Cumberland Island's potential as a national seashore.[33]

On November 4 Udall first stopped at Brunswick Junior College and then at Fort Frederica National Monument on St. Simons Island. There he met Alfred W. "Bill" Jones, chairman of a resort organization called the Sea Island Company, who had been instrumental in the state's acquisition of Jekyll Island and was a tireless supporter of the Cumberland Island National Seashore proposal. From there Udall journeyed on to Cumberland Island accompanied by Congressman Tuten, Georgia state parks director Horace Caldwell, several area mayors and chamber of commerce officials, and a host of newsmen. On the island Lucy Ferguson gave the party an automobile tour, and Margaret Wright hosted lunch at Plum Orchard mansion. The secretary then continued on to St. Catherine's Island, which had also begun to interest the Park Service.[34]

On Cumberland, Secretary Udall noted that the island as a national seashore park would be a "tremendous asset for Georgia and the region." He added that his dream was to see state and national seashore preserves all along the Atlantic coast, and he praised Georgia for acquiring Jekyll Island. He warned, however, that development of national parks "does not come easily" and that the normal "gestation" period for acquiring park sites is two or three years, as had been the cases at Cape Cod and the recently acquired Fire Island National Seashore.[35] The secretary added: "All of us who are conservationists are indebted to these people [the Carnegies and Candlers] who have preserved this island . . . in a wonderful and unspoiled condition. Now there is a possibility of getting together on a conservation plan which would preserve this for all time." Udall promised to give life estates to those owners who wished to sell and to allow those who wished to retain their homes to do so provided they agreed to give the National Park Service scenic easements to prevent subdivision or property development. As for Park Service development, the secretary promised to "leave the maximum amount unspoiled that is possible to do. We would probably develop a ferry, have beach areas for intensive use, nature trails, camp sites, places for boating and fishing. Our main objective would be not to develop the island so intensively that we would spoil it."[36]

Trouble in Paradise

In spite of this highly successful meeting, the possibility of a Cumberland Island National Seashore in two or three years immediately faded. First, one heir sold a strip of land in segment 2S (see map 2.5) to developer Robert Davis, who subdivided and began selling home lots. Then, in a grievous blow to plans for an islandwide national seashore, Andrew Carnegie III, Thomas Carnegie IV, and Henry Carter Carnegie sold their lands in segments 5N and 4S to Charles Fraser, developer and owner of a major resort complex on Hilton Head in South Carolina. The price was $1,550,000, or about $500 per acre. Later, other family members stressed that "the Carnegie boys" had not wanted to sell but were forced to in order to secure money for education purposes.[37]

Fraser arrived at a fluid time in Cumberland Island's history. Of the four remaining family branches, the Johnston group vigorously supported the seashore idea, the Perkins group expressed interest, and Nancy C. Rockefeller and her sister, Lucy Rice, had adopted a friendly but "wait and see" attitude. The Ferguson group, however, insisted on maintaining their holdings as private property, although they were unresolved about the merits of a national seashore on the rest of the island. Little Cumberland Island was under development and apparently unattainable. Finally, the Candlers, like the Rockefellers, were interested in conservation but only if the rest of the island became a national seashore and if they could maintain their vacation presence on the island.

The sale of one-fifth of the Carnegie holdings to Charles Fraser disrupted the slow, gentle rhythm of negotiations. Fraser arrived on the island armed with a grand plan not just for his tracts but for the entire island. He was aggressive and had a take-charge attitude and a vision for Cumberland Island. Earlier he had developed the Sea Pines Plantation resort on Hilton Head as a low-density residential retreat for the well-to-do. The resort included areas of natural landscapes, carefully designed golf courses, and a variety of other recreational facilities. It was conspicuous for its suppression of tawdry shops and amusements, signs, and other unsightly features typical of most public beach areas.

The sensitivity of Fraser's development was widely lauded by the press and government officials. However, Sea Pines Plantation only commanded the southern 5,000 acres of Hilton Head. Subsequently, other entrepreneurs introduced those undesirable features to other portions of the island. Fraser vowed not to let this happen to Cumberland Island.[38] He came to Cumberland armed with an islandwide plan designed to eliminate any possibility of later intrusions by "billboards, large trailer parks, summer hoards of litterbugs, commercial strip developments, Ferris wheels, etc."[39]

He proposed two entities to manage the island. The first, Cumberland Island Holding Company, would be a profit-making corporation owned and operated by his Sea Pines Plantation Company. Its purpose would be to purchase land and construct unobtrusive and sophisticated private recreation communities. In his introductory letter to the island owners, he promised that the company would also conduct "master planning, . . . control architectural standards, engage in ecological research, and maintain support facilities such as a ferry, medical center, fire protection, airport, etc."[40] Later Fraser suggested that 150 homes might be appropriate for the island. The holding company would seek conservation leases from those owners unwilling to sell their land that would prevent them or later purchasers from developing lower-standard facilities. Fraser also promised to stabilize the ruins at Dungeness, rehabilitate the Recreation House, preserve other significant historic buildings and sites, and maintain some natural areas in an undisturbed state. For advice in the latter undertaking, the Cumberland Island Holding Company would consult with the Marine Institute on Sapelo Island, the Institute of Ecology at the University of Georgia, and the National Park Service.

The second entity in Fraser's grand plan was a tax-exempt, nonprofit corporation to be called The Cumberland Island Conservancy. Its purpose would be to accept either gifts of land in fee simple for permanent natural areas or open space easements from owners who wanted to ensure permanently that their land would not be developed. Fraser pointed out the sizable tax benefits to any land donor but stipulated that if the owner could not realize any tax benefits, the Cumberland Island Holding Company would donate $100 per acre to the Island Conservancy for acquisition and preservation of such lands.

Fraser clearly had evaluated the island and its owners carefully before proposing this grand scheme. He readily recognized their disdain for a causeway to the mainland by stipulating that transport to his resort would be by air or automobile ferry. He offered minority ownership in the two companies to the heirs and secured an agreement with Robert Davis to purchase his subdivision. Over and over he stressed the high quality of his operation and the facility with which the Carnegie heirs could maintain their island lifestyle under his plan. Finally, he stipulated that he would not develop his resort unless all the owners of land plots larger than 60 acres signed an agreement for scenic easements.[41]

At the same time he dangled these inducements in front of the heirs, Fraser vigorously sought approval from officials and the public in both Camden County and the state of Georgia. His aggressive campaign repeatedly reminded the public that his resort would be open to them and not just to a privileged few. Initially, Fraser ignored the evolving National Park Service campaign, but he later incorporated it into his own. He espoused a national park on part of the island with his resort forming a necessary and complementary partner. His development would serve the public as people arrived to see the beauties of the new seashore.

The various parties interested in Cumberland Island reacted in different ways to Charles Fraser and his island plan. The National Park Service could do nothing but pursue park status for as much of the island as it could get. Director George Hartzog met Fraser, and the two tried to accommodate each other. Officials in Camden County were ecstatic. They envisioned a substantial increase in tax revenues as well as vastly increased tourism in the county. Some years earlier the army had built a major supply base at Kings Bay and promised a monetary windfall for local governments and businesses. However, the military had not developed the base, and revenues for the county were minimal. Locals were pleased that a private corporation, not another untrustworthy federal agency, wanted to develop Cumberland Island.[42]

Environmental groups, led by the Georgia Conservancy and the local chapter of the Sierra Club, were slow to join the evolving debate over the future of Cumberland Island. However, after Fraser's plan became public, the Sierra Club actively opposed it. In January 1969 club representatives promised Vincent Ellis, the Park Service's lead agent, that they would do all they could to see that the island became a national seashore. True to their promise, the Sierra Club held a meeting in Atlanta called "Crisis at Cumberland" on February 13. Its newsletter, the *Georgia Sierran,* began to report routinely on the island negotiations and to call for support for a national seashore.[43]

Of greater importance, the remaining Carnegie heirs and the Candlers, owners of more than 85 percent of the island, were frightened and dismayed at this sudden external competition for an island they had held for more than eighty years. Coleman Perkins urged the Carnegie family to pull together and resist this most serious threat to their island. The heirs reacted with outrage to both Charles Fraser and his plans. Most responded that they wanted to be left alone and did not want the island developed. They called Fraser's plan a profit-oriented scheme that would ultimately destroy the fabric of nature and the solitude they enjoyed. Fraser's orderly and all-encompassing plan offended their proprietary sense. Putnam McDowell, son-in-law and business manager for Lucy Ferguson, found that the Fraser plan would "effectively eliminate any possibilities of selling the property except to Hilton Head."[44] They deeply resented his take-charge approach. A furious Nancy Carnegie Rockefeller summed up the Carnegie opinion in an emotional letter to Fraser after the developer had publicly suggested that the selfish heirs were unworthy to keep the island to themselves:

> What right have you to criticize the Carnegies and their 84 year stewartship of Cumberland Island. If it had not been for the four generations of Carnegies, the Island would not be in its 100% natural state. Not one of them destroyed a single thing in 84 years, nor allowed anyone else to, and *we are not going to*. Now, because of the almighty dollar, you want to desecrate this last outpost under the heading "Why should a few enjoy what belong[s] to the people . . . ." You were ill advised to have written as you did about our family to which each of us, naturally, takes exception. Who are you to twist untruths to benefit your plan?[45]

Candler heir William C. Warren III went a step further and published his evaluation of the proposals for Cumberland Island in the *Atlanta Journal.* After reviewing Fraser's plan and explaining that poor communication had allowed segments 5N and 4S to be sold to the developer rather than another island landowner, he delivered the bristling indignation of the island owners:

> Now, all at once, Camden County officials, Messrs. Fraser, [Congressman] Stuckey and others are pricking up their ears and all have the right answers. Some want a bridge, which I might add, would ruin the island. Some want cars and parking—another destroyer of nature. Who knows the answers?
>
> Past experience should be a consideration; present owners should be considered; and present owners should be heard above the voices of those who would put themselves in positions of attempting to dictate to those property owners over which they have no right or authority.
>
> We purchased our land years ago, own it and its meager improvements outright and now we are confronted with those who tell us what we should do and must do with our land in order to best serve our fellow man.
>
> This smells Red and makes one see red especially when, in effect, we haven't put up the billboards, hot dog stands, service stations and other eyesores that these so-called Cumberland Island experts are fearful of, nor do we intend at any time to ever erect any such man-made horrors.
>
> Let the sleeping beauty sleep![46]

All this antagonism from the island owners frustrated Charles Fraser, and he sought any and all means to gain control of the island. With strong local and commercial backing, he tried to run over the owners and develop his resort in spite of them. However, new-money Charles Fraser had not reckoned with the power that older money could bring to bear. A few years earlier Robert W. Ferguson had hired a young attorney, Thornton Morris, to represent the Ferguson interests on the island. By 1969, as the conflict became heated, Morris also represented the Perkins branch. In time he would become an important counselor for most of the Carnegie heirs and later even the Candlers. Morris coordinated the heirs' efforts to combat Fraser. He began by investigating Fraser's financial status and the backing he had for the Cumberland Island project. He informed the heirs that Fraser was stretched very thin and might not have sufficient funds to carry out the development he proposed.[47]

In the meantime, the property owners formed the Cumberland Island Conservation Association to handle the increasingly complex threats to their island stewardship. Morris reported any and all developments in the Cumberland Island affair to its executive committee. On April 18, 1969, he exposed Fraser's desire to reroute or widen the Main Road. The Carnegies and Candlers quickly used the 1964 court document dividing the island to block Fraser's road plan. Morris carefully followed the public campaign Fraser mounted through spring 1969. Fraser hired Dr. Hugh B. Masters, a specialist in outdoor recreation and administration, who lectured throughout southeast Georgia on the recreation potential of Cumberland Island. Masters cited a wide variety of possible sports and activities that would stem from Fraser's resort development.[48]

In mid-March 1969 Morris evaluated a series of statements by Fraser about the taxes paid to Camden County by Cumberland Island owners. He suggested that the developer's ploy might be to let his purchase of Cumberland Island land establish a new land value, pay the higher tax on his land, and then use his many allies on the mainland to bring legal action over the discrepancy between his tax payments and those of the other owners. This would have the effect of bringing a large tax increase and, possibly, bills for back taxes to the island owners and might force them to sell.[49]

Not all of the Georgia officials favored Fraser's plan for the island. The Georgia State Tourism Division, the Georgia Planning Department, the State Recreation Council, and the Coastal Area Planning and Development Commission all backed the national seashore plan. The *Atlanta Journal and Constitution* was solidly behind the national seashore project, as were most newspapers outside Camden County. The Sierra Club offered to help island owners withstand Fraser, and the Georgia Conservancy roundly criticized Fraser and his plans in the press and other public venues.[50]

Amid all the public and private acrimony over Cumberland Island's future, Charles Fraser dealt his most powerful card. Camden County's representative to the state legislature was Robert Harrison, a local attorney. On March 4, 1969, he introduced a bill in the Georgia House of Representatives that called for the establishment of a Camden County Recreational Authority. Among its other functions, it would be empowered to acquire land on Cumberland Island by condemnation. A sympathetic colleague submitted an identical bill to the Georgia Senate. Initially there seemed to be substantial support for the bills and for state rather than federal control of Cumberland Island.[51] Suddenly, the future of Cumberland Island as a retreat for the Carnegie and Candler heirs, as a national seashore and as a de facto wilderness, was in serious jeopardy. A week later another bill sought to expand the power of eminent domain of the North Georgia Mountains Authority to cover the entire state. The director of that authority was none other than Fraser's publicist, Dr. Hugh B. Masters.[52]

The National Park Service could do nothing but continue its steady correspondence with the owners and campaign for a national seashore. The environmental groups called in all their contacts and favors to defeat the bills. However, it was the island owners who beat this most serious challenge. Robert W. Ferguson had served in the Georgia legislature, and he and Lucy were well known and well liked by current members. Although Lucy had misgivings about a national seashore, she recognized that this bill could force the Fergusons off the island. Thornton Morris lobbied to defeat the pro-Fraser bills. Other members of the Carnegie family and especially Sam Candler, who was deeply involved in conservation and planning in Georgia, also gave battle.[53]

Only days after the initial Robert Harrison bill, Herbert Johnson, an attorney for T. M. C. Johnston, called a meeting with Sam Candler, William Warren III, Thornton Morris, four members of the Georgia House, two members of the Georgia Senate, the head of the State Game and Fish Department, several conservationists, and others interested in blocking Fraser. They decided to introduce a substitute bill that would authorize a study commission for the island. The commission would include members of the Carnegie and Candler families.[54] Ultimately this plan and the efforts of the island families and their friends and allies defeated the condemnation bills.

The state legislature established the study commission, and negotiations with the National Park Service continued. Charles Fraser, his plan for the island scuttled, tried to make the best use possible of his segments of the island. He informed the *Savannah Morning News* that he would sell his options on the island to the Park Service contingent upon their purchase of the Carnegie and Candler lands as well. He suggested that this might take some years and that in the meantime he would coordinate his planning with the Park Service and develop his land. He envisioned lodging and camping facilities and expected to be named as concessioner for the seashore when it was finally established.[55]

Working toward the National Seashore

As the battle against Charles Fraser unfolded, the National Park Service continued to negotiate and plan for a national seashore on the island. It needed the approval of the island owners, Camden County officials, local congressman Williamson S. "Bill" Stuckey, the environmental community, the state of Georgia, and, ultimately, Congress and the president. Each of these parties had different constituencies and different desires. Support from the environmental groups was, of course, continuous and vigorous. By the summer of 1969, most state agencies supported the national seashore concept. The other parties, however, still required convincing. Negotiations settled on five interlinked issues. These included: (1) land acquisition from the Candler and Carnegie families and the retained rights to be afforded them in the proposed seashore, (2) the embarkation point or points and means of access from the mainland, (3) the level of economic development proposed for the new park unit, (4) the fate of small landholders on both Cumberland Island and Little Cumberland Island, and (5) the source for money to acquire the island.

First, and ultimately most important, was support from the Carnegie and Candler families. One of the branches of the Carnegie family had removed itself from the debate by selling to Charles Fraser. The remaining four, the Johnston, Perkins, Ferguson, and Rockefeller/Rice branches, as well as the Candlers, had three primary provisions in mind: if they refused to sell, they would not have their land condemned; if they did sell, they would receive generous retained rights preserving their traditional uses of the island; and if the Park Service took over, it would work with them to maintain Cumberland in its existing state of nature and historic preservation. Each of these required delicate negotiations, and each became a sore point in the subsequent management of the national seashore.

Negotiations got off to a poor start because of the Park Service's penchant for secrecy. Through most of its long investigation of Cumberland Island, the agency had studied the island quietly, hoping to control how and when news might be released to the wider public about their interest. Most of the agency's contact had been with the Johnston group, and other Carnegie heirs felt they were being ignored. Island owners repeatedly complained about this to Park Service representatives.

After a couple of visits in late 1968, Vincent Ellis reported to his superiors that this approach was backfiring. He noted the particular concern of the Ferguson family and recommended that in the future any Park Service representative should openly "make a personal approach to each member of the five branches of the family and explain our Seashore Proposal." He added that the island grapevine let everyone know when a Park Service official was on the island anyway. Furthermore, he claimed that secretive attempts to tour Plum Orchard and the island had created an aura of fear and mystery. He concluded that the agency should rent rooms and vehicles at the Fergusons' Greyfield Inn and operate openly. Fortunately, Ellis had contacted the Fergusons before his October 15 and November 1 visits. He wryly pointed out to the regional director that his "trip could have turned into a fiasco" if he had not done so.[56]

The concern of the Fergusons through all the bickering and negotiations over Cumberland Island's future was that they be allowed to keep their land at least through the lives of Lucy Ferguson and her children. She expressed this wish to Stewart Udall during his 1965 visit and later when the Park Service asked her to help pass the seashore legislation. She kept a steady drumbeat of opposition to any measure that threatened her private land. Through her son-in-law Putnam McDowell, attorney Thornton Morris, and friends in the Georgia legislature, she never strayed from the themes of private ownership and her long life on the island. When Park Service official George Sandberg came to negotiate for land sales and Congressman Stuckey came to ask her support for the seashore bill, she reminded them that she would never relinquish her land. In response, Park Service director George Hartzog and Interior officials promised that they would not even approach her to discuss her lands. Still, she widely reported her mistrust of the federal agency. Finally, Fraser's clumsy attempt to have a state authority condemn Cumberland lands forced her to make a decision. McDowell wrote her on June 23, 1969, in hard pragmatic terms:

> You speak (I thought disparagingly) of "Knuckling under to the Park" and about the younger generation not standing up to fight. Well, this isn't a moral or ethical issue for which one should be willing to fight to the death. This is a question of a piece of idle real estate in a rapidly changing and crowded world. Fighting to freeze its status quo is nothing more than windmill tilting which is a game for the rich or idle. Not only that, but it is a piece of real estate which has brought more grief than happiness to those who have clung to it. The forces of taxes, politics, and changing circumstances among the owners are going to force the owners of Cumberland to do something more constructive with it. This is the fact with which you must reckon. . . . If we are not going to go along with the approach [the Park Service option] that has been described in recent reports from Thornton, then we must tell the others and tell Udall that they cannot count on the Fergusons. I have said everything that I can say on this subject. Now it is up to you to decide what you want to do.[57]

McDowell certainly underestimated the emotional bond that attached his in-laws to Cumberland Island, but his analysis of the situation was undeniable. Lucy Ferguson recognized the reality expressed in his words and ultimately came to grudging support for a bill to establish Cumberland Island National Seashore.

Support from the rest of the Carnegie heirs depended on the nature and duration of the "retained rights" they might enjoy after selling the island to the government. Traditionally, retained holdings in national parks meant continued use of a residence within a new park and a small parcel of land around it for either twenty-five years or the lives of the owners. In rare cases the agreements ran through the lives of their children.[58] However, for the Carnegie and Candler heirs, their proprietary sense of the island and their desire to be co-stewards moved them to seek more.

As Stewart Udall and Thornton Morris drafted a seashore bill for Congressman Stuckey, they received a letter from Andrew Rockefeller that outlined the owners' desires. First, they wanted the right to build a residence on their land in the future without any time limit. Second, Rockefeller suggested that the family's bargaining position was strong enough to secure a right through the lives of their grandchildren. Finally, he allowed that 40 acres would be enough for each retained estate.[59] Ultimately the island owners would win concessions on most of these requests in the initial

Cumberland Island bill. With retained rights settled and the Ferguson privacy protected, the Carnegies and Candlers fully supported the national seashore campaign.

A more difficult task for the Park Service was convincing Camden County officials and the local public to accept the seashore. The rights of the Carnegies and Candlers were unimportant to these groups. They focused instead on access to the island, the level of economic development, the tax revenue to be gained by approving the seashore bill, and the fate of small landholders on both Cumberland Island and Little Cumberland Island. The most pernicious issue in the campaign was whether a causeway should be built to the island. Most local officials steadfastly demanded one all the way through the legislative process. Only with a causeway, they reasoned, could enough people journey to the island to give measurable economic benefit to Camden County.[60]

During Stewart Udall's visit in 1965, Congressman Russell Tuten supported the idea of a national seashore but insisted on a causeway. Later the Park Service, island residents, and environmentalists convinced him to change his mind. For this apparent betrayal of Camden County, Stuckey vilified him in the next congressional election. After winning Tuten's seat in Congress, Stuckey flatly stated, "Nobody wants to wait for a ferry." Subsequently, the new congressman told the *Atlanta Constitution* he would oppose any seashore that did not satisfy local demands.[61]

However, the Park Service had already begun its campaign to sway his opinion. Assistant to the Secretary of the Interior Max Edwards wrote to him, "Even if causeway access were provided, at a great expense, and terminated at a parking area on the island, the pressure for further circulation would be too great to contain and with the mounting increase in vehicle numbers, it would only be a matter of a short time that the resource would be overrun—creating just another intensive use area without regard to its other valuable assets."[62]

Eventually, Stuckey too came to oppose a causeway, although he insisted that the Camden County Commission must agree to back any bill he would introduce. Through 1968 and 1969 intense negotiations with local officials continued. Eventually, Camden commissioners were convinced to back a seashore bill that allowed consideration of a causeway at a later date. At one delicate point in the negotiations, Stuckey invited the commissioners to the Greyfield Inn to secure their support for his introduction of a seashore bill to Congress. At the time, two commissioners favored the bill, one adamantly opposed it, and two wavered. Clerk of the commission J. E. "Fats" Godley took the seashore opponent down to the bar and "got him drunk" while Stuckey and others convinced the fence-sitters to support the bill. With four of the five commissioners nominally in favor of a national seashore, Stuckey introduced the bill.[63] Camden officials did exact a promise from National Park Service agent and future superintendent Sam Weems that all access to the island would be through Camden County.[64]

Local officials also worried about a repeat of the disappointment they suffered with the Kings Bay Army Terminal. In order to placate these concerns, the Park Service planned for an extensive recreation presence on the island. In October 1967 the agency released a development plan to the public that promised seven development areas on the island and one on the mainland at Cabin Bluff. Later the agency commissioned the Bureau of Business and Economic Research at the University of Georgia to estimate the economic impact of the proposed seashore. The university researchers projected that if the Park Service carried out its plan, the island would receive 225,000 visitors in the first year, escalating to more than a million during its fifteenth year. Over that period they expected visitors to spend nearly $70 million in Camden and adjacent Glynn Counties. They added that the Park Service expected to spend more than $12 million to carry out its development plan and ultimately would employ 915 people.[65]

Later, as Stuckey's seashore bill wound through Congress, the National Park Service released a master plan for the proposed unit in June 1971. It promised a fleet of twelve 100-passenger ferries, 300 picnic sites, a jitney service, 150 campsites, and a variety of interpretive and conference centers. At the mainland embarkation point, the agency planned to develop extensive camping and lodging facilities. A daytime visitor capacity of 10,000 persons was routinely mentioned. Now these were the kinds of numbers Camden County could accept! And it did so. Although these visitor projections must have appalled the island owners, they remained quiet, preferring perhaps to deal with the more immediate questions concerning Charles Fraser and their place in any future seashore.[66]

Another issue of local concern was the fate of small landowners on Cumberland and Little Cumberland Islands. During the years since Udall's visit, the Little Cumberland Island Association had grown to sixty members who hailed from twenty states. Most had small homes on the island or plans to build them. The group was a nonprofit organization that promoted limited residential development, preservation of natural resources, and scientific study. A number of the members were also active in the Nature Conservancy, including five past or present governors of that organization.

Initially the Park Service bungled communications with this group too. Association president Ingram H. Richardson finally wrote to Udall in October 9, 1967, asking for a meeting. The agency initially excluded the smaller island from the seashore but eventually worked out an arrangement whereby Little Cumberland Island would be included in the seashore boundary but not acquired. Instead, it would be left to its present owners as long as they continued to operate a trust that prevented extensive residential or any commercial development.[67]

On the larger island owners of small parcels were divided into two groups: a few north end residents who had owned land and homes for decades and those who had purchased land from Robert Davis. The latter group vehemently opposed the seashore, and when it became apparent that they could not block it, they tried to keep their properties. However, this potential settlement in the middle of an important visitor area was unacceptable to the Park Service, to most of the larger landowners, and to the environmental organizations. In spite of the fact that many county officials owned land in the new subdivision, they faced a losing battle to stay on the island. Yet the promise of extensive economic development still led most of them to support the national seashore legislation. Later, when the legislative process reformulated the national seashore bill, they were powerless to stop it.[68]

One final issue confronted the National Park Service and other seashore proponents: where to find money to pay for the land and development. The congressional process was a slow one, and some owners wanted to sell immediately. The threat of losing the island to Charles Fraser or to a state recreation area was still very real.

Initially, island owners and the Park Service, assisted by Sea Island Company owner Alfred W. Jones and members of the Little Cumberland Island Association, approached the Nature Conservancy. However, during a meeting in Washington, D.C., in April 1969, the Nature Conservancy warned island owners and the Park Service that it might have a problem securing funds for the purchases. Conservancy representative Tom Richards explained that his group had a line of credit from the Ford Foundation for purchase of future parklands. However, it could not be used for Cumberland Island because the Ford group required a definite time limit after which the Conservancy could reclaim its money with a sale to the government.

Despite this problem, the Conservancy planned to seek options on land-holdings with funds raised by "a committee composed of nationally known figures and prominent Georgians."[69] After Congressman Stuckey introduced a national seashore bill in February 1970, the Nature Conservancy did attempt to work out agreements with Fraser and other island landholders. However, by April of that year, the environmental group relinquished its role in Cumberland Island, presumably due to monetary restrictions.[70]

Fortunately the Park Service secured another familiar beneficiary. Over the years National Park Service director George Hartzog had discussed the difficulties of land acquisition on Cumberland Island with representatives of the Old Dominion Foundation. The Mellon-funded organization was hesitant to acquire the island because it had no land management capabilities. However, on December 18, 1967, Congress established the National Park Foundation, which replaced the thirty-two-year-old National Park Trust Fund Board. The new foundation was authorized to "accept, receive, solicit, hold, administer, and use any gifts, devises, or bequests, either absolutely or in trust of real or personal property or any income therefrom" for the benefit of the National Park Service. Its primary duties were to accept gifts and purchase property to hold during the glacial congressional process of creating a new national park. Once the new park was established, it would sell or donate the land to the Park Service.[71] If the Mellon group could be convinced to fund the National Park Foundation, the land could be acquired, possibly at no cost to the government.

In August 1968 Hartzog wrote to Ernest Brooke Jr., president of the Old Dominion Foundation, and explained the new National Park Foundation. He asked that the foundation Brooke directed take "direct action" to help with Cumberland Island.[72] However, Old Dominion did not immediately step forward, perhaps because it and the Avalon Foundation were restructuring to become the Andrew Mellon Foundation. By early 1970 the Mellon group still resisted the pleas from the Park Service.

What happened next is one of those serendipitous sequences that seem to occur on Cumberland Island. One of the senior trustees of the Mellon Foundation, New York attorney Stoddard Stevens, visited the Cloister Hotel at Alfred W. Jones's Sea Island Resort. During his stay hotel manager Richard A. Everett gave an illustrated talk on the Georgia Sea Islands and mentioned that Cumberland Island would soon be developed. Everett apparently gave a stirring narrative about the losses that would result from the intrusion of resort construction on his favorite of the islands. Stevens requested a visit to Cumberland and was immediately impressed with its beauty and pristine character. He reported back to the Mellon Foundation, and negotiations for their support immediately intensified. On April 8, 1970, representatives of the foundation, including Paul Mellon, met with George Hartzog, Thornton Morris, and others at Sea Island to discuss the issue. The Mellon Foundation agreed to work through the National Park Foundation and supply at least $6 million for land acquisition. Once again Mellon money enabled the federal government to save a choice piece of coastal property for the American public.[73]

The Congressional Process

As this patchwork of agreements, alliances, and promised moneys coalesced, the actual design of a seashore bill proceeded. Island landowners shaped this process by hiring former secretary of the interior Stewart Udall to represent their Cumberland Island Conservation Association and to draft a suitable bill for Congressman Stuckey. Udall's national prominence, compelling personality, and understanding of Congress were critical assets. He and Thornton Morris designed a bill that Stuckey introduced to the House of Representatives on February 3, 1970, two weeks into the second session of the 91st Congress.[74]

This initial bill, H.R. 15686, clearly reflected the difficult compromises worked out over the preceding two years. The proposed seashore would encompass all of Cumberland Island, including islands and marsh areas in Cumberland Sound. Conspicuously excluded was Little Cumberland Island. The bill promised longtime island owners that no "improved residential property," including any structures started before the government acquired 50 percent of the island's land, would be condemned. It proposed that retained estates would last forty years or the lifetime of the last child of the owner, that owners could rent their estates for vacation or "year-round use," and that rights holders could initiate building on their property up to ten years after the bill became law. Finally, it limited all retained estates to 40 acres except for 100- to 200-acre parcels around High Point, Plum Orchard, Stafford, and Fraser's Sea Camp, plus an undefined fourth of segment 5N near Lake Whitney, and the area west of the Main Road from segment 5N to the northern tip of the island.[75]

H.R. 15686 also offered much to Camden County. It deferred a final decision on a causeway but required that ferries capable of moving at least 300

vehicles per day to the island be started within three years of initial land acquisition. Furthermore, an advisory commission composed of one Camden County commissioner, three other Georgia officials, four scientists
from the state's Ocean Science Center, and only two Department of the
Interior officials would be required to revisit the causeway question every
two years. As for development, 15 percent of the island would be devoted to
privately operated lodgings, campgrounds, marinas, and dining facilities.
Those areas also would include state-run education centers. Private developments as well as the retained estates would remain on the tax rolls for the
county. Charles Fraser, with the specific allowance of a 200-acre retained
right at Sea Camp and the explicit language allowing these activities, stood
to gain the concession operation he sought.

Beyond those benefits to Camden County, the state of Georgia could exact state taxes from the estates, dominate the advisory commission, and require the National Park Service to consult with its Ocean Science Center in
questions of planning and management. At the time Stuckey introduced
H.R. 15686, no foundation had agreed to pay for land acquisition, so the bill
placated Congress with a plan to acquire property options on Cumberland
lands. This would allow delay of final purchase until June 1975. Finally, the
bill provided $1 million for the options with ceilings of $12 million for land
acquisition and $10 million for development. Seashore ally and Georgia
Coastal Island Study Committee member Hal Webster reported that
Stuckey's bill allowed for only 58 percent of the island to be set aside for
public use. It reserved 19 percent of the land for the owners, 15 percent for
development, and 7 percent for roads.[76]

Reaction to the bill was mixed. Margaret Wright phoned Theodore
Swem at National Park Service headquarters in Washington, D.C., and expressed her pleasure with the bill and her commitment to rallying other island owners behind it.[92] Camden County officials seemed satisfied. The
Park Service also had learned its lesson about keeping communications
open. During March it announced a series of visits by agency officials who
would be charged with the development of the island. Woodbine mayor
Clarence Haskins called the bill a compromise of all individual interests and
pointed out that whatever locals did not like they could get changed later
anyway.[77]

However, not everyone was pleased. State legislators, spurred by Governor Lester Maddox, reopened the question of developing a state park on
Cumberland Island similar to the one on Jekyll Island. This resurrection of

state interest came with the final report of the Georgia Coastal Islands Study Committee, the agency established in lieu of Charles Fraser's Camden County Recreational Authority. The study committee recommended that another body, the Georgia Coastal Islands and Marshlands Planning Commission, be established to "devise, implement and enforce a comprehensive land use plan which would control and regulate the future use and development of Georgia's coastal islands and marshlands." That commission would have the power to approve or disapprove acquisition of island lands. Once again, island owners' representative Thornton Morris worked to diminish the threat and managed to obtain a block of four of the twenty-one seats on the proposed commission for the Georgia Marshland and Island Foundation, a conservation group chaired by Sam Candler. In this way the owners hedged their bet by assuring they would have input if the state scheme replaced the national seashore plan. By mid-March, however, the state legislature rejected a bill to implement the study commission's recommendations. In the process, however, Georgia did claim all the marshlands in Cumberland Sound, a move that would significantly affect the national seashore's management in the coming decades.[78]

Meanwhile, Charles Fraser decided to anticipate the pending seashore bill. In early April 1970, with no warning, he suddenly began cutting two cross-island roads and a 500-foot airstrip on his parcel of land in segment 5N. In response to blistering criticism from Bob Hanie, director of the Georgia Natural Areas Council, and editorials in the Atlanta newspapers, his vice president responded: "We are going to great efforts to coordinate our plans with the National Park Service in the event the national seashore bill goes through. Mr. Fraser spent the entire week last week with Mr. George Hartzog (director of the National Park Service) and they know everything we are doing."[79]

It did not take long for Hartzog to reply. He flatly denied that any development on the island was approved or coordinated with the National Park Service, telling the *Atlanta Constitution:* "Our plans do not include an airport where Mr. Fraser is building his. Neither do we have cross-island roads like the ones I saw when I flew over the island yesterday." The director did allow that he knew that work was taking place on the island but added "my knowing about it and my approving it are two different things."[80]

Island residents and environmentalists reacted angrily. Sam Candler told reporters that residents favored Stuckey's seashore bill and were "very much against" Fraser's development. A few months later, to underscore their opposition, more than ten environmental organizations held a rally on Jekyll Island. Although it was a light-hearted affair, some 300 participants underscored the support for the Cumberland Island National Seashore bill among environmentalists across the state.[81]

Ultimately, Charles Fraser realized that this last effort to force his resort development onto the island would also fail. At the same time, his Sea Pines corporation ran into financial difficulties that necessitated raising quick cash. On April 24 he informed reporters that he would sell his holdings on Cumberland if a conservation organization bought the rest of the island for a national park. He stressed that he would insist such a park be for "very active public use" and not for a nature reserve. That summer Fraser signed an option to sell his property to the National Park Foundation. The following year Fraser finally did receive some positive press, albeit from distant New York. Author John McPhee published in the *New Yorker* magazine a segment of his forthcoming book about environmental activist David Brower. In it he portrayed Fraser as a dedicated conservationist whose sensitive plan to create something beautiful was defeated by social, political, and financial pressures and by a campaign of "ecological propaganda."[82]

The Final Legislation

Congressman Stuckey's popular bill was never reported from the Committee on Interior and Insular Affairs and died with the end of the 91st Congress on January 2, 1971. On July 15, 1971, he introduced another bill, H.R. 9589, in the House of Representatives. Two weeks later Georgia senators Herman Talmadge and David H. Gambrell submitted an identical one to the Senate.

Over the eighteen months since Stuckey introduced the first bill, many conditions on Cumberland Island had changed. Charles Fraser was gone except for a small retained estate. The National Park Foundation owned or had optioned nearly three-quarters of the island. The state was fully behind the seashore concept, as were both parties in Congress and the Nixon administration. The momentum of the campaign clearly favored the National Park Service.

Agency officials used that time to continue shaping the new bills. Gone was the exclusion of Little Cumberland Island. In its place was the stipulation for a conservation trust to operate it. Gone was the explicit prohibition of the condemnation process. Instead, the secretary could "acquire lands, waters, and interests therein by whatever legal method available to him."

The new bill modified the retained rights to last for forty years or the life of an owner or owner's spouse and vastly reduced the large retained areas around the older estates. However, it specified that any agreements already reached by the National Park Foundation would be honored.

Gone too was the stipulation that 15 percent of the island be privately developed. Replacing it was permission to acquire 100 acres on the mainland for a visitor center and headquarters and to construct a parkway from Interstate 95 to that site. The advisory commission remained but was limited to a ten-year term. Significantly, the explicit order to consider a causeway every two years was replaced by a softer statement that the "Secretary or his designee shall, from time to time, consult with the Commission" regarding ferry service and the desirability of a causeway. Hunting and fishing remained. The most significant addition was a redefinition of "improved property" to mean dwellings begun before February 1, 1970, two days before Stuckey introduced his original Cumberland Island bill.[83]

These changes not only reflected the desires of the National Park Service but also the growing influence of environmental organizations. Jane Yarn, an active conservationist in coastal Georgia, is credited with convincing Stuckey to alter the section demanding reconsideration of a causeway every two years. Yarn's argument was supported by a study released in May 1971 by the University of Georgia College of Business Administration. Authors Charles Clement and James Richardson bluntly stated: "A major ecological and recreational feature of the coastal islands [of Georgia] is the insularity provided by the wide expanses of water and salt marshes. Causeway construction that would destroy this insularity or interfere with the natural functioning of the marsh areas should be discouraged."[84] Although they had participated in meetings and events supporting a national park presence on Cumberland, conservation groups became far more vocal with the new bills and the obvious momentum of the campaign. The Georgia Conservancy lobbied state senators and representatives while the Sierra Club and other groups courted congressional allies nationally.

Local reaction was predictably unfavorable. Most of the Carnegie heirs had already sold to the National Park Foundation and, hence, were unaffected. However, the Fergusons, the Rockefellers, and the Candlers had not. The disappearance of the explicit ban on condemnation troubled them, especially Lucy Ferguson. Camden County officials and residents greeted the new bill as a betrayal. The Camden County Commission asked Stuckey to add a causeway and halt land acquisition in order to preserve some area for private development. Stuckey refused. State legislator Carl Drury, a proponent of the seashore, told reporters that the commission "raised the voice of parochialism and the effect of granting their requests would destroy what we want to accomplish."[85] Each chamber of Congress referred its new bill to its Committee on Interior and Insular Affairs. This time, because of overwhelming support for the bills, the committees acted promptly.

Four members of the House Subcommittee on National Parks and Recreation, including chairman and Colorado Democrat Wayne Aspinall, visited the island in early November. There they conferred with Stuckey, various pro-seashore state officials, and island owners. They received and ignored a resolution from the Camden commissioners requesting a causeway and rejecting both land acquisition by condemnation and the February 1, 1970, construction cutoff for retained estate eligibility. The congressmen toured the island and, like everyone else, admired its beauty and pristine character.[86]

The subcommittee held its public hearings on April 20 and 21, 1972. Speakers included representatives of the National Audubon Society, the Georgia Conservancy, the Sierra Club, the National Parks and Conservation Association, and Save America's Vital Environment (SAVE). All favored establishment of the seashore, prohibition of a causeway, and explicit language ordering the National Park Service to emphasize preservation of the island's natural environment. A number of island residents spoke, including Nancy Carnegie Rockefeller and her daughter Georgia Rose, Lucy Ferguson with son-in-law Putnam McDowell, Franklin Foster, and Nancy McFadden. Although they all favored the seashore proposal and prohibition of a causeway, they wanted aspects of the bill changed. Ferguson wanted a more explicit statement of retained-rights eligibility. McFadden suggested that recreation should go to Jekyll and other islands while Cumberland remain a nature preserve. Nancy Rockefeller urged the addition of a ban on condemnation and submitted her draft of an agreement by which continuing residents would promise to manage their lands in accordance with national seashore purposes. Foster complained of difficulties in donating land for the future seashore.[87]

Camden County and the state of Georgia were also well represented. County Commissioner J. E. Godley, the same one who distracted the anti-seashore commissioner in the Greyfield Inn bar, spoke for both Camden's local government and small landholders who had recently purchased property on the island. He urged the same changes as the county resolution of November 2, 1971. Robert Davis, developer of one of the new subdivisions on Cumberland Island, took a harsher view of the proposed legislation. He opposed anything that interfered with his right to sell and develop lots. The subcommittee gave him one or two minutes to speak before it shifted to questioning Franklin Foster again. Another of Davis's purchasers, Kenneth Harrison, echoed his objections to the February 1, 1970, cutoff for land retention. One former resident of Camden County, William Voight Jr., also saw problems with H.R. 9859, but for very different reasons. He wanted the proposed unit to be renamed Cumberland Island National Park and devoted to much more stringent nature preservation.[88]

Other speakers, including Congressman Stuckey, Georgia Department of Natural Resources commissioner Joe D. Tanner, Assistant Secretary of the Interior Nathaniel Reed, and George Hartzog, spoke glowingly about the bill. Reed suggested that the congressmen delete the provision for a Cumberland Island Advisory Commission, citing the existence of the nationwide Advisory Board for National Parks, Historic Sites, Buildings, and Monuments as adequate.[89] George Hartzog explained the development program, repeatedly citing a figure for maximum daily visitation of 10,000 per day during the peak tourist season. Several congressmen expressed concern over the Park Service's ability to handle so many visitors and maintain the quality of environment on the island.

The hearing clearly demonstrated the rising influence of the environmental lobby and the futility of local resistance to the bill. Two weeks later the Senate subcommittee held its hearing. The same organizations and most of the same speakers testified. Nothing had changed the variety of opinions expressed. Local pleas for a causeway and some private development met even less sympathy than the House subcommittee had afforded. Chairman Alan Bible of Nevada proved to be especially attuned to protecting the island from overdevelopment. His subcommittee's report to the full Senate suggested adding language that mandated preservation of the island's "wild state," citing a similar provision in the act that had authorized Cape Hatteras National Seashore. Specifically, the committee proposed an amendment to the bill which would prohibit construction of a causeway. In an interesting departure, however, the senators moved the cutoff date of residences eligible for a retained right to August 3, 1971, the date that the first Senate bill was introduced.[90]

In late May 1972 legislators from the Senate and the House met to hammer out the differences between the two bills as amended by their respective subcommittees. The *Atlanta Constitution* reported that passage during this congressional session was assured.[91] The Senate acted first, passing its version on July 25. On October 10 the House passed a version that amended the Senate bill and returned the estate eligibility date to February 1, 1970. Senator Bible urged his colleagues in the full Senate to pass the House version, which they did by voice vote. On October 23, 1972, President Nixon signed Public Law 92-536 establishing Cumberland Island National Seashore.

What did the final act say? Essentially H.R. 9859 remained as introduced except for four notable changes. First, lawmakers granted Assistant Secretary Reed's request and dropped the requirement for a Cumberland Island Advisory Committee. Its deletion eliminated vital roles for Georgia and Camden County. Second, Congress strengthened the language directing the National Park Service to preserve the island "in its primitive state." Third, at Senator Bible's urging, legislators inserted a specific ban on a causeway to the island. Finally, in a very late addition, Congress added a requirement that the Department of the Interior carry out a study of wilderness feasibility for the island and report back in three years. This final section once again demonstrated how fear of Park Service overdevelopment troubled many seashore supporters as well as key congressmen.

The tone of development plans for the island also subtly changed. Although the Park Service still suggested up to 10,000 visitors per day, no cars would be ferried to the island. Hans Neuhauser of the Georgia Conservancy told reporters that without a causeway the Park Service would have a problem getting that many people to the island. Other environmental organizations, including the Sierra Club and the Nature Conservancy, also expressed satisfaction with the new law's measures to "preserve the natural integrity of the island."[92]

As the weeks passed, Camden County, the state of Georgia, and the country became acquainted with their new national park unit, at least through the press. No visitors would be allowed for another two years. Newspapers and magazines featured stories on the island's history and attractions as well as the difficulty encountered in negotiating, introducing, and passing the national seashore bill. Two weeks after passage of the Cumberland act, the Department of the Interior awarded Alfred W. Jones of Sea Island a conservation service award for his role in aiding the Cumberland Island campaign as well as his support for Fort Frederica National Monument, Jekyll Island State Park, and the Sapelo Island Foundation.[93]

However, as the National Park Service moved in to take over negotiations for the remaining private lands and management of the complex human ecology of Cumberland Island, the cost of land acquisition and the reality of their legal agreements would become painfully apparent. Congress, meanwhile, adjourned without providing any funds to administer the new unit.

4

Land
Acquisition
and Retained
Rights

Land issues form the central theme of the Cumberland Island National Seashore story. Acquisition of Carnegie and Candler land, defense against subdivision, and the negotiation of unusual and complex retained-use estates have politicized and complicated all other management issues. Hence an understanding of how the National Park Service acquired the island lands and the sacrifices it made to do so is critical. This chapter describes the government acquisition of more than 140 individual tracts of land on the island, where and why retained estates were granted, and the characteristics of the rights associated with those estates.

National Park Service land acquisition policy developed over many decades. During the early twentieth century, Congress and the presidents created parks from the public domain, and land acquisition was simple. The overwhelming majority of the property was already in federal hands. In summer 1959 the Park Service fielded 188 parks totaling more than 20 million acres. The inholdings (private lands) within those parks amounted to only 32,000 acres. The agency chose to ignore inholdings unless their owners threatened park resources or carried out activities contrary to park purposes.

For parks established after July 1959, however, the agency changed its land policy to prompt acquisition of all private lands. This change coincided with its aggressive pursuit of new recreation parks in already settled areas. Public land law expert Joseph Sax attributes this new policy to Park Service officials who "view ultimate fee acquisition as a faithful response to congressional policy and National Park history."[1] In some areas the Park Service suspended land acquisition as long as local zoning remained restrictive. At Cumberland Island, however, because developers sought sub-division, such a suspension was never an option.

During the primary period of land acquisition at Cumberland Island, 1970-84, the Park Service initially avoided condemnation in deference to the original owners' willingness to support the national seashore and sell their lands cheaply. However, by the late 1970s the agency actively used this un-popular tool. A June 1977 memorandum from Park Service acting director Ira Hutchison explained the philosophy behind such actions:

> More and more over the past several years there has been a notable reluc-tance on the part of some field managers to pursue the acquisition of certain inholding properties and to recommend and support legislation needed to adjust park boundaries, based primarily on their view that this would create an undesirable public relations problem with affected landowners and/or community relationships.
>
> We recognize that we cannot be totally unmindful of these relationships, but we must also recognize that our primary responsibility is for the long-range protection of park resources and park environment. We *cannot* com-promise this duty and responsibility of our stewardship for the generations yet unborn simply to make our job easier on a day-to-day basis.[2]

Thus land acquisition by any means, including condemnation, was both policy and moral charge for the National Park Service as it approached Cumberland Island.

Ultimately, the zealousness of this singular approach would be blunted by considerations for the generosity of the Carnegie heirs, by vast price in-creases, and by public antipathy to eminent domain and to the federal gov-ernment in general. Indeed, by the late 1980s that antipathy made condem-nation unacceptable to the public across the nation and to its congressional representatives as well.

When the National Park Service began shopping for Cumberland Island, its task seemed a simple one. A 1958 appraisal by Robert M. McKey showed only nine owners of 23,683 acres of upland and marsh (the area now held by the State of Georgia was not included). After a careful study of lands and structures, McKey suggested a total value of $2,148,000. By far the largest owner was the Trust for Lucy C. Carnegie, with 91 percent of the land and 86 percent of the appraisal value. In 1965, after the trust ended and the in-dependent ownership of a tract by the Olsen family was recognized, the agency still faced only twenty-five owners. A majority of them favored Park

Service acquisition. Optimistic officials believed that standard retained-estate contracts would see the seashore in fee simple government ownership by the end of the century.[3]

Over the next decade land acquisition would become vastly more complicated. The number of landowners swelled by more than 100. Six subdivisions would occur, including Fraser's grand project and three by Carnegie family members. Three of these would see residential construction (map 4.1). At the same time, the National Park Foundation, the organization charged with acquiring lands for the future park with donated funds, set a precedent by granting liberal but highly variable and complex retained estates to the original owners. When the National Park Service assumed negotiations, both large and small landowners drastically raised their prices and demanded equally generous retained rights.

Land Acquisition before 1973

When Congress established the National Park Foundation on December 18, 1967, its first major project was the acquisition of Cumberland Island with Andrew Mellon Foundation money. Even as the new organization struggled to organize, negotiations with Carnegie family landowners began. In order to carry out these negotiations, the Park Service lent experienced land officer George Sandberg to the project. The intense desire of all parties to make this seashore a reality led Sandberg to offer more generous retained rights than normal. Standard retained-rights agreements were for small parcels of land to be held for either twenty-five years or the lives of the sellers. At Cumberland Island typical terms included larger parcels for either forty years or for the lives of the sellers' children. Each Carnegie or Candler family landholder could secure one or more parcels for his or her piece of the island.[4]

Ironically the first landholder to sell was Charles Fraser. Embittered by the rejection of his project, he lashed out at the Carnegies and Candlers "who have this arrogant hostility toward the common man" and who destroyed what he regarded as an environmentally sensitive island plan.[5] On September 29, 1970, Fraser sold his land in segments 5N and 4S to the National Park Foundation for $799,500.[6] He then took this money and developed a resort on nearby Amelia Island in Florida. This figure roughly equaled his cumulative cost for land acquisition and development on Cumberland up to that time.

However, his sale came with a number of strings attached. First, he kept

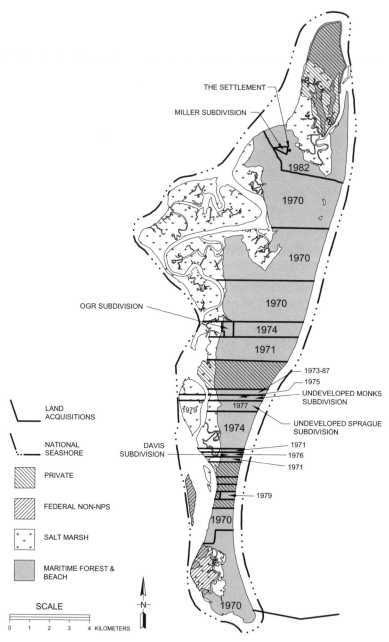

THE SETTLEMENT

MILLER SUBDIVISION

1982

1970

1970

1970

OGR SUBDIVISION

1974

1971

1973-87
1975
UNDEVELOPED MONKS
SUBDIVISION

1979
1977

1974

UNDEVELOPED SPRAGUE
SUBDIVISION

DAVIS
SUBDIVISION

1971
1976
1971

1979

1970

1970

LAND
ACQUISITIONS

NATIONAL
SEASHORE

PRIVATE

FEDERAL NON-NPS

SALT MARSH

MARITIME FOREST &
BEACH

SCALE

0   1   2   3   4 KILOMETERS

N

Map 4.1. National Park Service land acquisition and partially constructed and proposed
subdivisions through 1984

his two development sites, the 2-acre parcel at the present Sea Camp Visitor Center and the nearby 38-acre parcel containing a campground he had built, as retained forty-year estates. Stipulations in the agreement allowed for construction of seven houses, selection of another 40 acres in place of the above parcels if Fraser so desired, and the right to lease the houses for not less then one week at a time.

Second, earlier agreements between Fraser and the heirs of Thomas Carnegie II gave each of them a retained estate as well. Henry C. Carnegie and friend James Bratton formed Cumberland Island Properties, claiming two parcels totaling 10 and 5 acres, respectively, in segment 4S. These carried the rights to build three houses. Gertrude Schwartz, as trustee for Andrew Carnegie III, received a 7.5-acre estate in the same area that she later traded for an equal-sized parcel of segment 1S near Old House Creek. Not really interested in the island, she rented the property to a Ben Jenkins for many years. Finally, Thomas Carnegie IV received a 7.5-acre estate in segment 4S on the shore of Cumberland Sound.[7]

The most avidly pro-seashore branch of the family, the heirs of Nancy Carnegie Johnston, provided the second National Park Foundation acquisition and set the tone for negotiations and retained-rights agreements thereafter. Five heirs, Coleman, Thomas, and Marius Johnston, Margaret Wright, and Lucy Graves held in common segments 3N and 5S. Thomas Johnston's portion was divided in thirds among his children, Margaret Richards, Nancy Butler, and Thomas Johnston Jr. Segment 3N included the Plum Orchard mansion. Over a span of three months in the fall of 1970, these four grandchildren and three great-grandchildren of Lucy Carnegie sold their land to the foundation for a combined $1,615,000. At the same time, they donated Plum Orchard mansion outright along with $50,000 as seed money for a fund to maintain it. This would have major management implications for the future seashore.[8]

Sandberg negotiated a bewildering array of retained-right estates for the Johnston group, including a 12-acre, forty-year estate along the main road in segment 3N for Margaret Wright; small estates for the lives of Nancy Butler, Thomas Johnston Jr., and their spouses and "issue" adjacent to Plum Orchard; and a similar arrangement for Margaret Richards and her family near the ocean dunes on Duck House Road, also in segment 3N. Lucy Graves negotiated two forty-year estates of 5 and 10 acres in segment 5S. Later she too swapped the larger one for a plot in segment 1S between the ocean dunes and the Main Road. Her smaller parcel contains The Grange.

Coleman Johnston traded for a 40-acre parcel on the Brickhill River in segment 4N. Finally, Marius Johnston negotiated for $20,000 more than his siblings received in lieu of a retained estate.[9] All parties maintained rights to drive the Main Road, occupy existing structures or build a new one within ten years, gain access to and use the family cemetery, and debark at various docks on the island. In 1979 Margaret Wright sold her 12-acre retained estate with improvements to the National Park Service for $16,000.

As the Fraser and Johnston sales proceeded, pressure on the foundation to acquire at least 60 percent of the island increased. The seashore bill had reached a critical phase in Congress, and legislators wanted proof that the land would be available for the new park unit. Coleman Perkins, owner of 4,100-acre segment 4N, was initially reluctant to sell. Perkins and his immediate family, plus attorney and friend Thornton Morris, had formed Table Point Company as a business operation. In April 1970 Table Point had sold its timber rights to the Georgia-Pacific Corporation.

However, the promise of a huge 186-acre retained estate on Brickhill River and Malkintoos (MacIntosh) Creek throughout the lives of all the shareholders' children induced Table Point to sell to the foundation for $1 million on November 2, 1970. Several weeks earlier Georgia-Pacific had accepted the inevitable and sold its timber rights to the National Park Foundation for $19,500. Along with the many retained rights stipulated in the contract, such as exclusive hunting, dock privileges, and Main Road and cemetery use, Sandberg verbally promised the Table Point Company rights to use Charles Fraser's South Cut Road for beach access.[10]

The following spring the National Park Foundation culminated its major purchases by concluding agreements with Mary Bullard and Margaret Sprague for their thin strips of land in segments 2N and 2S. The land totaled 1,500 acres, and the two granddaughters of Margaret Carnegie Ricketson received approximately $455,000 each. Margaret Sprague retained a 15-acre estate north of Greyfield in 2S while Mary Bullard chose a 10-acre estate on former Perkins land between Plum Orchard and Table Point. Each retained-right estate will last through the lives of the respective women's children.

The final acquisition by the National Park Foundation came after establishment of Cumberland Island National Seashore. In July 1973 Lucy Foster, daughter of Lucy Carnegie Sprague Rice, donated 21 acres of land in two parcels of segment 1N containing the Stafford house and The Chimneys. She maintained exclusive, lifetime rights of use on both properties.[11]

In February 1973 the National Park Foundation donated 3,460 acres to

Cumberland Island National Seashore and followed in December with 9,229 more. Negotiations were well under way with the Rockefellers for segment 3S, with Margaret Laughlin and her daughter Cynthia Cooper for 1S, and with the Candlers for their High Point tract. As agreed, no one approached Lucy Ferguson. The first major project of the National Park Foundation was a resounding success. It had acquired all of tracts 3N, 4N, 5N, 4S, and 5S plus substantial portions of 2N and 2S.

These acquisitions, however, were not unqualified. The retained rights were confusing and variable and would deeply affect management. Still, the land was in Park Service hands, and the seashore was a political reality. At this point the federal government was responsible for continuing land acquisition in the young seashore.[12]

A Rash of Subdivisions

Despite the success of the National Park Foundation, the land acquisition picture at Cumberland Island had a dark side. Well before the foundation began negotiating and even before Charles Fraser entered the scene, cracks appeared in the Carnegie family's resolve to protect Cumberland Island from development. Oliver Ricketson III, nephew of Lucy Ferguson, lived in New Mexico and had little use for his two thin parcels in segments 2N and 2S. In what must have been a controversial move, he sold his 102-acre tract in segment 2S on September 26, 1967, for $60,000.[13]

The buyer was one Robert Davis, sometime jockey and cowhand and close personal friend of the Fergusons. Davis was no ordinary employee. He is devotedly remembered by those who knew him. He was so close to Lucy Ferguson that some family members resented his status as "adopted son." And he was an astute businessman. Friend and attorney Robert Harrison later described his canniness for making money and said Davis "could buy a piece of land and the minute he bought it, somebody else wanted it and wanted it for more than he did."[14] So it was on Cumberland Island.

In 1968 he formed the Davis Land Company, subdivided his property west of the Main Road into seventy-one lots and offered them on the market for $4,000 each. It quickly became apparent that the desire for a sheltered getaway on the legendary rich man's island was intense. Only tax considerations forced Davis to slow his sales in the face of fierce demand. Prices that rose to $6,000 in 1970 and $7,500 in 1972 had no effect on demand. By July 1973 Davis held only four unsold tracts.[15]

Meanwhile his land east of the Main Road also brought money. In July

1970 real estate broker Henry A. Crawford introduced Davis to a group of Atlanta businessmen and doctors whom he represented. These investors sought a ninety-nine-year lease to build and operate a motel. Like the small landowners in the Davisville subdivision, they did not believe the seashore legislation would pass, or if it did, that the provision allowing retained estates only on structures initiated before February 1, 1970, would hold. The lease was a curious one that showed Robert Davis's business savvy. Lease payments for the 14.5-acre parcel were set at $560 per month for the first two years, $1,112 per month for the next three, and $2,224 per month for the remaining period of July 1975 through June 2069. Furthermore, lease increases based on the Consumer Price Index could be added in the years 2000, 2025, and 2050. Thus the lease was affordable for the period during which the fate of the seashore would be decided but worth much more if the Park Service condemned the leased land.[16]

The great success enjoyed by Robert Davis spurred Oliver Ricketson III to try the same scheme with his northern strip of land. In 1971 he formed OGR, a New Mexico corporation, and began offering lots on both sides of the main road. Although small individual purchases were made, most of the land went to a relative few parties who in turn further divided their portions and resold to friends or relatives. For example, the clerk of the Camden County Commission, James E. Godley, bought 7.5 acres in September 1971 for $2,000 per acre. He then sold pieces to other county commissioners at cost. That same month R. R. McCollum bought 5 acres for $10,000, and he sold three parcels totaling half of the acreage to Robert Van Cleve, Martin Gillette, and Riley Harrell Jr. for a total of $10,500 six months later.[17]

Unlike Davis, Ricketson's company, OGR, also sold large parcels of land to several individuals at $1,000 per acre. Florida land surveyor and speculator Louis McKee bought the largest parcels, 30 acres of upland and 96 acres of marsh. Originally employed by the Candlers, McKee became a familiar figure in the Camden County land office, buying and selling parcels in all the subdivided portions of Cumberland Island. Three weeks after buying these two tracts, he conveyed them to three other people, Elinor Giobbi, Heloise McKee, and Carnegie family member Nancy R. Copp. Each obtained an undivided one-third of the marsh and sole ownership of a 10-acre tract of upland.[18]

Lucy Ferguson also bought a large tract of 28 acres. She then sold 16 acres at a slight profit to friend Wilbur Readdick and 12 acres at a loss to longtime employee and close confidant J. B. Peeples. Both Readdick and Peeples sub-

divided their land among family members and resold various portions after the seashore had been established.[19]

At the north end of the island, Laurence Miller and his aunt, Mary Miller,
developed a third subdivision on lands surrounding the Settlement. In the
years preceding creation of the seashore, the Millers had quietly added lands
from nonresident owners to their holdings. By 1972 they had accumulated
more than 100 acres. After passage of the seashore's establishing act, Laurence Miller platted fifty-four lots on 30 acres, all upland but with no coastal
access. Over the next two years, they sold clusters of lots plus other Miller
land to businessmen from Brunswick, Atlanta, and Camden County. Among
the buyers were Louis McKee and attorney Grover Henderson. The latter,
son of one of the original Davis subdivision buyers, bought one parcel for
himself and another with law partner Robert Harrison, also a Davisville
owner. Most of the lots measured one-half to two-thirds of an acre and sold
for $6,000. Grover Henderson also bought another tract independently,
traveling to Miami to buy a lot containing a small ramshackle "house" from
Thomas Lee, heir to one of the Settlement's freed slaves.[20] All of these purchases created dismay and some panic among both the Park Service and
the Carnegie and Candler owners who had worked hard for the seashore's
creation.

The National Park Service Takes Over

In 1973, as the National Park Service took up land negotiations, it faced this
vigorous subdivision as well as large tracts of unacquired Carnegie and
Candler property. The enactment of the seashore law had come too late for
Cumberland Island to receive an official budget for fiscal year 1973. To overcome this setback, the Park Service reprogrammed approximately $500,000
for land acquisition and another $94,700 for administration from other
1973 projects nationwide. The Southeast Regional Office in Atlanta appointed its land specialist, William Kriz, to continue negotiating for Cumberland Island land.[21]

Immediately controversy arose over the Park Service's land acquisition
priorities. With acquisition funds limited to $10.5 million by Congress, the
agency needed to choose its purchases wisely. Park Service officials decided
to negotiate options on the large parcels while construction continued to
take place in the subdivisions. The option procedure had been authorized
in an amendment to the Land and Water Conservation Fund Act on July 15,

1968. With this method the buyer and seller agree on a price. The buyer pays a portion of it to prevent sale of the land to anyone else and then later exercises the option by paying the remainder of the price. The law required that an option last a minimum of two years and that the option amount be credited to the total price. During 1973 the entire $500,000 land budget for Cumberland Island was devoted to options on large tracts of Carnegie land.[22]

This approach drew immediate criticism from a variety of sources including Georgia governor Jimmy Carter and various newspapers in the state. On October 12, 1973, Carter wrote to Secretary of the Interior Rogers Morton that developers had increased the price of the land from the $500-per-acre price that the Carnegie group offered to the $5,000 for less than one-half acre being charged by subdivision developers. He pointed out that "since the Act places a spending ceiling of $10 million on land acquisition for the Seashore based on current appraisals of value, the actions of these developers are not only likely to raise the price of certain parcels, but also endanger the establishment of the entire island as a National Seashore." Carter also noted that large landholders, including the Candler, Ferguson, and Rockefeller families, "are irate at what the others are doing to the island. They refuse to negotiate with the National Park Service until the building ceases." He concluded by urging Morton to seek more acquisition funds and to expedite land acquisition.[23]

As the governor continued to monitor the situation, more alarming information came to him. As the Park Service appraised and bought lots for $14,000 per quarter-acre unit, Franklin Foster sold 5 acres in segment 1N to developer Robert Davis for $63,000. Carter wrote this time to Georgia senator Sam Nunn, stating: "This five-acre tract is at least as valuable and, if acquired by the lot, would cost in excess of $175,000. This one transaction could cost the taxpayers over $112,000 more than originally anticipated by the Park Service." Meanwhile the *Atlanta Journal and Constitution* issued an inflammatory article describing extensive construction.[24] Outraged letters from the public and environmental groups quickly added to the governor's complaints.

The National Park Service answered Senator Nunn with a detailed explanation of its strategy. Acting associate director Lawrence C. Hadley wrote that there were some seventy-five landowners at that time, but only three owned "over 1,000 acres, two over 500, three over 300, and six in the 60 to 80 acre range." The government, he reported, was negotiating on all the larger parcels. Hadley then explained the logic of the Park Service approach: "If a lot holder builds, for example, a $10,000 improvement on his property, this will in all likelihood not cost the United States any more than what he has spent on it, whereas, if $100 per acre in value through inexorable escalation is added to one of the larger tracts, there can easily be from $300,000 to over a million dollars added to our cost. So it is really escalation in land values that is skyrocketing, not the building on small lots. If 50 lots were built on, less cost escalation would result than the appreciation in land values on the larger tracts."[25]

Secretary Morton, responding directly to Governor Carter, added that larger owners were eligible for retained rights while these later landholders were not. He explained that the structures on the small lots would be removed and that the governor should not worry because "the land heals quickly." Furthermore, noted the secretary, the newspaper articles reported "more activity is occurring than is actually the case." The National Park Service had observed only two new construction starts.[26]

In essence the Park Service strategy was a sound one, but land acquisition proved more difficult and vastly more expensive than anyone had expected. Although 1974 saw acquisition of Laughlin-Cooper segment 1S for $2.4 million and the unsold OGR lands for $830,000, other large landholders including the Rockefellers and the Candlers backed away. Even worse, two Carnegie family members belatedly attempted to subdivide and sell their lands. Both were located in segment 1N.

Robert Monks, son-in-law of Lucy C. S. Rice, engineered the first attempt to subdivide. Late in 1972 millionaire Monks, a failed Republican candidate for the Senate from Maine, advertised lots for sale at a price the *Atlanta Journal and Constitution* called "twenty times" the amount being asked by owners selling to the National Park Foundation.[27]

The first to respond to Monks's scheme was Democratic governor Carter. Expressing his outrage, he warned buyers not to be gullible and fall for vastly inflated prices for land that they would ultimately lose to the federal government at "fair market prices." In addition, he asked state officials to investigate possible fraud charges, calling Monks's sales "an overt attempt" to escalate land values by trying to create a subdivision without roads, utilities, or any access to individual lots.[28]

Jekyll Island real estate agent Douglas Adamson, representing Monks, retorted that the millionaire had the law on his side and, furthermore, that land had sold on the island for $25,000 per acre. Where he got such a figure is unknown. Certainly at that time no Cumberland Island land had sold for anywhere near that amount. Adamson also told potential buyers that the National Park Service would acquire only 70 percent of the island. Governor Carter called that assertion "a false statement" and added that state health officials "probably will not approve septic tanks on such tiny lots." Meanwhile the Park Service warned that whoever bought land could not build on it and would lose it eventually. Apparently Monks thought better of the plan, for he never sold any lots. In January 1975 the Park Service bought his 324 acres for $754,400. Monks retained a twenty-five-year right to the "Stafford Beach House" but conveyed it to his nephew and niece, Franklin W. Foster and Lucy Carnegie Sprague Foster, shortly thereafter.[29]

Phineas Sprague, another heir of Lucy C. S. Rice, executed the last attempt to subdivide a portion of the island in 1977. The preceding September, after negotiations reached an impasse, Superintendent Paul McCrary recommended to Regional Director David Thompson that Sprague's land, along with more than a dozen smaller parcels, be condemned.[30] Sprague responded by selling some land to Robert Davis, who began subdividing and selling lots near the ocean beach. One member of the Carnegie family speculated that Davis was deliberately provoking the Park Service and worried that it might lead to condemnation of all remaining private land. This time the Park Service reacted with alacrity. On February 18, 1977, the agency issued a "declaration of taking" against Sprague and the four purchasers of his lots. This form of condemnation allows the government to assume control of the disputed property at once. Subsequently, the agency paid $3,648,499 to Sprague and another $84,405 to the others. No retained rights were granted.[31]

In addition to these large and expensive land purchases, Charles Fraser approached the Park Service in 1974 and asked to sell his two retained rights at Sea Camp. The agency hired South Carolina appraiser H. Philip Troy to appraise the estates. In his January 1975 report, Troy suggested, not very subtly, that the government was being had. It had executed a transaction "inversely comparable to the fabled purchase of Manhattan Island" by selling land that it could not use for forty years. Now, he added, the government was preparing to pay a second time for the same 40 acres of land plus the steep rise in their value over the previous four years. After getting this off his chest, Troy assessed the retained-right contract as a forty-year leasehold and appraised the two parcels at $570,850.[32]

When the agency reacted slowly, Fraser wrote to National Park Service director Gary Everhardt and threatened extensive development. He pointed out his rights to choose an alternative 40-acre site and to build seven houses. He described his plan to "'stretch' our 40 acres along the entire oceanfront of tract 4 South, building each of the seven houses on a very wide, shallow oceanfront site." He and "a group of corporations" planned to commence their construction within sixty days and to lease these properties on a weekly basis. Finally, he wrote, "should our perception of Park Service lack of interest not be correct, and if, in fact, the Park Service has the funds available to make such an acquisition and desires to do so at a fair price, we would be willing to defer a bit longer." In July 1975 the Park Service paid Fraser's Cumberland Island Holding Company $600,000 for relinquishing its rights to the original holdings and granted another 2-acre retained right on Cumberland Sound just south of Sea Camp.[33]

A postscript to this activity occurred in 1988. By that time Charles Fraser had sold the Cumberland Island Holding Company to a group of investors led by Richard Goodsell and attorney S. Larry Phillips. The latter contacted the Park Service to complain about heavy visitor intrusion through the company's retained estate, which straddled the popular River Trail between Sea Camp and Dungeness Dock. He suggested a swap of retained rights again, this time to two existing houses plus 5 acres in the Davis subdivision in segment 2S. After he threatened to exercise the company's right to build a "lodge" along the River Trail, the government again acquiesced and again relocated that retained estate.[34]

By 1980 most of the remaining interests of the Thomas Carnegie II heirs who originally had sold to Fraser ended. First, Thomas IV sold his retained right back to the government in 1976 for a paltry $4,900. Meanwhile, Cumberland Island Properties, of Henry C. Carnegie and James Bratton, neglected to build any residences on its 15 acres. Its agreement, like most on Cumberland Island, required the retained-right holder to start construction of a residential structure within ten years of the National Park Foundation purchase or lose the estate. A cryptic handwritten note jotted on the Southeast Regional Office's copy of the retained-right agreement states "not used, not built, see next page, 10-year building requirement." The Park Service thus recognized the partnership's failure to comply with the requirements of the retained right. Nevertheless, the seashore continued to list Henry Carnegie and James Bratton as retained-rights holders until 1999 when the error was exposed.[35]

During these expensive acquisitions Lucy and Franklin Foster Sr. tried to work out an arrangement to donate more land for the national seashore, but even this became complicated. In 1973 the Fosters offered more than 200 acres in installments over a period of years, obviously to benefit from tax considerations. To their consternation the government insisted on negotiating rather than simply accepting the gifts.

Franklin Foster Sr.'s complaints to the *Atlanta Journal* forced Park Service land agent William Kriz to respond. In the first year of the Foster plan, Kriz explained, they would donate 40 acres paralleling the ocean but the beach itself would remain in their possession. But to the Park Service anything less than its fee simple ownership of the beach was unacceptable even though Foster promised access to the public. Eventually negotiations calmed ruffled feathers, and the Fosters donated 212 acres in seven separate gifts between 1973 and 1987, retaining rights of use on most. Ultimately, Lucy Foster was the only Carnegie heir to donate land for the national seashore other than a few acres that accompanied the Johnstons' Plum Orchard mansion.[36]

Subdivisions Revisited

While the agency emphasized acquisition of large holdings, it also pursued appraisals and negotiations with "new" owners in the three ongoing subdivisions. The Davisville group, first to buy onto the island and most active in construction, drew the initial attention of Kriz and the National Park Service. During and immediately after the congressional campaign to create the national seashore, various lot owners maneuvered to cement advantages wherever possible. Those who wished to keep their island property built homes with hopes they could convince the government to relent and let them stay.

Attorney J. Grover Henderson, who co-owned a Davisville property with his mother, became their spokesman. He waged a media campaign blasting the Park Service for discrimination against the small owner while allowing the larger ones, the Carnegie and Candler families, to negotiate for retained estates. He neglected to mention the provision in the seashore's enabling act that allowed owners who began construction of houses before February 1, 1970, to have such a right. The older owners had such structures and hence the rights. Most subdivision occupants had no such construction by that date and, therefore, no rights. The *Florida Times-Union* also quoted a Henderson statement calling for conversion of the area to a national wildlife refuge so the owners would be left alone and a "limited number" of visitors could still see the island.[37] This campaign to protect new, local landowners sold well in Camden County where people already harbored a growing dislike for the National Park Service.

Ultimately, the Davis subdivision went quietly, but by no means cheaply, to the government. Acquisition began in earnest during 1974, and most properties were in Park Service hands by early 1976. Prior to and during this period, however, both local owners—most descended from families who had known each other for generations—and speculators like Louis McKee and Charles Fraser engaged in a blizzard of lot sales between themselves and with other confederates on the mainland. For example, Davis sold three lots to Richard Brazell for $18,000. In July 1972 Brazell sold lot 5 back to Davis for $15,000. Three months later the lot was back in Brazell's hands, and he executed a sale and buyback with Jiles Hamilton. Finally, in July 1974 Brazell sold lot 5 and another lot for $28,000 to the National Park Service. The government rewarded Brazell's initial investment of $6,000 per lot with $14,000 per lot based on "current market value." So many sales took place, usually upping the price along the way, that some county residents passed through the ownership histories of several lots and never negotiated with the government.[38]

Most Davisville owners received $14,000 for their .39-acre lots. Those with more land received equivalent per-acre settlements. Those who had built homes were able to demand much more. However, the most important factor in the eventual settlement price was the ability of the owner to stall negotiations and resist condemnation. The longer an owner kept his land, all other factors being equal, the more elevated its price became.

Park Service appraiser Finis Rayburn calculated that the value of small lots on the island rose 92 percent between May 1970 and October 1972, an increase of 3.17 percent per month. He posited that the rate had slowed to 2 percent per month from October 1972 to his October 1973 appraisal of William Rogers's Davisville holding. Rayburn concluded that Rogers's three lots, plus improvements, were worth $65,000. Four months later the Park Service paid William Rogers $76,350 for his 1.3 acres.[39]

In addition to land price inflation, a delay in selling gave an owner the chance to increase the value by building a house. For example, Camden County tax commissioner George Law rebuffed the Park Service on two lots until March 1976. He received $15,000 per lot at that time. On another Davisville lot, purchased from Louis McKee, Law built a home and refused to sell at all. The Park Service filed for condemnation, and Law answered by demanding a jury trial. Local jurors awarded him $120,000 for his .39-acre parcel and small house. The government unsuccessfully appealed before paying the full amount to Law in June 1979.[40]

In the meantime, Robert Davis offered to sell his own remaining lands within and adjacent to the subdivision, including the leased tract. In 1975 he held 28.99 acres of upland and claimed 46.22 acres of marsh. Henry Crawford had somehow become sole holder of the lease on the eastern parcel. Thus, the sale required three separate components. First, Robert Davis held fee title to the land. Second, Davis required reimbursement for the projected lease payments for the rest of the ninety-nine-year lease. Finally, Crawford required payment in lieu of the money he expected to earn on the commercial property.

The Park Service appraised the land at $385,000. Davis and Crawford countered with a 56 percent higher figure of $602,000, three-fourths to Davis and one-fourth to Crawford. On September 23, 1976, the government paid $602,000 to the Davis Land Company for 29 acres of upland and a quit-claim on 46 acres of marsh. When added to the nearly $350,000 he realized from lot sales west of the road, Davis reaped at least $800,000 on land he had bought for $60,000 nine years earlier.[41]

National Park Service officials faced the same pattern of land swapping and recalcitrant owners at the OGR and Miller subdivisions as well. In a later interview Laurence Miller Jr. admitted that "we knew the National Park was comin[g] and we had no value on the land. You know land is valued according to what is sold in the area. And nothin[g] had ever been sold in the area [north end]. So in order to establish the value of the land, I subdivided some of it and I sold seven lots, I think. And by so doin[g] established the value of the land."[42]

At OGR, Louis McKee, various lawyers and real estate agents, former Davisville owners, and what Superintendent Bert Roberts called "a 'Who's Who' of county government" waged a buying and selling spree like that in the Davis subdivision.[43] From Ricketson's five original sales, new owners subdivided into twenty-one lots. A month after county clerk James Godley purchased 7.5 acres from OGR for $15,000 in September 1971, he sold 2.7 acres to County Commissioner George Hannaford for $5,480. Before the end of the year, the latter then conveyed .82 acre to M. W. Hannaford for $4,000. In January 1977 M. W. Hannaford sold his .82 acre to the United States for $20,000. George Hannaford sold .72 acre to the Park Service for

$13,000 during that same winter. Other lands resold by both Godley and George Hannaford enjoyed similar increases. The six-year price rise from Godley's purchase to the 1977 sales exceeded 800 percent.[44]

New Appropriations and Condemnation

Of course, these rapidly escalating land prices for both large and small tracts forced the Park Service to reach the enabling act's limit on land acquisition funds very quickly. As early as 1974 agency officials and Governor Carter began urging Congress to raise the spending limit for Cumberland Island property. Eventually, Congress included an increase in the land acquisition ceiling to $28.5 million in the National Parks Omnibus Bill of 1978 passed during Jimmy Carter's presidency.[45] As is common with omnibus bills, the near tripling of Cumberland Island's land budget escaped close scrutiny and criticism by legislators.

Armed with new funds, the Park Service became more aggressive in land acquisition. Between 1978 and 1982 the agency condemned forty-one tracts of land, primarily in the OGR and Miller subdivisions.[46] Many owners resisted, and the court costs and land prices were very high. A medical doctor, Robert Van Cleve, unsuccessfully attempted to maintain his holding in the OGR area by claiming he had built a clinic to serve island residents and visitors.[47] County politicians Robert Read and James Bruce and their relatives lost their land. So, too, did Lucy Ferguson's friends in the Peeples and Readdick families. However, most realized profits ranging from 800 to 1,400 percent for their six to eight years of ownership.[48]

On the north end the Park Service not only condemned all the parcels near the Settlement sold by Laurence Miller but all the Miller land as well. However, in deference to her long life and considerably longer heritage on the island, National Park Service land officials worked out a reserved estate for Mary Miller and her family. Rather than paying the court-decreed $125,000 for her remaining 84 acres, the agency paid $63,753 and allowed her a retained estate on Fader's Creek.[49]

The Park Service also pursued the lands held by the heirs of Olaf Olsen and the Bunkleys. The Olsen lands were divided into two tracts. Virginia Olsen Horton and family sold their 12.5 acres to the government in July 1977 for $136,930. Her brother, O. H. "Bubba" Olsen, held a second tract that he refused to sell. The government condemned his property in late 1980, and he received $120,000 for 10.8 acres. A long list of Bunkley heirs split $90,000 for their 7-acre parcel after the agency condemned it in 1980. Only the Horton group, which included Bubba Olsen, received a retained right.[50]

At that same time a 1.1-acre tract on Fader's Creek with a preexisting structure purchased from the Millers by Louis McKee and the lot in the Settlement owned by Grover Henderson became embroiled in a remarkable series of events and exchanges that resulted in two more retained rights on the north end of the island and another in the Davis subdivision. It began with the arrival of a biology student and lifelong naturalist Carol Wharton (née Ruckdeschel) on the island. For a time she managed to stay on the island by working for the Candlers at their High Point estate. Around that time she became romantically involved with surveyor Louis McKee. Subsequently McKee made her co-owner of some of his lands and heir to his estate. On March 31, 1977, McKee and Wharton conveyed the tract on Fader Creek to the Park Service, reserving life estates on it for each of them. A month later Wharton quitclaimed her interest in that retained right to McKee for $6,000.[51]

Events took a turn for the bizarre in late 1979. McKee and Carol Ruckdeschel (having retaken her maiden name) still held joint ownership of another property within the Settlement that McKee had purchased six years earlier for $6,000. McKee quitclaimed his interest in this land to Ruckdeschel and her parents during the Christmas holidays after she optioned this land to the Park Service. In March 1980 Ruckdeschel sold this property to the agency for $45,000 plus a retained right. Barely a month later, Carol Ruckdeschel shot and killed Louis McKee.[52]

The episode is still clouded in mystery and some controversy. Apparently Ruckdeschel became estranged from McKee. Robert Coram of the *Atlantic Weekly* later reported that McKee physically abused her in the weeks before the shooting. On April 17, 1980, while Ruckdeschel sat inside her Settlement house with Peter DiLorenzo, a visiting hiker, McKee appeared and demanded entry. Ruckdeschel later stated that she feared for her safety and would not allow him to enter. As he tried to break through the closed door, she shot him in the chest. By the time rangers responded to her phone call, McKee had died. The Camden County sheriff and park rangers took Ruckdeschel and DiLorenzo to the sheriff's office in Woodbine, where the hiker corroborated her statement. A coroner's jury cleared and released her without charges the following day.[53]

Upon the death of McKee, the National Park Service presumed that his life estate on the Fader Creek property had ended. In an effort to remove retained-rights holders from the historic settlement, the agency traded rights with Grover Henderson and gave him rights to the Fader Creek parcel in August 1982. One year later Betty Johns, representative of the estate of Louis McKee, quitclaimed all rights to the Fader Creek property back to Ruckdeschel at no cost. On April 30, 1984, Ruckdeschel filed suit against Henderson, claiming he had no right to the property and should never have occupied it. The federal government, in turn, sued Ruckdeschel to quiet her claim against the tract.[54]

The various legal maneuvers ultimately led to an agreement between the three parties whereby Ruckdeschel maintained a retained right in the Settlement adjacent to the old church and secured a right for herself and her parents on the Fader Creek acre. Grover Henderson in turn relinquished the latter property for a 1-acre plot in the Davis subdivision with a much better house built a few years earlier by original Davisville buyer Alvin Dickey before the government condemned his property. Henderson also demanded and received a better package of retained rights. The federal government wound up losing a good deal of money and carrying three—not two—retained-right estates.[55]

The Park Service also condemned one last large tract on the southern end of the island during this period. Negotiations with all five Rockefeller landholders reached an impasse by 1978. Government appraisers insisted that an 80-acre parcel was worth about $250,000. Nancy Rockefeller and her four children demanded a great deal more for their land in segment 3S near the Sea Camp development area. Ultimately, the Park Service and the Rockefellers agreed to proceed to condemnation and let the court decide on the land's value. In a 1996 interview James Stillman Rockefeller Jr. stated that he and his siblings "don't know for sure why my piece was picked on," but the Park Service proceeded apace with prosecution of his case. Both Rockefeller and Southeast Region lands specialist Tom Piehl speculate that his was a "test case" to gauge the expense of further condemnation of the parcels of the Rockefellers and other large landholders.[56]

In June 1979 the court condemned J. S. Rockefeller's 62.6-acre tract but ordered the Park Service to pay $1,126,080, five times its offer for the land three years earlier. This huge settlement, coupled with the $3.65 million court award granted to Phineas Sprague, scared the agency away from condemning the other parcels of Rockefeller land. Even with a new, much higher land acquisition ceiling, these per-acre prices threatened the Park

Service's ability to afford the remaining private land on the island. J. S. Rockefeller's tract remains the only piece of original Carnegie land taken from a family member who did not threaten subdivision or development.[57]

One reason the Park Service hesitated to spend much more money on condemnation was the optimistic turn negotiations for the huge High Point property had taken. It actually ranked third in a 1980 land acquisition prioritization of the agency's desired lands. However, the first two were unattainable. First on the list was a mainland headquarters parcel at Point Peter near St. Marys. But its purchase was mired in the controversial general management planning under way at that time. Listed second were the two Greyfield tracts. This served no purpose other than unnerving the Fergusons because the agency continued to honor its promise not to bother Lucy Ferguson.[58]

The High Point property consisted of nearly 2,200 acres. Negotiations had proceeded fitfully since before the seashore's establishment. Finally, the Candlers signaled their interest in pursuing the sale. Congress allocated $10 million for the purchase, and the Trust for Public Lands took over direct negotiations. As the discussion proceeded under the glare of media attention, however, a new factor entered the scene. With legislation for wilderness pending in Congress, environmental organizations questioned the terms of the proposed acquisition. The Wilderness Society opposed granting the Candlers a "50 acre" retained estate around their compound in proposed wilderness with rights to exclude the public. In response, Regional Director Joe Brown pointed out that there were already eleven retained estates north of Plum Orchard and that "it is unrealistic to expect that we can achieve pure wilderness on the island until these rights have expired."[59] Presumably this was meant to suggest that another retained right was a small price to pay for acquiring such an important piece of land.

Both negotiations and the wilderness legislation proceeded into 1982 with the Candlers insisting on the exclusion of their estate from the wilderness, environmentalists pushing for at least "potential wilderness" status, and the Park Service looking for some middle ground. Eventually the agency agreed to a complex package of retained rights including a 38-acre estate, exclusive rights to adjacent docks, roads and beach access, the right to post "No Trespassing" signs at the compound, and a price of $9.6 million. These rights are to last until the death of the last of twenty-eight named individuals, some of them small children at the time of the sale. On January 20, 1982, High Point conveyed the lands to the Trust for Public Lands, which passed them to the National Park Service five days later. Nine months after that, Congress established wilderness on the northern portion of the island, setting up a whole new slate of management problems.[60]

Active private land acquisition halted after the January 1984 purchase of a .12-acre plot in the Settlement from George Merrow, a descendant of a former slave. In twelve years the National Park Service had accepted, bought, or condemned 149 tracts of private land totaling 18,687 acres for a combined price of slightly over $23,843,700, well over double the original congressional appropriation. Lucy Foster would donate three more parcels totaling 8 acres over the next three years. Five private owners, four Rockefellers and Lucy Ferguson, still held title to more than 1,650 acres in seven tracts. Six of those tracts spanned the island from sound to ocean. In acquiring land for the seashore, the National Park Foundation and later the National Park Service granted twenty-one retained estates ranging from twenty-five years to the life of unborn children.[61]

Three other owners of large Cumberland Island properties, all public entities, also continued to reject National Park Service overtures. The Department of the Navy owned Drum Point Island, a low mound of dredge spoil west of the Stafford property. Over the years vegetation had covered the island, creating a visual buffer between Cumberland Island and the Kings Bay military facility. The Park Service sought the 139-acre island in order to prevent further dumping and maintain the vegetation screen.[62]

A similar but more troublesome threat faced the agency at Raccoon Keys at the southwest end of Cumberland Island itself. There the U.S. Army Corps of Engineers rebuffed the Park Service and continued to hold the 518-acre tract as a potential site for future dredge spoils from the Intracoastal Waterway. Both these properties remain in military hands today.[63]

The third government property owner was the state of Georgia. The state owned all land below the mean high-tide line, all saltwater creeks, and extensive marshes west of the northern half of Cumberland Island. The enabling act required that the National Park Service could only obtain these lands through donation. During the early years of the seashore, the agency confidently expected the donation at any time. However, as the wilderness planning proceeded, it became clear that the Park Service (in reality the Wilderness Act) would not allow motorboats to access the beaches and creeks. In February 1978 the Georgia Department of Natural Resources, responding to loud complaints from local mainland residents, announced it would not turn over any lands unless the Park Service struck that requirement from the proposed plan. Four years later Congress passed a wilderness bill for the island. The state still maintains control of 13,820 acres of beaches and inland waters.[64]

The Retained Rights

Twenty-one persons or parties received retained rights to twenty-four pieces of property during the active land acquisition years. Margaret Wright and Thomas Carnegie IV sold theirs back to the government. Charles Fraser returned his two and accepted a new one and a large amount of money. Robert Monks conveyed his property to the Fosters where it became part of a single bundle of rights on all their donations. Finally, Cumberland Island Properties, of Henry Carnegie and James Bratton, failed to satisfy contract requirements and became a ghost entry on Park Service land records. By 1984, therefore, these arrangements had distilled down to seventeen parties holding rights to eighteen pieces of land, with Marius Johnston having no land while Lucy Graves and Carol Ruckdeschel each held two parcels (map 4.2). Each and every agreement differed in contractual obligations and permissions.[65]

In July 1975 Martin Baumgaertner of the Southeast Regional Office summarized the retained rights on Cumberland Island granted by the National Park Foundation.[66] The description of retained rights below is drawn from his document plus the deeds and retained-rights files of other island residents. It gives only a general sense of the rights and requirements mentioned because many variations of language and specifics exist within each contract. Through the history of the national seashore, retained-rights holders, the National Park Service, and other interested parties have questioned, challenged, or ignored many contract stipulations. The most common issues can be grouped into five areas: structures, docks, roads, beach driving, and damage to park resources on or off various estates. Arguments and contrasting interpretations have arisen over other specific stipulations, but these five have dogged management of the seashore consistently for more than thirty years.

The use of buildings for noncommercial, residential purposes is the most fundamental element of a retained-right contract. Two types of stipulations exist: the right to build one or more new dwellings on a piece of property and the right to use and perhaps modify an existing one. Some rights holders have both. Those who could build usually had a time limit

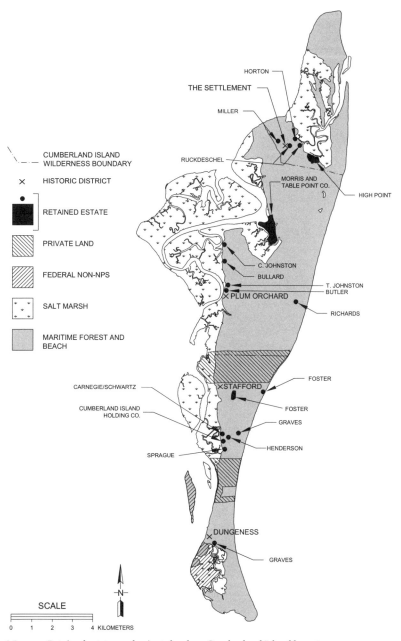

CUMBERLAND ISLAND
WILDERNESS BOUNDARY

✕ HISTORIC DISTRICT

● RETAINED ESTATE

PRIVATE LAND

FEDERAL NON-NPS

SALT MARSH

MARITIME FOREST AND
BEACH

HORTON

THE SETTLEMENT

MILLER

RUCKDESCHEL

MORRIS AND
TABLE POINT CO.

HIGH POINT

C. JOHNSTON

BULLARD

T. JOHNSTON
BUTLER

✕ PLUM ORCHARD

RICHARDS

FOSTER

CARNEGIE/SCHWARTZ

✕STAFFORD

CUMBERLAND ISLAND
HOLDING CO.

FOSTER

GRAVES

HENDERSON

SPRAGUE

DUNGENESS

GRAVES

SCALE

0   1   2   3   4 KILOMETERS

N

Map 4.2. Retained estates and private land on Cumberland Island by 1987

within which to do so, most often ten years. A number of controls on new home construction are included in the contracts. First, each holder was limited to a maximum number of new dwellings. Second, some were limited by dwelling size restrictions. Third, nearly all had height restrictions of forty feet. Finally, some deeds specify the right to add support structures, including sheds, fences, and corrals.

Those with structures in existence on February 1, 1970, can use them for residential purposes. Indeed, the presence of these houses allowed the only three outsiders, Henderson, Ruckdeschel, and the Goodsell-Phillips team, to retain estates. The age and value of these dwellings varied from the century-old Stafford mansion, to the ramshackle 1930s Trimmings house in the Settlement (also on the National Register), to the Sea Camp house built by Fraser in the late 1960s. Also governed by contract stipulations for the use of existing buildings are various Davisville houses for which retained-rights holders traded their original retained estates. The contracts usually, but not always, stipulate that the homes cannot be substantially altered or expanded, especially those in historic districts. In some cases the contracts define who should pay for maintenance; in other cases they are silent.

One further provision, present in all but three of the contracts, states that the holders can lease their "non-commercial, residential" property to anyone they choose. In most cases leases must last at least one week. A few require ninety-day leases. This very commercial-sounding noncommercial use means that most but not all of the holders can temporarily transfer most of their other rights to anyone who rents their estate.

The matter of docks and island access is one of the most diverse and confusing issues. Some contracts stipulate that the holder may use the "Main Dock" at Dungeness. Others specifically deny its use. Still others are silent. Many refer vaguely to "National Park Service docks," others to assorted extant docks from one end of the island to the other. Some grant permission to build a dock. Some are exclusive. Some are not. Many include a combination of these provisions. Furthermore, a couple of the contracts allow their holders to "maintain navigability" in streams that flow through or alongside their properties. The matter of use of the three existing airfields plus any built by the Park Service in the future is equally diverse.

Cumberland Island contains a number of roads and a few trails that once functioned as roads. The right to drive a vehicle over any or all of these forms another highly variable issue. Use of the Main Road is universal although the Schwartz holding allows use only from the estate to an unspecified dock. Other roads are more problematic. Contracts with the Johnston branch usually stipulate freedom to use the Plum Orchard road. A few specify use of the road leading to the Duck House area (Richards estate). Most are vague. The contract for Mary Bullard, in deference to her physical disability, allows her to drive a vehicle on any and all roads and trails. One or two others seem to hint at similar rights. Most contracts are silent on road use other than the Main Road.

Intimately associated with driving the roads and trails of Cumberland Island are two specific issues: access to the beach and driving along the beach. Unfortunately for later seashore managers, most contracts are silent on both. Only Mary Bullard's contract specifically stipulates the right to drive on the beach. For the others one is left to decide whether the beach is, in fact, a traditional "road."

The final group of issues surrounding the retained-rights agreements concerns the matter of resource protection. The National Park Service owns this property. It has a right to expect that estate holders will protect the valuable natural and historic resources for which the seashore was established. Nearly all contracts specify that residents or their assigns should not "commit waste" or cut timber. In a few cases the minimum size of a living, off-limits tree is specified. On the other hand, several contracts forbid cutting trees except for "residential purposes," but that phrase is left unexplained. Some contracts invest their holders with rights to stabilize sand dunes and develop a pathway to the beach but also forbid driving over nonvegetated sand dunes.

Finally, some agreements call for maintenance of estate grounds in a "neat" or "tidy" condition. Coleman Johnston's contract carries a particular 1975 addition for protection of archaeological resources on his Table Point estate. He agreed to check with Park Service officials and allow a professional survey of this rich Indian site before building his new dwelling and support structures.

The sum of these general and specific stipulations, their wildly diverse applications between seventeen parties and eighteen estates, and the confusion of meaning in the very words used in the contracts set the stage for conflict among parties interested in the national seashore. Furthermore, seashore managers soon realized that what these terse legal contracts state and what the Carnegie and Candler heirs claim they were promised are two very different things.

Looking at the saga of land acquisition at Cumberland Island, one is

struck by the extraordinary prices paid for some land and the generous retained rights granted, at least by Park Service standards. Essentially it was a
matter of desire. That desire began with the first seashore surveys in the
1930s. It swelled as the agency frantically assessed surviving coastal recreation opportunities during the 1950s. Desire sharpened further with each
Park Service visit to the idyllic island.

After Congress established the seashore, desire drove the agency to seek
fee simple ownership of the entire island. Most Carnegie descendants remained true to their pledges to sell at low cost to the government. However,
a few, coupled with many latecomers, sought to make money on the new
park. That desire for the island betrayed the Park Service. Cagey owners, old
and new, recognized that the agency would pay high prices to gain control
of paradise. That desire also led to exacting retained-rights agreements, far
more complicated and permissive than those present in other parks. Those
agreements quickly became and remain today the major management issue
of Cumberland Island National Seashore.

Planning and
Operating in
the 1970s

In November 1972 National Park Service officials turned from celebrating the establishment of Cumberland Island National Seashore to the reality of managing it. The legislative history of the new unit stressed its importance in all three of the national park system's raisons d'être—recreation, cultural resource protection, and natural resource preservation. From the beginning, Park Service planners at Cumberland faced uncertainty and conflict over how to prioritize them. Budget limitations and conflicting laws and policies demanded decisions unpopular among some agency personnel and segments of the public. The presence of a highly vocal and politically powerful group of island residents further complicated management.

In addition to the ongoing program of land acquisition, seashore officials faced a daunting array of tasks. First, they needed to evaluate the natural and cultural resources, which required research on the ecology, archaeology, history, and historical architecture of the island. Assessment of the impact of decades of vacation use by the Carnegie and Candler heirs was also necessary. Decisions had to be made on what to preserve, what to ignore, and what to eliminate (map 5.1).

Extensive planning would also be required. The National Park Service had to locate and acquire a permanent headquarters site on the mainland, develop recreation and interpretation facilities on the island, and devise a transportation plan for the new national seashore. Planners also had to decide upon the levels of visitation and tourism infrastructure. At least four island areas—Dungeness, Stafford, Plum Orchard, and High Point–Half Moon Bluff—required historic resource, landscape, and public use plans. Adaptive use of existing historic structures had to be developed. Even historic interpretation for visitors required more than the usual amount of

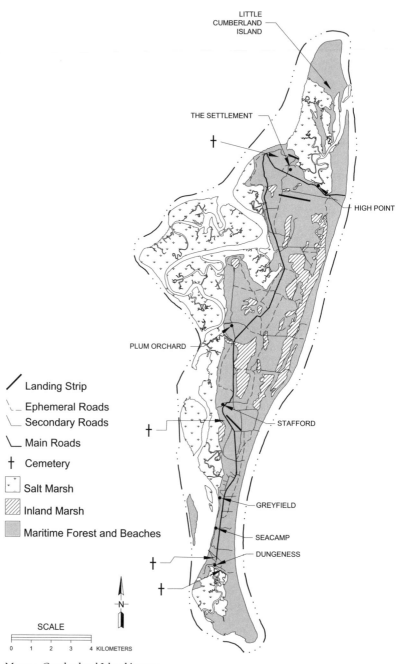

LITTLE
CUMBERLAND
ISLAND

THE SETTLEMENT

✝

HIGH POINT

PLUM ORCHARD

/ Landing Strip

＼－ Ephemeral Roads

＼ Secondary Roads

＼ Main Roads

✝ Cemetery

Salt Marsh

Inland Marsh

Maritime Forest and Beaches

✝ STAFFORD

GREYFIELD

SEACAMP

DUNGENESS

✝

✝

N

SCALE

0   1   2   3   4 KILOMETERS

Map 5.1. Cumberland Island in 1972

planning. What historic period should be emphasized: that of the Indians, the cotton planters, or the rich man's retreat? In addition, section 9 of the act that established the seashore required the Park Service to report on the suitability of any part of the island for wilderness designation.[1] Given the agency's responsibility for tourism development on a designated recreation area, as well as its traditional reluctance to pursue such restrictive management zoning, this would be a particularly controversial task.

Finally, while Park Service officials grappled with these fundamental, long-term issues, they had to open the seashore to visitors and manage the day-to-day operations. These duties required a staff and a regular budget plus special funds for planning, development, and resource management. The agency had to locate and open a temporary mainland visitor contact station, develop interim transportation to and on the island, provide a few minimal facilities on Cumberland for visitor use, and interpret the natural and cultural features in a very short time. These immediate needs inevitably led to another question. What privileges and limits did federal law and the retained-rights contracts impose on the island's residents, both contract holders and fee simple owners? Long-term planning was expected to settle most management issues, but satisfying a public anxious to visit its new recreation reserve and protecting the resources demanded immediate decisions.

During the 1970s the National Park Service tackled all of these issues and programs with varying degrees of success. Seashore officials met shocking legal setbacks, extensive disagreement with other government agencies, and criticism from every quarter of the public. By 1980 the Park Service realized that it was in the middle of a management quagmire.

What Did the National Park Service Receive?

The National Park Service presence on Cumberland Island and the nearby mainland was initially minimal. For the first year Superintendent Walter Bruce of Fort Frederica National Monument conducted the affairs of the new unit. Other than land acquisition, which was being handled by officials from Washington, D.C., and Atlanta, Park Service activities consisted of periodic visits to the island and information-gathering studies. Personnel from the Southeast Regional Office in Atlanta undertook many of these initial inspections, while the agency's Technical Information Center in Denver, Colorado, or specialists outside the agency conducted the more in-depth studies. No budget for Cumberland Island existed in fiscal year 1973, so the regional office diverted $94,700 for operations from other parks. By April 1973 the Park Service had two employees, two boats, and two jeeps on the island. Because the latter were in a constant state of disrepair, much of the inspection work on the island had to be done in vehicles borrowed from the state of Georgia.[2]

Almost immediately an issue arose that presaged future management problems. The nine decades of Carnegie presence on the island had deposited a substantial amount of trash in dumps near the various estates. Added to this were waste materials left by Charles Fraser and those residents constructing new homes or altering old ones on their estates. Numerous old automobiles in varying states of decomposition lay scattered around the island. The *Atlanta Journal and Constitution* described one waste pile east of Dungeness as a "rancid open garbage dump." Another observer said it was "900 feet square." Carnegie in-law and sometime island resident Landon Butler told the newspaper, "When we wanted the Park Service to take over the island we thought they would take care of things like this" (fig. 5.1).[3]

National Park Service officials responded that their hands were tied. Jim Bainbridge, associate regional director for operations, told the *Atlanta Journal* that because Congress had not authorized funds for the new seashore, it was illegal for them to carry out any activity on the island. He added, "The residents are responsible for putting the garbage there and now they are the ones complaining about it." The islanders wanted the Park Service to bury the trash with a rented bulldozer, but Thompson responded that the agency could not do that because "we might find that burying it is going to pollute an underground water system."[4] Eventually, after an Environmental Protection Agency inspection, the Park Service removed most of the waste material. Members of the Youth Conservation Corps (YCC) accomplished much of the work during summer 1974. Nevertheless, simple cleanup work continued for nearly a decade on the island.

The removal of old cars was complicated by questions of ownership. As late as 1983 the agency was still looking up ownership of identifiable vehicles and notifying the last known titleholder either to remove it or sign a release giving the Park Service the power to do so.[5] Relations with the residents soon became inflamed over the disposition of some automobiles. When Bert Roberts was named superintendent of the seashore in November 1974, a number of older cars in very good exterior condition were stored in the Dungeness carriage house, which the Park Service planned to convert to a maintenance facility. Roberts ordered rangers to tow the cars outside and

Fig. 5.1. The wrecked remnants of automobiles littered the island in 1972 when the Park Service took over.

place them in a row along a nearby trail. This relocation added fuel to the residents' suspicion that the Park Service had no desire to preserve historic resources. In the 1980s the cars still formed an attractive curiosity for passing visitors, but currently they are rapidly becoming an oxidized soil layer.[6]

Three things were notable about the incidents with the trash and abandoned automobiles, and they set the tone for the incessant management conflict in later years. First, the Park Service faced the fact that it owned an island heavily changed by centuries of human use. The agency's desire to make it resemble an ideal national park would become controversial and burdensome.

Second, the retained-estate holders expected to be comanagers of the island. They had very well established ideas of how the natural and historical resources should be handled. These ideas would repeatedly clash with standard Park Service policy and the personalities of some agency officials.

Fig. 5.2. The dock at Dungeness was unusable in 1972.

Finally, the National Park Service was seriously underfunded for the tasks at hand. The difficulty of accomplishing the relatively simple chores of trash and abandoned automobile removal took years to solve. Monitoring the resources of Cumberland Island National Seashore, maintaining them, especially the more than 100 structures on the island, and interpreting them for visitors would prove enormously expensive. At no time did the national seashore have anywhere near the money necessary to do all these tasks well. As resources conspicuously decayed, many island residents, visitors, and the local public refused to accept that explanation (fig. 5.2).

Early Operations

The national seashore opened its first mainland office on land provided by the Brunswick Pulp and Paper Company near Shellbine Creek. As the date for opening of the seashore to the public approached, however, the Park Service decided that it needed a site more accessible to deep water and handy for visitors. In addition, the Brunswick Company lost interest in donating land to the government. With a final decision on a mainland site mired in the planning process, the Park Service elected to move to St. Marys on a temporary basis. Initially it intended to occupy a trailer in front of the riverfront MacDonnell Building until office space in the Century Theater Building across the road (now a submarine museum) became available. However, in May 1975 seashore officials moved into the MacDonnell Building on a month-by-month rental basis.[7]

Public pressure on the National Park Service to open the seashore for visitation began in November 1974. Although park rangers had received periodic questions about public access, the national seashore's second birthday ignited criticism that the Park Service had "locked up" the island. In response, the agency began planning to open to visitors during the summer of 1975. Public hearings and meetings with Camden County officials and businesses followed. Director Gary Everhardt and other senior Park Service officials visited to demonstrate the agency's commitment. The national seashore chartered a boat to conduct two round-trips per day from St. Marys to the dock at Sea Camp. The seashore opened on June 5 with plans to run through the end of August. Because the boat was small, reservations were required.[8]

On the island the Park Service, believing that a walk from Sea Camp to Dungeness and back again was too strenuous for some visitors, decided to provide transportation. Initially it used three trams to convey visitors (fig. 5.3). Later the General Services Administration provided a forty-passenger bus. The bus turned out to be one of those ideas that look good on paper but do not work well on the ground. On its maiden trip down the Main Road, it struck a low-hanging oak branch and ripped off part of the roof. This left only the trams, which repeatedly bogged down in the sand. Subsequently the Park Service abandoned the transportation system when seashore maintenance crews repaired the dock at Dungeness and the passenger boat was able to moor at both docks on the island (fig. 5.4).[9]

During the first three months, the island received 3,482 visitors, prompting the Park Service to continue boat runs through the off-season on fewer days the week. The following full year saw 17,480 visitors. More than 3,400 stayed overnight on the island, primarily at Charles Fraser's old campground near Sea Camp. By 1978 visitation topped 37,000 and included nearly 13,000 overnight campers and backpackers. Most visitors to Cumberland walked to and through the Dungeness complex, paying special attention to the ruined mansion. They then returned by foot or tram to Sea Camp, often by way of the nearby beach. Nearly all island visitors reported that they greatly enjoyed their time on the island.[10]

Fig. 5.3. In the 1970s the Park Service moved visitors between Dungeness and Sea Camp by tram. Once both areas had ferry service this expensive practice was discontinued.

Fig. 5.4. Visitor access is still by two or three passenger ferries each day.

One reason for their enjoyment was the elaborate interpretive program developed by the seashore's rangers. They offered such activities as seining in the ocean to study sea life, sunrise beach walks, evening history talks, "marsh tromps," and education about waves, currents, and the basics of surfing with the use of "boogie boards." Rangers also provided orientation talks on both the mainland and the island. Interpretive signs and a park brochure were available by 1977. The Park Service planned even more programs for the seashore's permanent mainland base once that was in place. In 1979 the Park Service added tours of the Plum Orchard mansion, which proved immensely popular. Unfortunately, the constraints of transportation limited the number of visitors who could enjoy this additional feature.[11]

Opening the seashore to visitors blunted much of Camden County's criticism of the National Park Service. However, the intense demand to enjoy the island soon raised the issue of access by private boat. Docking facilities on the island were limited, and many were private. Before 1979 Sea Camp and Plum Orchard were two potential landing sites for the public. However, public ferries monopolized the former while the latter was very small. During the first two months that the seashore was open, National Park Service officials turned down several hundred requests to land private boats on the island from "public bodies, corporations, organizations, groups, and individuals."[12]

The private landing issue drew Park Service attention to a potential public relations nightmare. During the seashore's first two years, the Brunswick Pulp and Paper Company had periodically brought guests to the island for tours. Seashore rangers facilitated these visits by providing logistical support and guide services. The policy of allowing these tours developed under Sam Weems during the time when the Park Service expected to have its headquarters and a parkway on Brunswick land. On occasion the company used a boat provided by the Georgia Game and Fish Department to bring vehicles in order to drive around the island. The Park Service anticipated legal problems when public visitation began and planned to issue a special-use permit to the Brunswick group. However, once the requests for landing private boats began pouring in, agency officials quickly changed their minds. Regional Director Thompson informed Brunswick officials that public scrutiny would certainly call this arrangement favoritism, and therefore it must stop. He cited the pressure coming not only from Camden County but also from tour companies in Fernandina. Thompson suggested that the paper company work out an arrangement with the Greyfield Inn. However, as the Brunswick Company's interest in hosting the seashore headquarters waned, so did its desire for a special-use permit.[13]

Questions about Retained Rights

Island residents began to test the limits of their retained rights as soon as Cumberland Island National Seashore became a reality. Questions of new construction, extension of residents' rights to renters and guests, employment of off-duty national seashore personnel, auto use, airport landing rights, and docking privileges all surfaced during the first seven years. Some were resolved, but most were not. National Park Service officials soon found that the broad agreements negotiated by the National Park Foundation, as well as by the agency itself, left a lot of room for differences of opinion.

For example, most retained-right agreements allowed docking on estate lands or Park Service facilities. One of the first conflicts to arise concerned the dock at Dungeness. Although the Park Service did not use it for passenger embarkation until 1979, it did unload construction materials there. The agency turned down several requests to use the dock by those without specific rights in their retained-right agreements.

Another dock in dispute was at Old House Creek. A man named Ben Jenkins secured a long-term lease from Gertrude Schwartz, trustee of Andrew Carnegie III. His attempt to build a new dock 400 feet from the Park Service's structure resulted in a brief, furious conflict. It was settled when seashore officials allowed him to adapt their dock for his use. At Hawkins Creek on the north end of the island, the Candlers used Brickhill Dock, which actually was on property formerly owned by the Carnegies and Charles Fraser. Decades of tacit permission to use it had engendered a proprietary feeling among the Candlers. When others began using the dock, they complained and succeeded in gaining a special-use permit that excluded other residents. Questions of who had rights to do what and where they could do it also extended to parking privileges, beach-crossing areas, airfields, and private roads.[14]

Other retained-rights issues also arose. The Park Service repeatedly turned down one resident's request to hunt and trap all over the island at any time ignoring state game laws. Conflict arose over some construction projects on retained estates and the waste they caused. Residents complained about trespass by both visitors and Lucy Ferguson's cattle. The rights to rent or sublease property had to be clarified repeatedly. Law enforcement jurisdiction on both federal and retained-estate lands remained in limbo until 1982 when Georgia established concurrent authority with the Park Service. Establishing rights of use and residency was a process of the government and the residents feeling out each other. Debate persisted about the letter of the law, the exact wording of the retained-rights agreements, and the spirits of both.[15]

Driving on the island was the most serious retained-rights issue to surface during these early years. Although seashore officials did not challenge anyone's right to drive at this time, they attempted to close the South Cut Road that Charles Fraser had bulldozed in 1970. The road lay well within the area the Park Service expected to be designated wilderness. It was not a historic road, and the Park Service did not anticipate a negative reaction to its announcement. However, the Perkins family and Thornton Morris held retained estates adjacent to the western end of the road and used it to access the beach. They even had renamed it "Perkins' Beach Road." Their reaction and the effort they made to retain use of the road demonstrated the residents' fierce determination to protect their right to drive all over the island.[16]

In September 1978 Thornton Morris anticipated the problem and began laying the groundwork for retention of his rights and those of the Perkins family. He explained that National Park Foundation representative George Sandberg had assured him "that the Perkins retainees would be able to use the Perkins' Beach Road to get to and from the beach, and that this use right would continue until such time as the National Park Service provided an adequate alternate means of transportation from the reserved area to the beach." He also reminded Park Service regional director Joseph Brown that Coleman Perkins had not wanted to sell his land to the government. He had done so because without his sale the Andrew Mellon Foundation would have refused to provide money for Cumberland land acquisition.[17]

The National Park Service rejected Morris's argument after securing an opinion from the regional solicitor. Agency director William Whalen told local congressman Bo Ginn, who inquired after receiving complaints from Morris, that in the hearings for wilderness planning, the public had expressed an intense desire to establish the maximum amount of wilderness and to limit areas of "potential wilderness." The latter would include areas of nonconforming uses like roads. As a result, park planners decided to limit access from the Main Road to the beach to a single route, the Duck House Road. That area had to remain open anyway because of a specific retained-right agreement. Whalen stated that the Perkins family and Morris were being given "the same right of access to the beach as others."[18]

By April 1980 environmental organizations became involved in the controversy. The Park Service and Congressman Ginn received letters from the Sierra Club, the Wilderness Society, and the Georgia Conservancy urging closure of the road in order to protect wilderness values. Congressman Ginn defended Morris and the Perkins family explaining that while they did not have a specific legal commitment to use South Cut Road, they did have "sound documentation to support a route of 'convenient access' to the beach." This communication to the Wilderness Society followed a summit meeting on the issue at Ginn's office on April 30. Representatives of nine congressmen and Senators Nunn and Talmadge, as well as Park Service representatives, Thornton Morris, and the Georgia Conservancy, attended. The conservation group saw the writing on the wall as the lawmakers sympathized with Morris.

An agreement was reached whereby the Park Service would seek a "potential wilderness" designation for the road, and Morris and the Perkins group would be issued twenty-year special-use permits. The level of congressional pressure that could be brought to bear by the island residents swept away Park Service resistance. It would not be the last time that the Carnegie and Candler heirs used their powerful contacts to influence seashore policy. On September 22, 1980, Joe Brown sent the special-use permits to Thornton Morris. He stated that nothing in these permits automatically assured their renewal. However, he added that "based on the information now available," he anticipated that the permit would be renewed.[19]

Into the Fray: Long-Term Planning at Cumberland

The National Park Service developed a steady management operation and a popular program for visitors on Cumberland Island during the years 1972 to 1980. However, the agency never intended for many of those practices to become permanent. The Park Service began trying to determine the level of recreation development and resource management even before legislators established the seashore. In addition, Congress awaited a report on the wilderness suitability of the island. Two interconnected planning processes, one for a "general management plan" and another for a "wilderness recommendation," became the most consistently controversial actions for the agency in its early years on Cumberland.

During the congressional hearings to establish Cumberland Island National Seashore, the National Park Service promised extensive recreation development for visitors over much of the island. Officials explained that the designation "national seashore" meant a high capacity for recreation and promised up to 10,000 visitors per day for Cumberland. Indeed, the Park Service could not have secured enough local and state backing to pursue establishment of the new seashore park without such a goal. After Cumberland Island National Seashore became a reality, agency planners concluded that they were bound by congressional intent to move ahead with the extensive development.[20]

What they experienced, however, was a backlash against large-scale development that was sweeping through the park system across the country. The rise of the environmental movement in the late 1960s and early 1970s changed the rules by which park officials could manage their units. The staffs of nearly all of the fourteen national seashores and lakeshores endured a similar cycle of promising extensive tourist development for avowedly recreation-oriented units only to have it thrown back in their faces during the planning process.[21] These public reactions were facilitated by passage of the National Environmental Policy Act of 1969 (NEPA). This critical environmental law orders federal agencies to absolutely minimize the effects of their actions on the natural environment. In addition, NEPA requires the federal government to conduct research on the potential effects of any action and make the results public. Thereafter, citizens and organizations can comment on the plans, and their ideas must be noted in the final decision. At Cumberland Island environmentalists, residents, historic preservationists, and others continually used the NEPA process to reject or modify Park Service plans.[22]

The contentious struggle for an overall "general management plan" and a "wilderness recommendation" focused on five primary issues. The first concerned the location of a headquarters, visitor center, and ferry embarkation point on the mainland. A second issue was the amount of tourism infrastructure to be developed on the island. Third, the maximum number of visitors allowed onto Cumberland also had to be determined. A fourth question concerned what form of transportation should be furnished to visitors on the island, if any. Finally, the amount of acreage of the young national seashore, if any, that should be placed under the protective but highly restrictive designation of wilderness had to be decided. Each of these issues spawned controversy.

Devising a General Management Plan

In 1971 the Park Service released a master plan that outlined its ideas for recreation on Cumberland (map 5.2). The mainland base would be near Cabin Bluff and connected by a new road to Harrietts Bluff Road or directly to Interstate 95. Roughly 300 acres in size, it would accommodate 1,500 cars, employee housing, and maintenance facilities. The plan envisioned private hotels nearby where more parking would be available. A fleet of twelve 100-passenger ferries would connect to the island at Brickhill Bluff (Brickkiln), Plum Orchard, and Dungeness. Each of these sites would have interpretive facilities, a jitney terminal for island transportation, and concession facilities to rent bicycles and camping equipment. The jitney service would run the length of the island, connecting the three docks as well as three beach areas, six campgrounds capable of accommodating a total of 400 overnight visitors, and high-interest areas like Half Moon Bluff, High Point, Terrapin Point (near the old Cumberland Wharf), and Stafford. The concessions would sell picnic supplies also.[23]

For the visitors the Park Service planned stables for eighty horses and many miles of new trails across the island to be divided into horse, bike, and hiker categories. Interpretive facilities would be located at all recreation and concession sites and adjacent to historic or archaeological resources. The British forts and the Spanish mission would also receive interpretation should they ever be found. Three areas of intense beach use would accommodate 7,000 sunbathers and swimmers at one time. Each would have bathhouses, shelters, comfort stations, and concessions to sell beach equipment and "light refreshments."

Seashore officials planned to turn the Plum Orchard and Stafford mansions into environmental conference and study centers operated by a concessioner. They also suggested that the Recreation House at Dungeness could be used as an interpretive center "if restoration proves feasible." The planners offered no specific upper limit of visitors, but the level of development and the capacity of the intensive-use beach zones suggested that 10,000 visitors per day could be easily accommodated.[24]

A year later, as Congress considered the Cumberland bill, the Park Ser-

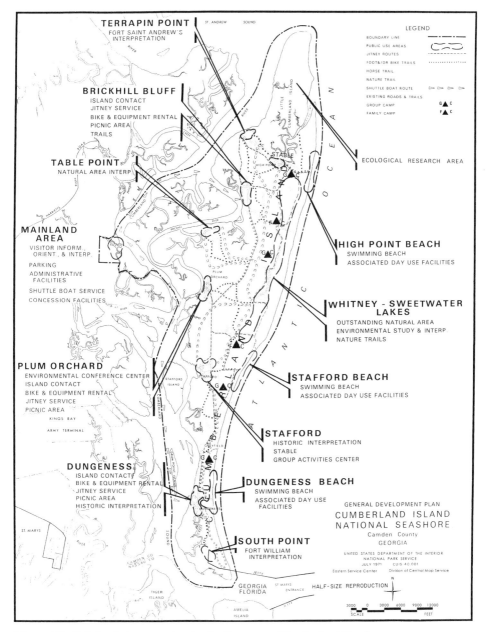

Map 5.2. The Park Service's 1971 master plan for the proposed Cumberland Island National Seashore. (National Park Service, Oct. 1971, *Master Plan, Proposed Cumberland Island National Seashore*, CINS Library)

vice released a draft environmental statement on the proposed plan.[25] The environmental statement supported the development concepts of the master plan. Seashore planners projected that the majority of the negative environmental effects would occur at the mainland headquarters site. They called island impacts such as dredging near docks, changes to the undisturbed character of parts of the beach, and control of noxious insects insignificant. The master plan and the environmental statement helped convince Camden County and the state of Georgia to back the national seashore legislation.

Any reservations about the propriety of these development levels seemed to be answered by a series of in-depth studies conducted by the highly respected Conservation Foundation, an environmental study and promotion group based in Washington, D.C. In 1972 the foundation published a document entitled *National Parks for the Future*.[26] The organization's researchers made numerous recommendations for future national park management. Among them were suggestions that the Park Service should involve citizens in the planning process, a procedure already mandated by NEPA, base its planning on natural resource impacts, also ordered by NEPA, conduct carrying-capacity studies to determine the appropriate number of visitors, and manage resources according to established guidelines.

In the case of the latter recommendation, a 1970 law, often called the General Authorities Act, already provided the foundation.[27] The statute discontinued the Park Service's practices of categorizing units as predominantly historical, natural, or recreational and managing resources in accordingly hierarchical levels of importance. The law ordered that the agency manage the entire park system as one complex unit and afford all resources equal protection.[28]

The Conservation Foundation study had many specific recommendations, some of them fairly theoretical. Subsequently the organization acquired a grant to study these proposals on the ground and selected Cumberland Island as its test case because it was a new unit. Foundation staff members and Hans Neuhauser, a leader of the Georgia Conservancy, took charge of the project to study and plan for Cumberland's management program. Four teams studied natural resource management, the mainland embarkation options, priorities for interpretation, and the visitor carrying capacity of the island.[29]

Albert Ike and James Richardson of the University of Georgia Institute of Community and Area Development conducted the carrying-capacity study. In a 1974 draft they gave figures based on recreation and ecology literature for each of the development areas of the Park Service's master plan. They estimated the total capacity of the island at nearly 16,000 visitors per day.[30] Eighteen months later Ike and Richardson released a revised report that dropped the maximum carrying capacity to just under 14,400 visitors.[31] The Park Service argued that these figures, coming from a reputable conservation organization, supported its master plan.

However, the foundation's authors offered a number of conceptual and concrete recommendations that urged caution in developing infrastructure and setting visitor limits. First, they emphasized the need to determine the psychological carrying capacity of each area. The figures given in their report were based purely on ecological tolerance. Further studies would be needed to determine whether visitor enjoyment would be compromised at these visitation levels. Second, they suggested that the Park Service should start with small areas devoted to activities like swimming, horseback riding, and camping in order to measure their actual carrying capacities. Thereafter, it might expand the development areas or activities to the plan's proposed levels as test results warranted. Ike and Richardson also recommended that the Park Service make visitor education the top priority. They prioritized the acceptability of various outdoor activities, rating nature walks, beachcombing, and hiking highest and swimming and cycling somewhat lower. At the bottom of the list were fishing, hunting, horseback riding, and marine boating. This ranking took into account both the psychological and environmental impacts as well as the availability of those activities on the mainland.[32]

Finally, the report made specific development recommendations. The authors proposed two rather than three intensive-use beach areas. They suggested dropping the High Point area because of its importance as a turtle-nesting site. Ike and Richardson also recommended substantial research on soil and biotic tolerance before any development took place. They proposed no more than twenty sixty-passenger jitneys and forty horses for guided tours. They endorsed camping for up to 1,300 persons in fifteen areas but insisted that these visitors should bring their own equipment with them rather than rent it from a concession on the island. Of greatest importance, they recommended that the Park Service plan for 10,000 visitors per day rather than 14,400 to allow flexibility. The general tone of the report urged slow and cautious incremental development in order to test both carrying capacities and visitor demand.[33]

With the apparent sanction of this respected conservation organization, the National Park Service forged ahead with plans similar to those suggested in the 1971 master plan. However, the levels of visitation and development envisioned by the agency drew immediate criticism. Five months before the final seashore legislation passed, one advocate wrote: "The NPS still has a lot of people on its staff who are imbued with the idea that 'parks are for people' and to hell with the basic beauties and wonders that attract people in the first place. Some of them seem to forget that if the resource is degraded, the experience of the visit is degraded."[34]

A noisy portion of the criticism leveled at the Park Service came from Lucy Ferguson and her guests at the Greyfield Inn. Consumed by fears that the government would condemn her land, she railed against the Park Service and virtually any action it took. After Congress established the seashore, she even hired former attorney general William Ruckelshaus to manage a political campaign to redesignate the island as a national wildlife refuge or as a national park with a strict emphasis on natural resource protection.

Son-in-law Putnam McDowell repeatedly sought to tone down her virulent antagonism and save her from what he saw as a pointless and expensive campaign. In May 1974 he wrote to her: "The biggest thing you are up against is that the legislation passed by the Congress, under which about 80% of the Island was acquired, calls for a National Seashore Recreational Area and nothing else. This classification has a certain meaning under the law and it is hard to see how the Park Service can move very far away from the recreational precedent set in the case of Hatteras and Cape Cod."[35] Later, McDowell confessed to Thornton Morris: "If I recommend to her that she stop now—which is what I think she probably should do—I think she is going to have some fun with the idea that she's a real fighter, but I'm a quitter, and wouldn't support her. I usually face things like this head on; but I'm inclined in this case to let Mr. Ruckelshaus and Mrs. Ferguson drift hand-in-hand into the sunset until money separates them."[36]

McDowell sensed that many factors would contribute to a Park Service decision to back away from its most extreme plans. He believed that Ruckelshaus's influence was very limited and that other members of the Carnegie heirs were unwilling to contribute funds for this battle. Eventually, Lucy Ferguson shifted her campaign to encouraging the establishment of a designated wilderness over as much of the island as possible. In this she found broader support from other island residents. In her campaign against the

Park Service, Ferguson lost no opportunity to enlist her guests' support. The Park Service and the secretary of the interior received a number of complaints about agency plans to remove feral hogs, transport visitors to the north end of the island, develop beach recreation, and receive 10,000 visitors per day. In some cases considerable disinformation came from Greyfield, which the Park Service had difficulty correcting.[37]

Lucy Ferguson's antagonism was but a small part of the public unhappiness with National Park Service plans for Cumberland Island. Under the glare of public attention, the park planners began to reduce the level of planned visitor development for the island. While they did not abandon the idea of 10,000 visitors as a maximum, they stressed that this was not a goal for every day but a limit for a few days of the year. Furthermore, they suggested that this figure represented the visitation to the entire seashore unit. Perhaps half of those visitors would only stop at the mainland center and not take the time-consuming trip to the island. Finally, they assured congressmen and the public that any visitor development would take place slowly with constant monitoring to detect any adverse environmental effects.[38]

Environmental organizations responded angrily to the Park Service's apparent intransigence and brought unexpected pressure to bear. Lucy Smethhurst of the Georgia Conservancy persuaded her cousin, Deputy Assistant Secretary of the Interior E. U. Curtis Bohlen, to visit the island, listen to Conservancy representative Hans Neuhauser, and then order the Park Service to carry out additional visitor carrying capacity and ecological studies before developing the island.[39]

In February 1975 the Park Service held a hearing in Woodbine, the seat of Camden County, to receive input from the public on its proposals. Approximately 200 people attended, primarily members of various environmental organizations. The Park Service presented an array of options for each element of a general management plan. For example, ten different locations for a mainland visitor center were offered, including four on Brunswick Paper Company land, two at the Kings Bay Army Terminal, one at Crooked River State Park, one at St. Marys, one at Point Peter just east of that town, and one at Fernandina Beach, Florida (map 5.3). Seashore planners also presented a wide variety of options on the island for camping, beach use, embarkation points, historic interpretation programs, and visitor limits. In addition, the Park Service offered several different proposals for wilderness areas totaling from none to 29,300 acres.

The *Atlanta Journal* reported that the most hotly debated issue was the

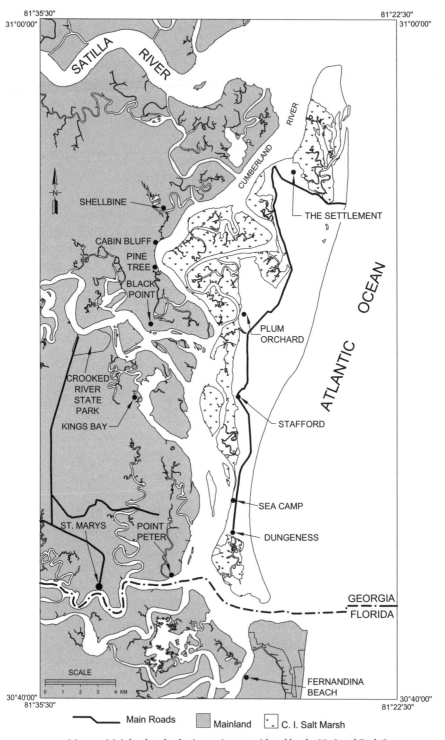

Map 5.3. Mainland embarkation points considered by the National Park System

number of visitors to be allowed on the island each day. Superintendent Bert Roberts suggested that a realistic figure might be 5,000 to 6,000 per day, but most speakers demanded still lower limits. The only exception was a representative of Charles Fraser's Sea Pines Company who insisted that its contract with the National Park Foundation required the Park Service to develop extensive recreation facilities and high visitation levels. Possibly Fraser still envisioned a role as island concessioner.[40]

A second contentious issue was the amount of land to be nominated for wilderness status. Congress initially passed the Wilderness Act of 1964 in response to the U.S. Forest Service's single-minded pursuit of logging and other consumptive uses at the expense of recreation and environmental preservation. However, the law ordered all federal land agencies to review every roadless area of 5,000 acres or more for suitability as a "wilderness area." Within such a preserve automobiles and other mechanical devices, buildings, and nearly all other modifications of the natural environment are prohibited. Later judicial interpretations and congressional additions have strengthened wilderness status to exclude bicycles and even baby carriages. At least initially, the National Park Service opposed the legislation on the grounds that it would constrain its dual mandate to protect park resources but also to provide for their use by the public.

Designation of each new wilderness area requires an act of Congress. At both this hearing and a follow-up one in East Point, Georgia, the participants predictably sought the maximum amount of wilderness designation. Mildred Frazier of the Georgia Coastal Audubon Society proposed that the entire island be classified as wilderness.[41]

After the hearings Park Service planners returned to the job of designing draft versions of the general management plan and wilderness recommendation. The vast difference between the original plans and what the public at the hearings demanded demonstrated the gap between the Park Service's concept of its mandate and the opinions of environmentalists and island residents. Somehow, the Park Service had to forget promises made during the legislative battle and its own experience in developing mass seashore recreation. In early September 1975 Assistant Secretary Bohlen informed Congress that the Park Service would not have a wilderness recommendation ready by the October 23 deadline.[42]

As the Park Service planners toiled, another element of the original master plan changed. Brunswick Pulp and Paper's decision to rescind its land donation to the seashore eliminated four of the potential visitor center sites.

At the same time, a resurgence in activity by the U.S. Army at Kings Bay removed two more possibilities.[43] Finally, the feasibility of embarking from Fernandina Beach, Florida, evaporated. Bert Roberts gave the resort town hope that it could become a secondary embarkation point during a visit in February 1975. However, the Conservation Foundation team studying the issue of mainland access included an official from Camden County. He convinced the rest of the team to recommend that all island access originate within the county in order to compensate it for lost tax revenue on Cumberland.

Over the ensuing decades Fernandina Beach promoters periodically mounted campaigns to secure a second embarkation point with applications to the Park Service and letters to their congressional delegation. Typically, this occurred when a new superintendent arrived at the seashore. In some ways Fernandina Beach does offer better facilities, but the Park Service has continued to honor its commitment to Camden County.[44]

A New Plan Fails

In October 1976 the Park Service released an "environmental review for the general management plan and wilderness recommendation." In every area the new plan sharply contrasted with the 1971 master plan. The only options left for the mainland base were St. Marys and Point Peter. Seashore planners recommended that the headquarters stay in St. Marys, but the visitor center and primary ferry terminal be at Point Peter. A single owner held the historic peninsula, and the agency saw no reason to doubt that it could obtain the land.

Park Service planners had drastically reduced the level of development on the island. The plan called for two day-use beaches at Sea Camp and at Nightingale Avenue northeast of Dungeness. Instead of up to 7,000 persons total, each would accommodate only 200. It also proposed five wilderness campgrounds with a total limit of 80 campers and three improved ones south of the wilderness with a combined maximum of 180. Concession services were reduced to some vending machines in the developed areas. No horseback riding or bicycling concessions were mentioned.[45]

The greatest change came in the daily visitation limit. The Park Service included a breakdown of its reduced recreation facilities at maximum use. A total of 1,060 persons could be accommodated on the island at one time. Assuming a 50 percent turnover for the day-use activities, the planners added another 400 for a daily total of 1,460. This was a far cry from 10,000 to 16,000 per day. The jitney service remained in the plan to operate primarily between Stafford and Dungeness. Occasional jitney tours to Plum Orchard and the north end of the island would be scheduled on a reservations-only basis. Seashore planners had decided to continue the jitney service based on its success at Yosemite National Park, where the agency used it to eliminate automobiles from part of Yosemite Valley.[46]

The Park Service recommended wilderness designation for slightly over 20,000 acres north of Stafford (map 5.4). Planners excluded the Main Road, which they labeled a "historic trace," the High Point and Half Moon Bluff complexes, and the retained estates. In addition to the upland acreage, the wilderness would include the marshes adjacent to both Cumberland and Little Cumberland. In July 1977 the Park Service issued two more documents for public review, a draft general management plan–wilderness study and a draft environmental statement. These documents reiterated the visitor development plans of the 1976 review and clarified the nature of the proposed wilderness. The exclusions divided the wilderness into three separate sections amounting to a total of 8,851 acres for immediate designation and another 11,794 acres to be established as potential wilderness. The latter would automatically become wilderness when conflicting uses stopped.[47]

In September 1977 the Park Service confidently presented its wilderness plan at hearings in St. Marys and Atlanta.[48] Once again the reaction of the audiences shocked veteran agency officials. For the second time park planners heard themselves characterized as an irresponsible agency rushing heedlessly toward destruction of the island. Hans Neuhauser set the tone of public response:

> The Park Service's wilderness plan indicates that the area north of this line [at Stafford] would be wilderness, except for the road to Plum Orchard; except for the road to Terrapin Point; except for the road to High Point; except for the road to the beach at the north end; except for the reserved estates; except for the north end of Cumberland from Terrapin Point to High Point; except for the powerline corridors from the mainland to Cumberland; except for the powerline corridors from Cumberland to Little Cumberland; except for major tidal creeks; except for Christmas Creek; except for jeep or motorcycle patrols on the beach; except for pig hunts; except for turtle management; except for septic systems; except for drainfields and land use around them; except for reserved rights of access; and except for the submerged lands.

There appears to be more exclusions of wilderness than there are inclu-

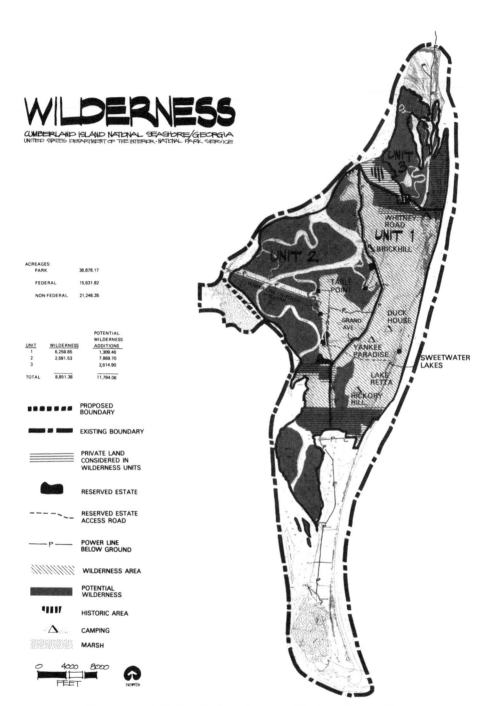

Map 5.4. An early National Park Service proposal for three discrete wilderness areas on Cumberland Island. (National Park Service, June 1977, *Draft Environmental Statement, General Management Plan, [and] Wilderness Study,* Denver Service Center Map 640/ 20030B)

sions. In fact, the Park Service's wilderness plan consists of just the scraps left over from their development plan. Whatever the Park Service feels that it cannot conveniently develop is relegated to wilderness.[49]

Every environmentalist challenged the idea of running jitney tours to the north end. John Crawford of Wilderness Southeast used an analogy: "If you can drive to the top, why climb the mountain? Why climb the mountain if people are driving to the top?" Carol Ruckdeschel added:

> In the environmental impact statement it says, more or less, that the via-jitney visitors, even though not physically capable of a true wilderness experience, will be able to get a taste of wilderness by being driven to the edge of it. That's more or less a quote. In other words, a taste of wilderness when you get off the boat and get on the wheels and look at it from your seat, I assume, and that's a taste of wilderness.
>
> I think that right there we ought to stop and analyze this planning of wilderness, because it doesn't seem to reflect an understanding of wilderness, of what it's all about.[50]

Some environmentalists challenged other aspects of the plan. Several rejected the agency's proposal to develop campgrounds near the island's lakes. Another opposed hunting on the island. Still another pointed out that one of the three units of wilderness proposed by the Park Service did not meet the 5,000-acre minimum required by the Wilderness Act.[51]

Presumably the National Park Service expected some antagonism to the plan from environmentalists. However, the opposition included Georgia mainlanders as well as island residents. In addition to Carol Ruckdeschel, four other retained-rights holders spoke at one of the two hearings. All but Mary Miller favored including the roads in the wilderness. Thornton Morris, representing himself, Lucy Ferguson, and other residents, explained that the requirement to propose wilderness was an addition made by a Congress that did not trust the National Park Service to protect the island environment. He challenged the agency to "finalize the commitments which have been made to me and the family members [Carnegies] and owners of Cumberland Island by the national park officials for the last ten years regarding Cumberland Island, and to keep it as it is today, protected under the wilderness act, so that our grandchildren and their grandchildren can enjoy the same experience that I do."[52]

A few comments did not follow the general trend. A representative of the Peachtree Sportsmen's Club ridiculed the idea that Cumberland Island was a true wilderness and threatened to sue the Park Service if it did not open the island to hunting as the establishing legislation mandated. Several Camden County people agreed.

Alternatively, William Voigt Jr., former executive director of the Izaak Walton League, explained the history of the Wilderness Act of 1964, the Eastern Wilderness Act of 1975, and other legislation in order to support the Park Service position. He bemoaned the existence of retained rights but agreed that those commitments should be honored. He concluded by urging the Park Service to do everything in its power to manage the proposed wilderness for ecological preservation and to add the excluded areas as soon as their retained uses expired.[53]

The Park Service also received a number of public responses to the 1977 plans through the mail. The review by the Georgia Department of Natural Resources (DNR) was one of the most extensive and important. The DNR agreed with most of the other oral and written respondents in opposing federal proposals to run a jitney to the north end of the island and to exclude a 490-acre tract along the entire northern coast. The National Park Service had proposed the exclusion to "provide a setting for the historic zones at High Point and Half Moon Bluff." State reviewers answered that wilderness was "the primary focus of the Department's concern." And they rejected the exclusions as well as a concurrent recommendation to designate the entire island a historic zone. As an afterthought they also opposed the Park Service's proposal for a mainland center at Point Peter.[54]

The Georgia officials then made a number of recommendations. The National Park Service should abandon the idea of jitney travel north of Stafford except for visits to Plum Orchard. These could be halted when boat service to the mansion became available and retained rights of driving the Main Road expired. All retained estates, roads, and Little Cumberland Island should be included as potential wilderness. Thus, when incompatible uses ended, these would be immediately reclassified as wilderness. Interpretation of the north end historic features could be accomplished at the mainland visitor center or the south end of Cumberland for those unable to hike through the wilderness. Finally, wilderness legislation for the island should be structured so that the requirements of historic preservation legislation were met.[55]

Once again Park Service planners regrouped to study their proposals. However, the draft plans and environmental statement had touched off a continuing public debate. The conflict over jitney travel to the north end of

Cumberland eventually drew Director Whalen to the island. In late fall 1978 he inspected the road and historic north end and listened to Cumberland Island officials. He then returned to Washington, D.C., and ordered them to drop plans for jitney travel north of Plum Orchard. One year later the agency discontinued all motorized transportation for visitors at the south end of the island.[56]

In the meantime, the exclusion of 490 acres around High Point and Half Moon Bluff as a historic district continued to draw fire. Carol Ruckdeschel, who was living in and redesigning one of the historic structures in the proposed district, waged a letter-writing campaign to the Park Service, its Interior Department superiors, and the press. She challenged the historical importance of the various structures. This time, however, island residents resisted the environmental juggernaut that had so dramatically influenced Park Service planning. Joe Graves of the Johnston branch of the Carnegie heirs complimented the agency on its exclusion and decried the latest attempts by environmentalists to add even more acreage to a wilderness proposal that already included nearly 75 percent of the island's uplands. He added:

> During the public hearings conducted by the National Park Service in Georgia and since then some Island residents and those who like to backpack and camp in wilderness areas have lobbied very effectively for the creation of a large wilderness area.
>
> However, the people who will constitute the great majority of Park visitors are not organized to lobby for reasonable access and visitation in the Cumberland Island National Seashore. I wish to speak for those people.[57]

Eventually Hans Neuhauser submitted a compromise plan that called for the Park Service to exclude 2 acres at the Settlement at Half Moon Bluff and 54 acres at High Point. This would allow 434 acres to be added to the potential wilderness area. In August 1979 acting Park Service director Ira Hutchison recommended to Interior officials that Neuhauser's plan be adopted.[58]

By December 1980 National Park Service planning for Cumberland Island had changed dramatically. The agency sought a single wilderness zone beginning north of Stafford and ending in the marshes west of Little Cumberland Island. It included the smaller island but not the state-owned marshes west of Cumberland. The plan excluded the Main Road as far north as Plum Orchard, the estate itself, and the two historic areas at the north end. The new acreage totals were 8,540 acres of wilderness and 11,480 acres of potential wilderness. Seashore planners still called for a mainland embarkation center at Point Peter but proposed that it would not be necessary until the number of visitors coming through St. Marys topped 600 per day for the four high-use months. They continued to insist on a visitor limit of 1,460 per day but reduced their proposed island transportation to "one or two units" that would move visitors from the Dungeness and Sea Camp Docks to campgrounds, beaches, and the Dungeness historic complex. In addition, the Park Service still considered a reservations-only tour to Plum Orchard as a requisite.[59]

Having come so far from the original master plan, the National Park Service felt confident that this new set of proposals would succeed. The agency had received massive and continuous input from the public. Dominated by environmentalists, the public had forced the agency to drop its development of three docks and three beaches spaced along the entire island to two much smaller sets of landing points and beaches on the south. Proposals for visitation limits fell from 10,000 to 1,460, beach use from 7,000 to 400, and camping from 400 to 260. The transportation operation shrank from virtually the entire island to a few established roads in the southern part of the island. Gone were horse, bicycle, and camping concessions. Also gone were interpretive operations at every conceivable historic and archaeological site. The only thing that increased between 1972 and 1980 was the size of the recommended wilderness. Under the 1971 master plan it would have been ludicrous to suggest any wilderness remained. Now the agency proposed more than 20,000 acres in a single block. Although more than half was rendered "potential" by incompatible uses and ownership, the security of the island's natural resources seemed assured.

Resource Management in the 1970s

Resource management on Cumberland Island is complicated by two realities. First, natural and cultural resources are interspersed throughout the island. Counting the private Greyfield estate, six discrete complexes of historic buildings lie from one end of the island to the other. Isolated structures and archaeological sites are also numerous and even more widely dispersed (map 6.1). Each falls within one of the delicate island ecosystems that together define the seashore.

A second factor in resource management is the presence of retained estates and the specific rights that accompany each. These too are scattered throughout the island. Particularly significant are the rights to drive roads and the beach held by approximately 300 individuals. These factors together force the Park Service to face legal contradictions, immense pressure from special-interest groups, and a chronic budget shortfall in managing the island's resources.

Natural Resource Management

The National Park Service spent much of the 1970s on Cumberland establishing baseline data and management programs for the seashore's natural resources. Even before Congress established the new unit, agency-sponsored studies of the flora and fauna were under way.[1] When lawmakers passed the seashore legislation, in-depth research began on the geology, soils, hydrology, flora, and fauna of the island. The resulting reports identified natural and human-induced processes and provided data for planning and, in some cases, corrective measures. A University of Georgia team led by Hilburn O. Hillestad conducted the benchmark among these studies. Research began in 1973 with the specific objectives to "inventory and describe

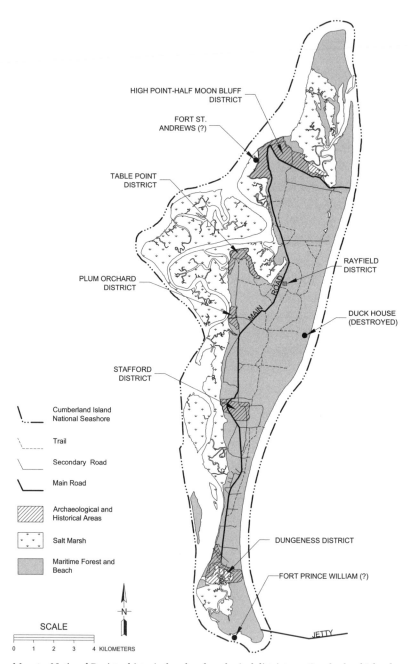

HIGH POINT-HALF MOON BLUFF
DISTRICT

FORT ST.
ANDREWS (?)

TABLE POINT
DISTRICT

RAYFIELD
DISTRICT

PLUM ORCHARD
DISTRICT

DUCK HOUSE
(DESTROYED)

MAIN ROAD

STAFFORD
DISTRICT

DUNGENESS DISTRICT

FORT PRINCE WILLIAM (?)

JETTY

Cumberland Island
National Seashore

Trail

Secondary Road

Main Road

Archaeological and
Historical Areas

Salt Marsh

Maritime Forest and
Beach

N

SCALE

0   1   2   3   4 KILOMETERS

Map 6.1. National Register historical and archaeological districts on Cumberland Island

the natural resources within the boundaries of the Cumberland Island National Seashore and to generally describe their functions and relationships." The scientists released their report in 1975 with chapters on geology, soils, water resources, vegetation, and fauna, as well as an opening summary of historical occupation and a concluding one on management issues. The latter pointed out many of the problems that would become critical in the ensuing twenty-five years.[2]

The first area of concern for Cumberland's new managers was the land itself. Coastal landforms are notoriously unstable because of long-shore currents and variations in sand supply, sea level, wind, storms, and a host of other interrelated physical processes. Beginning in 1963, Park Service policy ordered the agency's field personnel to foster and protect natural geological and ecological processes. Although precedents existed, this was a change in emphasis from earlier policies. It came as a result of a recommendation by an Advisory Board on Wildlife Management. Secretary of the Interior Stewart Udall formed the advisory board with chairman A. Starker Leopold and four other nationally recognized biologists to study overgrazing by elk in Yellowstone National Park.

In March 1963 the scientists issued their report, "Wildlife Management in the National Parks," more commonly known as the "Leopold Report." Board members went well beyond the task prescribed by Udall and offered a new, science-based philosophy for management of natural resources. They urged the Park Service to return parks to "vignettes of primitive America." This could be done with careful, ongoing research and management aimed at restoring natural processes regardless of what the public might aesthetically desire. Udall accepted the report and ordered all national parks to manage accordingly.[3]

In following this management philosophy, therefore, the Park Service should allow coastal processes to take place unless they are caused by human action. When Park Service researchers studied the geology of Cumberland Island, they found two problems. First, the edges of the island were eroding or accreting in different places. The erosion threatened both historic and recreational resources. Second, the system of ocean-side dunes on the island was actively migrating westward into the maritime forest and freshwater lakes. Each of these issues would become much worse during the national seashore's first three decades.

Most of the attention to coastal erosion focused initially on the Atlantic side of the island (map 6.2). To stabilize the entrance to St. Marys River,

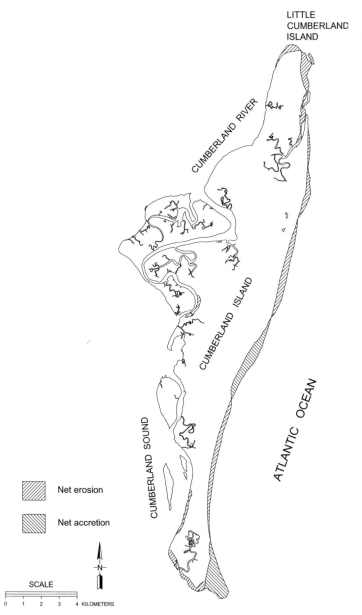

LITTLE
CUMBERLAND
ISLAND

CUMBERLAND RIVER

CUMBERLAND ISLAND

ATLANTIC OCEAN

CUMBERLAND SOUND

Net erosion

Net accretion

-N-

SCALE

0   1   2   3   4 KILOMETERS

Map 6.2. Shoreline changes on Cumberland Island since 1857. (Adapted from William H.
McLemore et al., 1981, *Geology as Applied to Land-Use Management on Cumberland
Island, Georgia,* prepared for the U.S. Department of the Interior by the Georgia
Geological Survey, contract no. CX5000-8-1563, reprinted in 1988 by the Cooperative
Park Studies Unit, University of Georgia)

Congress had approved construction of jetties seaward from the south end of Cumberland Island and the north end of Amelia Island in the late nineteenth century. Despite immediate and extensive erosion of the Florida island, the U.S. Army Corps of Engineers finished the jetties in 1905. It then began dredging the St. Marys River channel. In 1927 the corps lengthened the Cumberland jetty to more than two miles. It reinforced the jetty with concrete to prevent sand from passing southward into the channel. As a result, more than 500 acres of new land built up against the jetty on the southern Atlantic side of Cumberland Island. Partly in response to the altered ocean current, erosion has affected the northeastern and western sides, especially near Raccoon Keys on the southwestern portion of the island.[4]

In 1978 the Park Service established a monitoring program with twenty-eight stations spread the length of the island's Atlantic shore. After some initial data gathering, seashore personnel reduced the monitoring stations to twelve. As the years of monitoring accumulated, researchers were able to show seasonal changes in the beach profile, but they had insufficient data to reach a conclusion about long-term change.[5]

The Park Service's relative complacency about the Cumberland Sound side of the island was suddenly and unexpectedly shaken during the summer of 1974. Nearly twenty years earlier the U.S. Army had developed the Kings Bay Army Terminal across the sound. Army engineers also dredged a ship channel along the Intracoastal Waterway to a depth of thirty-four feet. Thereafter, the base had been used only as an occasional training facility. In 1973 Representative Bo Ginn threatened to have Congress take the facility away if the army did not increase its use. A year later the U.S. Army Command applied to the Corps of Engineers to dredge a turning basin for ships in Kings Bay and reestablish the thirty-four-foot depth of the channel.[6]

Although these plans were unsettling, the Park Service had little it could say to influence this activity and not much reason to do so. However, another part of the army's plan was intensely disturbing. During the original channel dredging from 1955 to 1957, the army had dumped spoils beside the channel, creating Drum Point Island, and on the southern end of Cumberland itself on lands taken with court order from the Carnegies. Now the army proposed to use those two sites again for additional spoils. The Park Service responded quickly, vigorously opposing the dredge spoil plan. Regional Director David Thompson pointed out that Congress intended to protect the area in the seashore's boundaries when it established the new unit. Park officials picked apart the army's initial environmental assessment and sought help from the secretary of the interior and Georgia's congressional delegation.[7]

In reality, the army may never have intended to use Cumberland Island for dredge spoils. Its true purpose was to use the seashore's land as a pawn to get its way elsewhere. In July 1974 Park Service official Anthony Rinck accompanied members of the Corps of Engineers and the Georgia Department of Natural Resources to inspect the projected sites for spoils deposition. The most important one was near Kings Bay itself and would have no significant effect on the seashore. Subsequently he reported that the army's representative told him that "the Cumberland Island spoils area will be used only if the environmentalists refuse to let the Corps of Engineers free dump on the west side of the channel."[8]

In August 1975 the Camden County Board of Commissioners added its opposition to the deposit of spoils within the seashore boundaries. In the meantime, the army drafted an environmental impact statement that received even more criticism from environmentalists, the Park Service, locals, and many others. Finally, Stephen Osvald of the corps called the Park Service in November to tell it that a certain amount of dredging was immediately necessary. To avoid delays, he stated the spoils would be deposited on an upland area within the Kings Bay Terminal and on a spoil island near the Kings Bay Wharf. In June 1976 the army's final environmental impact statement declared that as a result of "considerable opposition," it would not use Cumberland Island or Drum Point Island. However, it noted that this decision applied only on this "one-time basis."[9]

The Park Service did not have long to celebrate its victory over this military intrusion. On January 27, 1978, the U.S. Navy announced that it would take over the Kings Bay base and develop it for a fleet of nuclear submarines. Much more development on land and a substantial deepening of the channel would be necessary. Cumberland Island personnel and environmentalists worried that the deeper channel might increase erosion and that petroleum and chemical waste might enter the delicate marine ecosystem.[10]

As they awaited details on the navy's plan for Kings Bay and the Cumberland Sound channel, another threat arose. The S. C. Loveland Company of Philadelphia applied to the Corps of Engineers to install a mooring buoy a few hundred yards from the Dungeness Dock. What the company planned to moor to the buoy was unclear, but Park Service officials suspected it was barges and feared that they might contain chemicals or petroleum products. They adamantly opposed this idea on the basis that it would

endanger ferry operations, place a visual blight at one of the principal debarkation points for visitors, and potentially ruin ecological resources in
the event of a spill of dangerous chemicals. Ultimately, the vehement opposition from the Park Service and U.S. Fish and Wildlife Service successfully
stopped this intrusion. However, in the face of national defense needs, they
were powerless to stop the submarine project.[11]

In addition to the many threats to Cumberland's coastal fringe, the state
of its sand dunes also worried park officials. Dune vegetation was conspicuously missing from much of the island, and its absence allowed the dunes
to bury portions of the maritime forest and inland lakes. The Hillestad team
suggested that livestock introduced in the early 1900s was responsible for
this denudation and consequent dune instability. The team's report noted
that when livestock had been removed during the Civil War, it gave the island dunes a chance to revegetate. Although hurricanes had caused some
damage, most recently in 1964, cattle and feral horses were named as the primary culprits. By consuming the vegetation and trampling the dunes, they
threatened the entire ecosystem of the island.[12]

Pigs, Horses, and Cows

In the 1971 master plan for the proposed Cumberland Island National
Seashore, Park Service personnel were blunt about feral animals on the island: "The feral hogs must be eliminated from the island. They are too destructive to the island's vegetation and to the turtle eggs. . . . Although cattle
can be useful in keeping forest areas free from undergrowth and for improving wild turkey habitat, they are not natural and should also be eliminated." Interestingly, they neglected to mention the horses, which, because
of their size, are the most destructive to the dunes. Four years later the
Hillestad report agreed that pigs should be eliminated but also noted that
cattle and horses could be useful for vegetation control in certain open and
historic areas. However, the ecologists advised that they be greatly reduced
in number and absolutely restricted from dunes and beaches.[13]

Hence, despite centuries of human use and livestock grazing, the Park
Service insisted on removing these animals from all areas slated for management as natural ecological zones. In taking this position, the agency
based its decision on policies developed from the inception of the agency,
including the 1916 act to establish the National Park Service, Horace Albright's forestry and predator policy statements of 1931, the revolutionary

Fauna of the National Parks issued a year later, Director Newton B. Drury's arguments against emergency grazing in parks during World War II, the Leopold Report, and more than a decade of increasingly science-based resource management.[14]

In seeking to eliminate or reduce cattle, pigs, and horses, the agency's attention quickly focused on the island's only full-time resident and agriculturist, Lucy Ferguson. Cumberland's matriarch seemed to welcome that attention. She informed seashore officials that she had received permission from the other heirs to graze her cattle over the entire island. She expected the Park Service, as new owner of most of the land, to respect that commitment. Furthermore, she claimed all the feral pigs on the island were her property too. She vehemently opposed their removal, she said, because they kept down the population of poisonous snakes. She did not claim all the horses but only those on her land along with her pet burro that now ran with one of the horse herds.[15]

Of course the National Park Service rejected Ferguson's claims and asked her to fence her cattle on her own land. Superintendent Bert Roberts claimed the pigs were wild, had been periodically killed or trapped by former owners, and would be eliminated to the degree possible. Furthermore, he accused her of bringing more horses to the island and turning them loose after the national seashore's establishment.

This issue became quite inflamed in part because of the personalities of Bert Roberts and Lucy Ferguson. She had come to despise the National Park Service and lost no opportunity to say so to anyone who would listen. Government plans to develop the island for recreation, allow thousands of visitors, and restrict her use of the entire island infuriated her. Some of her family tried to assuage her and smooth over relations with the Park Service while others stridently supported her. Grandson Oliver Ferguson told the *Miami Herald,* "It hurts me to see the park rangers running around."[16]

Bert Roberts had succeeded unpopular Sam Weems as superintendent in November 1974. He promptly became even less popular with the island residents. Regional Director Thompson assigned the twenty-eight-year Park Service veteran to move Cumberland from a "project" status to an "operations" one—that is, to open the seashore to visitors. Thompson clearly expected a man who had previously been superintendent of Assateague Island and Cape Hatteras National Seashores to facilitate the island unit's opening. Roberts was a no-nonsense administrator who went by the book in his management activities.[17]

As the Park Service prepared for Cumberland Island's June 1975 opening, Roberts confronted Lucy Ferguson about the feral and domestic animals. After preliminary discussions, Roberts wrote an official letter stating the government's policy, which the island matriarch ignored. After a couple of weeks, Thornton Morris responded and noted that removal of the animals to Ferguson's land and completion of fencing around that land would be terribly expensive. He offered to meet Roberts to discuss the issue. Subsequently, Roberts offered Park Service help in removing cattle on its barge or rounding up and penning cattle and hogs on the island. Morris responded that Mrs. Ferguson no longer claimed ownership of any hogs not on her own land.[18]

When the Park Service opened Cumberland on schedule in June, cattle still roamed the island. A particularly large herd frequented the Dungeness estate. Thornton Morris, in a later interview, claimed that Lucy Ferguson knew she would have to remove her animals but that she defied Roberts in order to get him to offer help. Roberts evidently did not recognize the ploy. He wrote a lengthy and stern five-page letter to her on December 5, complaining that she had made no effort to control her cattle and that in fact their number on seashore land had increased. Furthermore, although Morris had stated that she did not claim pigs all over the island, some of her employees were seen feeding them, and another mentioned that he would trap her pigs on government land. Finally, Roberts accused her nominal employee and confidant J. B. Peeples of claiming part-ownership of the cattle and swine, threatening violence to federal officers and vandalism on federal land, and pledging to trespass whenever he wished. Robert's angry letter sparked an appeal by Ferguson to higher officials. Assistant Secretary of the Interior Nathaniel Reed sent Frank Masland, who promptly negotiated a ninety-day extension of the January 1, 1976, deadline set by the superintendent some months earlier. With that, Ferguson's employees began removing some animals.[19]

Bert Roberts relinquished the superintendency on December 31, 1975, although he stayed on to help the new superintendent, Paul McCrary, for a few weeks. He fired an aggrieved letter to the regional director on January 2, 1976, and welcomed removal of some of Ferguson's animals. However, he stated that family members expressed to him their fear that "Mrs. Ferguson is irrational where her thoughts about the National Park Service are concerned." He claimed that he decided on his very firm deadline after Thornton Morris, her son-in-law Putnam McDowell, a son, a daughter, several grandchildren, her personal secretary, and other Carnegies recommended he do so. He suggested that Assistant Secretary Reed and Masland "should be reminded of this."[20] Soon thereafter Lucy Ferguson's employees, with Park Service help, removed all the cattle to her land in segment 2N.

Pigs, however, were another matter. When she relinquished claim to the swine on government land, she handed the Park Service a nearly hopeless task that it still has not solved. Under benign circumstances a sow can give birth to nearly sixty young in one year. This level of reproduction means that a limited island ecosystem like Cumberland can see a nearly extirpated population of swine overtake the resources in a matter of a few years.

The only two options are complete elimination, a goal virtually impossible without dogs, or a never-ending program of hunting and trapping. The Park Service had sporadically hunted and trapped the pigs before 1976. That year it started an ambitious trapping program that removed 970 pigs from the island in two years. The *Florida Times-Union* reported that the Park Service then sold them to hunting reserves or butcher shops on the mainland and gave the proceeds to Lucy Ferguson. Despite this, the island matriarch told the reporters she was bitter about their removal and expected rattlesnakes to multiply "beyond reckoning." Park rangers estimated that another 500 hogs still roamed the island. Later that year a Park Service review of the trapping program estimated that it might require the removal of 250 pigs per year to hold the island population at about 600.[21]

Oddly enough, park officials paid little attention to the horses on the island despite their destructiveness. Undoubtedly their popularity with residents, visitors, and to some extent with park personnel preserved them from the harsh solutions visited on cattle and pigs. Nevertheless, they were exotic, and Park Service policy was clear. In its 1979 wilderness recommendation, the Park Service proposed to round up horses in the wilderness area and remove them. A representative herd could be maintained at the south end of the island. A 1978 study by ecologist D. L. Stoneburner supported this decision. He informed park ranger Zack Kirkland that the horses were having a deleterious impact on the island's freshwater lakes. Not only were they destabilizing dunes, but their "grazing and trafficking" along the lakeshores caused erosion (fig. 6.1). He suggested that studies of the horse population and their ecological impact should begin immediately. Five months later Superintendent McCrary sought aid from the Southeast Region for such research, citing concerns about inadequate horse management expressed in the recently released draft general man-

Fig. 6.1. Dune encroachment on the maritime oak forest near Sea Camp

agement plan. Nevertheless, several years passed before serious research on horses began.[22]

Native Animals: The Good, the Bad, and the Missing

The historic faunal assemblage of Cumberland Island was a rich one. Hillestad and his coauthors reported that the literature showed 26 terrestrial mammal, 7 marine mammal, 34 reptile, 18 amphibian, and 323 bird species lived or routinely visited the area encompassed by the seashore within the previous century. However, a number of species had disappeared while new arrivals competed for niches with those that remained. Six of the island's mammals were gone. One of them, the endemic Cumberland Island pocket gopher, had been extinct only since 1970. The other missing native mammals were the opossum, gray fox, eastern harvest mouse, bobcat, and black bear. A brief and unpublicized effort to reintroduce the bobcat in 1972–73 apparently failed.

On the other hand, the island had too many raccoon and white-tailed deer. The former preyed upon the nests of endangered sea turtles while the deer population, estimated in the thousands, had created a browse line in the forest over the entire island. One recent invader was the armadillo. No one was quite certain how it got there. Despite the armadillo's categorization as an exotic and the destruction of vegetation and archaeological sites it caused by digging for insects, the Hillestad team recommended no action to remove or control the armadillo population.[23]

In 1975 the Advisory Board on National Parks, Historic Sites, Buildings, and Monuments recommended that the Park Service find an eastern island on which to introduce and protect the endangered red wolf. Director Gary Everhardt responded that only Cumberland Island had the necessary size and resources to be a possibility. He suggested that a red wolf recovery team visit the island and further investigate the resources. Apparently, this went no further, for the red wolf never arrived on Cumberland.[24]

Several other animals required special protection. Manatees occasionally visited Cumberland's south end during the summer, a factor that had to be considered carefully in planning ferry and other boat traffic to the island. Least terns nested on the northern and southern ends of Cumberland Island in the interdune areas and on the ocean beach above the high-tide line. In August 1975 Assistant Secretary Nathaniel Reed ordered the Park Service to take extraordinary steps to protect tern nesting colonies in all national park units. At Cumberland Island park officials stopped all use of motor vehicles in the dunes and interdune areas but could not stop it below the mean high-tide line where the state had jurisdiction.[25]

Beach driving also affected another endangered species on Cumberland Island, the loggerhead sea turtle. This species drew the earliest and most aggressive research and protection activity on the island from the Park Service, U.S. Fish and Wildlife, and the Georgia Department of Natural Resources. The agencies were assisted by island resident Carol Ruckdeschel. She began a Park Service–supported study of the loggerhead in 1973 that lasted more than three years. With only occasional state or federal monetary support, she has continued her turtle and other biological research to the present time.[26]

In late November 1976 Superintendent McCrary replied to a request for information on sea turtle management from T. Destry Jarvis of the National Parks and Conservation Association. Using data from Ruckdeschel, as well as Jim Richardson of the University of Georgia, McCrary reported that between 50 and 100 turtles came ashore on the northern half of the island during the nesting season and that an equal number nested on Little Cumberland Island. He then reported on the dangers and management issues that affected the turtle population. These included turtles drowned by shrimp nets, nest predation by feral hogs, destruction of the protective foredunes by hogs, cattle, and horses, and motor vehicle use on the state portion of the beach. Deliberate human disturbance of nests was rare.[27]

The National Park Service initially saw little problem with the beach driving. In April 1979 Director William Whalen, land acquisition specialist George Sandberg, other agency officials, and John Bryant of the National Parks and Conservation Association met in Washington, D.C., to discuss the subject. Bryant reported that vehicular use of the beach was not a significant issue because the island residents conceded that Director George Hartzog had rejected it during land acquisition by the National Park Foundation. Only Mary Bullard had a right to drive there, he added. The Park Service leadership confidently awaited the imminent transfer of jurisdiction from Georgia at which time they would terminate nearly all beach driving. Like so many other issues, however, this would prove to be far more complicated than initially expected. Neither a halt in beach driving nor cession of the state's jurisdiction has come at this time.[28]

Protecting Archaeological and Historical Resources

National Park Service officials initially deemed the cultural resources of the new seashore as secondary in importance. In the 1971 master plan for the proposed unit, the agency seemed to relegate it to third in its list of responsibilities, noting that "although Cumberland Island is first thought of for its ecological significance and recreational potential, its archaeological and historical values are also considerable." After a disputatious inspection of the island's historic structures, Park Service architectural historian John Garner claimed that the legislative history of the seashore emphasized protection of cultural resources. Chief historian Robert Utley responded, "There appears to be no more compelling mandate for historic preservation at Cumberland Island than at any other natural or recreational area of the system."[29]

However, intense pressure from island residents and state historic preservationists would not suffer the island's historic and archaeological resources to be relegated to second place. In addition, a history of legislation and policy documents assured that the National Park Service would have to devote very large portions of its attention and money to these resources. When established in 1916, the agency inherited a variety of historic and archaeological monuments created by presidential proclamation under the Antiquities Act of 1906. Later, efforts by Director Horace Albright and his staff to add historical units in the eastern United States led to the Historic Sites Act of 1935. The relatively recent National Historic Preservation Act of 1966 demanded strict attention to these resources by the Park Service and all other federal agencies.[30]

In the act Congress created a National Register of Historic Places and ordered that the federal government grant funds to states so that they might identify historic resources, list them on the register, and preserve them. Legislators also created the Advisory Council on Historic Preservation to be chaired by the director of the National Park Service. This body must approve nominations to the register and actions taken with structures on the register. A majority of the members are from outside the federal government, usually experts from state and local governments.

In 1971 President Richard Nixon followed up with Executive Order 11593 entitled "Protection and Enhancement of the Cultural Environment."[31] This order required the federal government to assume a leadership role in historic preservation, to inventory eligible cultural resources on federal lands, and to initiate procedures to list them on the National Register.

Executive Order 11593 together with the National Historic Preservation Act created a body of restrictions and policies that shaped the historic preservation process. First, both the law and the executive order mandate that a nomination to the register from federal properties be cleared with the appropriate state historic preservation office. Second, an agency may dismantle a historic structure but only after qualified experts carefully record information and photograph it. After demolition, representative examples of style and materials are to be preserved. Finally, historic structures should be repaired or reconstructed only with appropriate historic materials.

The process of seeking state approval of actions pertaining to cultural resources came to be known as "section 106" after the clause in the National Historic Preservation Act where it appeared. The National Park Service evaluates properties for the National Register on the basis of significance in one or more of four areas. These are: (1) association with major events, (2) association with significant people in American history, (3) because they embody distinctive architectural or artistic values, or (4) because they may yield important information about history or prehistory.[32]

In order to enforce this new policy in the national park system, Director George Hartzog ordered in 1968 and again in 1972 that no structure fifty years or more old should be torn down without his permission. Furthermore, any park staff receiving permission should then follow a strict set of procedures. Nevertheless, in August 1974 Superintendent Sam Weems razed the dilapidated Plum Orchard laundry building without permission. Denver Service Center historian Edwin Bearss complained to Regional Director Thompson that Weems also planned to replace the Plum Orchard mansion's roof with one made of nonhistoric materials and to repair the nearby dock as he saw fit. Thompson balled out Weems and ordered him to take absolutely no action of any kind on any island structure without Thompson's written approval. A few weeks later Weems was replaced as superintendent.[33]

As the National Park Service moved to comply with the requirements for historic and archaeological preservation, it faced four tasks. First it had to conduct comprehensive surveys like the one Hillestad and his team were compiling for the natural resources. Second, the agency needed to evaluate each historic structure and decide which to preserve, which to reconstruct, and which to let disintegrate. Third, it had to actually maintain the buildings and convert some for use by the agency. Finally, it had to nominate appropriate archaeological and historic resources to the National Register.

Park Service specialists initiated two studies that would become the foundation of its cultural resource activities. John Ehrenhard of the agency's Southeast Archaeological Center led a team that surveyed the prehistoric and some of the historic resources of Cumberland Island. It investigated thirty-one areas with prehistoric resources and eleven specific historical sites as well as numerous roads, ditches, and canals. Ehrenhard issued his report in 1976. He proposed that seventeen prehistoric sites and two prehistoric "zones" were of sufficient importance and integrity to warrant National Register status. All of these were on the western edge of the island. Although a separate historic structures report would evaluate the more recent features, Ehrenhard recommended that seven specific historic sites be listed as well. These included the Tabby House, the Miller-Shaw cemetery, the ruins of the Deptford tabby house near Plum Orchard, the Stafford and Rayfield slave cabins, and the Half Moon Bluff cemetery. He also warned that dune encroachment and erosion were threatening some of the sites.[34]

Coincident with the archaeological research was the compilation of a his-

toric resource study by Denver Service Center historian Louis Torres. The resulting 1977 report consisted of two parts. The first portion detailed the history of the island, placing the various historical resources in narrative con-
text. The second portion was a "historic structure report" for the Dungeness
area. In this section Torres elaborated on the origin, characteristics, and current status of each structure at the old estate. His research was comprehensive, but he ran into trouble detailing the Carnegie era. He was unable to review most of the paper records of the estate because the heirs refused to
make them available to him. Torres did profusely thank Nancy Rockefeller
in the preface for her information and photographs. However, his attempts
to interview Lucy Ferguson were rebuffed, presumably due to her growing
distaste for the National Park Service and its actions on the island.[35]

During compilation of these two overviews, other more specific studies
were underway. In 1977 the Denver Service Center issued three lengthy reports by architectural historian David A. Henderson. They covered the
most architecturally significant standing structures on the island: the Plum
Orchard mansion and the Tabby House and Recreation House at Dungeness. Meanwhile, John Ehrenhard and island resident Mary Bullard carried
out an important archaeological investigation of the site of the Stafford
slave cabins, simply referred to as "The Stafford Chimneys." They found the
site to be of considerable significance, in some measure due to its preservation on an island generally inaccessible to the public.[36]

Mary Bullard proved to be a valuable ally in historical and archaeological research of Cumberland Island. Not only did she assist in a number of
investigations, but she later published her own research in a variety of venues. Some of the important ones were a monograph on the Settlement, a
Park Service report on Peter Bernardey, and a book on Robert Stafford and
the plantation era.[37]

The Park Service also conducted a number of inspections of the historical resources, often including experts from the state historical preservation
office and other interested agencies. Their reports, along with the more detailed historic structure studies, were necessary not only to evaluate the
worthiness of each building for listing on the National Register but to decide whether various buildings were worthy of preservation at all. To the
Park Service's dismay, its decisions at both the register and survival levels
led to immediate and prolonged controversy.

In November 1975 Park Service officials conducted a four-day "field review" of Cumberland's historic structures. Participating were senior offi-

cials from the Southeast Region and Cumberland Island National Seashore, several planners and historical architects, and historians Lenard Brown and Louis Torres. They focused on the Plum Orchard and Dungeness complexes and concluded that each area would be severely impaired if it was not preserved as a total building complex. Unfortunately at each site they found many outbuildings too far deteriorated to save.

At Dungeness, for example, they determined that the Recreation House, the woodworking shop, the poultry manager's house and its associated coops, a silo, the ice house, and the dairy barn had deteriorated too far to salvage (see map 2.1). The Recreation House was a particular disappointment because of its architectural significance. They decided to record the architectural data on each damaged building and either raze them or allow them to disintegrate on their own.

At Plum Orchard they found most of the support structures in "an advanced state of decay." Many of the Plum Orchard outbuildings were located on reserved estates where the Park Service had recently learned that it had no right to intrude and repair them. The field team concluded that "only several of the dozen or so structures will likely survive the term of the life estate (ending in 2013)." In its summary report the team suggested that the two complexes be interpreted overall as "ruins."[38]

Even those buildings to be preserved presented serious problems. Architectural historian John Garner wrote about the Dungeness complex: "If the unpreservable buildings are removed, the remaining ensemble will not portray the complex as it existed at any historic point in time. Retention of the support structures HS.67 [YCC kitchen], 68 [dairy manager's house], 69 [YCC dorm], 70 [washhouse], 72 [kitchen], and 75 [black servants' quarters] poses a serious management problem in that adaptive use, security, and ongoing maintenance are precluded by existing conditions of funding, staffing and isolation."[39] Here Garner predicted the problems that would continuously plague historic preservation on Cumberland Island.

The final part of the field review report listed the most significant salvageable structures in order of priority. Plum Orchard mansion ranked first. It needed restoration of the porch roof, balustrades, gutters, and rainwater disposal system (fig. 6.2). Next came the Tabby House, which needed work on the roof and exterior woodwork as well as treatment for insect infestation. Third was the Dungeness mansion ruins, which needed to be stabilized. The remaining structures on the list were the Dungeness carriage house, dock, and a storage building. Two months later, in January 1976, Robert Ut-

Fig. 6.2. The hot and humid climate of coastal Georgia took its toll on all the historic structures on Cumberland Island. This damage at Plum Orchard is an example of the costly repairs necessary to maintain the dozens of National Register structures on the island.

ley and Hank Judd from Washington surveyed the Plum Orchard and Dungeness complexes and concurred with the field team's recommendations.[40]

Almost immediately the Park Service list received criticism, much of it from the state historic preservation office. During a meeting a few days after the field review, state archaeologist Lewis Larsen demanded to know why so many buildings were in a terrible state of disrepair. He focused particularly on Plum Orchard mansion and the Dungeness Recreation House, and he blamed the Park Service for letting them decay. He brushed off Park Service replies that these buildings had not been maintained for years before federal acquisition.

Larsen then insisted that the historic resources were the primary value of the island, that other programs should be curtailed in favor of more funds for historic restoration, and that the Park Service should not have opened the seashore to visitors the previous summer when the money spent on transporting them could have been used for historic resources. He insisted that all structures should be preserved and that the entire island should be nominated to the National Register to prevent damage to its archaeological sites by the National Park Service.[41]

The Park Service team reacted to this tirade with relative equanimity. Its members explained their budgetary problems and asked Larsen to prioritize the structures he would save under those circumstances. The state man refused. They explained that Congressman Bo Ginn had pushed them to open the seashore to visitors but that the legislator was unlikely to push Congress for more historic preservation funds. Finally, they agreed to carefully coordinate all their plans with the state office.[42]

Much of the conflict over preservation of historic structures throughout the history of the seashore has focused on the two buildings Larsen identified. During an earlier inspection by Park Service personnel, an argument arose over the Plum Orchard mansion and over historic interpretation of the seashore in general. Superintendent Bert Roberts claimed that the important eras were the prehistoric and Spanish mission periods. He suggested that anything after 1800 rated little attention. Southeast Region chief of planning Frederick Ley Jr. added that "visiting specialists in and out of the Service, scientists, interpreters, etc.," held the same opinion. Furthermore, he called the Plum Orchard mansion a "white elephant" that the agency would regret accepting. Bill Everhart, one of the participants in the early 1950s surveys of the island, and others agreed. Southeast Region historian Lenard Brown disagreed with both the pejorative description of Plum Orchard and the relegation of post-1800 history to insignificance. Subsequently, Ley suggested that more research was necessary but that the matter of which history to emphasize was "not an issue for us to wade into, as the legislative history defines the historic elements."[43]

Despite the division between Park Service personnel about Plum Orchard, the agency sought to maintain and find a use for it. During 1976 contract specialists tented the entire structure and fumigated it to prevent further termite damage. Also, a park maintenance crew did minor work on the porch in 1976 and 1977. However, a series of investigations indicated that these actions were of little help in fully restoring the building. John Garner informed the Southeast Region director in November 1975 that various levels of improvement would range from $52,800 for minimal repairs to the porch and eaves to $1,193,280 for full restoration. Later David Henderson's detailed historic structure report recommended $399,000 as absolutely necessary to make the building usable. Unfortunately, because of budget constraints and attention to other historic structures, the mansion did not receive such serious attention until 1983.[44]

The search for a use for Plum Orchard mansion began even before Congress established the seashore. At the behest of the Johnston donors, the National Park Foundation proposed that it be used as a conference center. At the same time, the Park Service investigated using the house as a training center for historic preservationists. Seashore planners thought that the trainees' learning activities might offset part of the high cost of maintenance. Using the maintenance funds donated by the Johnstons, the foundation contracted with QRC Research Corporation to study the options. In August 1974 QRC researchers reported disquieting results. First, they demonstrated that it would be financially infeasible to use the mansion as an education center. Likewise, they discarded ideas of Plum Orchard as a museum or a "country inn." Then they reported that a conference center would be competitive only if the Park Service invested $1 million in repairs to the big house and $550,000 in building from twenty-two to twenty-seven additional bedrooms nearby. The QRC group added that the center supplemented by a full-size country inn next door would be even better.[45]

These options did not please the National Park Service. The costs were staggering—far more than it could wring from agency special funds for historic preservation. The idea of building a small hotel adjacent to the mansion was simply unacceptable. Nevertheless, it did not reject this or any other option outright. Instead, park officials and others continued to study and discuss the mansion's future for the rest of the decade. In the meantime, the $50,000 donated by the Johnstons for maintenance of the mansion ran out by early summer 1975.[46]

While many people were frustrated by apparent Park Service inactivity at Plum Orchard, the fate of the Dungeness Recreation House caused the most furious recriminations. Years later Nancy Rockefeller Copp expressed the Carnegie heirs' continuing anger at the fate of this unusual building: "It was so unique. . . . There was nothing like it. . . . And it was in perfect shape. All it needed was a new roof. And instead of that, they [the Park Service] took the $300,000 and stabilized the [Dungeness mansion] ruins. It was just a bad judgement."[47]

Copp's recollection of the structure's condition varied considerably from that of the Park Service's analysis. Architectural historian Henderson's 1977 report and photographs documented a building with severe roof damage from wind and storms and serious termite damage elsewhere. Exposure to sun and water had buckled the floor in many areas. Vandals had destroyed some of the columns in the portico and broken most windows. Nails, exposed by the buckling of wood shingles and siding, had rusted through, as had the metal eaves. Several trees had damaged walls and roof sections. Por-

tions of the complex had collapsed or sagged noticeably. Henderson estimated that the deterioration probably began in 1925 when Carnegie heirs
closed both this building and the Dungeness mansion.[48]

Henderson offered ten alternatives for the management of the Recreation House, ranging from construction of a steel railing for visitor safety as
they watched the building degenerate all the way to massive reconstruction
and use for interpretation and employee quarters. He suggested his sixth alternative as his preferred one. In this option the Park Service would restore
the exterior of the building, stabilize it, and allow the interior to continue to
decay. This would allow interpretation of its form and function for visitors
touring the Dungeness estate grounds. Henderson estimated the cost of this
work at $256,000.[49]

From the time of their arrival on the island, however, most Park Service
officials regarded the structure as unsalvageable. In answer to criticism
from the state historic preservation office, private preservation interests,
and the public, the Park Service invited a group of observers to inspect the
Recreation House as well as other sites on the island. The visiting party included David Sherman, the Georgia state historic preservation officer, his
staff, the Georgia state archaeologist, and two congressional aides. After
viewing the decaying structure, Sherman asked a number of questions
about the Park Service plans for it and expressed satisfaction with their decision. With apparent agreement from the state office, the Park Service
rejected Henderson's recommendations and allowed the once-wondrous
building to continue its rapid decay. A large portion of the center collapsed
in 1982, which accelerated the wind and water damage to the rest of the
building (fig. 6.3). Even after the state office gave up its opposition, the Camden County Historical Commission and island residents continued to chas-
tise the Park Service bitterly.[50]

The Plum Orchard mansion and the Recreation House elicited most of
the public attention, but the Park Service had many other cultural resources
to manage. Various surveys had identified 110 aboveground structures,
many of which could and would be preserved. For this work Cumberland
Island spent approximately $750,000 from its operating budget and
$400,000 from line-item appropriations between 1976 and 1979. In addi-
tion, the regional office added almost $800,000 from its cyclic preservation
program, close to one-third of the entire budget for more than 3,000 structures in fifty-three parks. Nearly all these funds were spent on restoration
of Dungeness outbuildings and stabilization of the Dungeness mansion
ruins.[51]

Fig. 6.3. During the 1980s and 1990s the magnificent Recreation House collapsed due to weather and termite damage.

The Park Service also spent a significant portion of the seashore's cultural resources budget on archaeological work. Here too the seashore's staff ran into trouble with state and some Park Service officials as well as some island residents. Ironically, the earliest archaeological reconnaissance of the new national seashore took place on the proposed mainland headquarters site on Brunswick Pulp and Paper Company land. Although this superficial survey found no discernible sites on the access road to Interstate 95, archaeologist Richard Faust did find materials from the Deptford period where the Park Service planned to build a large complex of administrative buildings, employee housing, and visitor campgrounds.[52]

On the island preliminary surveys turned up an immediate problem. The retained estate granted to Coleman Johnston was initially expected to be a 40-acre parcel in segment 5S at the southern end of the island. However, Johnston chose to relocate his estate to Table Point as his retained-right contract allowed. Unfortunately, the area proposed for his new estate included two important archaeological sites that contained both burial and house mounds. Archaeologist Donald Crusoe noted that the smaller of the sites appeared to be from the historic period and might have been the location of one of the Spanish missions. He suggested that the Park Service ask Johnston to move his retained estate southward and eastward by some 700 feet. However, Johnston refused to relocate. Eventually, archaeologists from the Southeast Archaeological Center conducted an extensive survey and recovery of artifacts while Johnston delayed his construction plans. The Carnegie heir also stipulated in the retained-estate exchange agreement that he would not destroy any cultural resources.[53]

A more serious problem arose on the retained estates of T. M. C. Johnston and Lucy Foster. Johnston had leased his estate at Plum Orchard to a Charles Hauser for the remainder of his reserved-estate right of forty years. The new lessee then began a series of modifications to two historic residences on the property. Letters to T. M. C. Johnston and visits by park officials to Hauser failed to stop the renovations. Cumberland rangers also reported that one of the life tenants on the Foster estate had threatened to remove the ruins of slave quarters at the Stafford Chimneys. Southeast Region director Thompson sought advice from the agency's regional solicitor on how to proceed.[54]

The answer was a shock to Cumberland Island officials. Solicitor Donald M. Spillman reported that the Park Service has the right to inspect property in retained estates but "only when there is belief that the remainder estate is endangered." Even worse, the agency could not legally stop alteration or demolition of historic resources, could not enter a site to conduct archaeological investigations, stabilize buildings, or salvage historical fixtures, and could not forbid any new construction. The only option left was to seek an injunction to stop any adverse action. In the case of the Stafford Chimneys, the Fosters never acted upon the threat. Hauser completed his alterations to the two buildings on his rental lot. Ironically, the Park Service could, with the owner's permission, enter a property and spend federal money to maintain a historical structure.[55]

The National Register

The Park Service faced one more cultural resource issue, the nomination of historical and archaeological sites to the National Register of Historic Places. Here too the agency experienced controversy. By the summer of 1975, the seashore had opened to visitors and was approaching its third anniversary. Agency historians had repeatedly inspected the island's cultural resources. Yet the Park Service had no nominations ready for the National Register when Director Everhardt asked for a list. Part of the reason was that agency officials still awaited the results of the major archaeological and historical overviews by Ehrenhard and Torres. However, more than research delays and the complexity of the nomination procedure lay behind their apparent inaction.[56]

During various meetings between Park Service and state historic preservation officials, a dispute arose over the amount of land and resources to nominate. The Park Service proposed several discrete and finite districts, each containing a number of buildings, prehistoric sites, or both. Among them were Dungeness, Plum Orchard, Stafford, and High Point–Half Moon Bluff. Agency officials planned to nominate other, more dispersed archaeological sites later (see map 6.1).[57]

Georgia historic preservation officers, on the other hand, wanted the entire island nominated as a multiple-resource district. This disagreement was a serious one that served as the catalyst for state archaeologist Lewis Larsen's diatribe against the Park Service during their November 1975 meeting. Larsen and other state officials regarded the Park Service's refusal to nominate the entire island as evidence that federal officers would carry out activities harmful to cultural resources on unlisted but potentially significant archaeological sites. Later his distrust spread to the historic preservation community, Cumberland Island residents, and some members of the general public.[58]

During the March 1976 inspection with Southeast Region and seashore personnel, state historic preservation officer David Sherman described the island as a "closed system of cultural development, a space-time continuum." Park Service resource managers argued that the island could not represent an entire system because so many structures and prehistoric sites had been damaged or allowed to deteriorate before the federal government took over. At the time, Park Service representatives believed they had convinced Sherman. For the rest of 1976 they submitted plans for renovation of various structures without any problems and continued to study new sites for potential nomination to the register.[59] However, state officials did not drop their plan to nominate the entire island. They continued to criticize the Park Service's cultural resources program vigorously.

In response, Park Service regional officials urged Denver Service Center specialists to hurry their review of nine potential districts. The acting regional director wrote in February 1977, "We seek, quite frankly, to place the burden of proof on the State Historic Preservation Officer to show why the entire island qualified rather than putting the National Park Service in the position of arguing against nomination of all of Cumberland." Later that year the Park Service declared four districts—Dungeness, Stafford, Plum Orchard, and the Main Road—eligible for immediate nomination. This step initiated the Executive Order 11593 requirement that federal agencies should manage properties declared as eligible as if they were already on the register.[60]

In the meantime, David Sherman and Lewis Larson prepared a letter explaining why the entire island should be nominated as one district. Directed at both the Park Service and the keeper of the National Register, it spelled out their idea of the island as a space-time continuum:

> Cumberland Island is a barrier island encompassing several ecosystems isolated from the mainland. In the absence of data to the contrary, we may assume that aboriginal populations operated on Cumberland Island as autonomous or nearly autonomous units. Any synchronic or diachronic research proposals concerning the anthropology or history of Cumberland Island would, of necessity, require that attention be directed to all site areas of the island as opposed to the district and site boundaries proposed by the National Park Service. The emphasis on preservation of the larger or less disturbed sites of the island to the exclusion of the small or damaged sites of the same temporal position, can only lead to biased interpretations of economic and social subsystems on the island.[61]

This was enough for the acting keeper of the National Register, Charles A. Herrington. He readily agreed with the Park Service historians that the four districts they had nominated were eligible. However, he endorsed the state's rationale of nominating the entire island. In addition, he agreed with criticism that the state preservationists leveled at the Ehrenhard report. They had strenuously criticized it because it did not explain in detail why some archaeological sites were excluded from the list of places to be nominated to the register or what would happen to them in the future.[62]

As the Park Service pondered this response, work continued on both renovation of structures and preparation of individual district nominations. Questions about the value of some historic resources arose. In his letter Herrington referred to the Stafford Chimneys as "one of the most important black history sites in the country." This level of significance was considerably higher than Torres had posited and called into question Park Service plans to grant retained estates in the Settlement to Carol Ruckde-schel and Grover Henderson. Now it was uncertain how or whether to proceed. However, Sarah Bridges, also of the National Register office, stated that the Settlement was a twentieth-century site and that a few retained rights there probably would not damage the resources.[63]

In 1978 the National Register issue came to a head. First, in its response to the 1977 draft general management plan and wilderness study for Cumberland Island, the Georgia Department of Natural Resources rejected the idea of one islandwide historic district as "neither practical nor feasible." Suddenly it was not simply a state versus federal conflict but a disagreement between state officials. Then, in early March regional historian Lenard Brown accused the state preservation office of refusing to process section 106 requests for structural rehabilitation work until the Park Service surveyed the entire island for archaeological resources and settled the matter of nominating all of it or a group of discrete districts.[64]

After a flurry of unsuccessful correspondence between federal and state officials, the Park Service fell back on an agreement it had with the Advisory Council on Historic Preservation. That policy provided that if a state office does not respond to a request for approval of a section 106 within forty-five days, the federal officers may assume that state officials have no comment. Using this approach, seashore officials resumed rehabilitation and development.[65]

The National Park Service never reconsidered its nomination of a series of separate districts and sites. In April 1978 the agency released a draft "historical resource management plan." The authors again proposed historic districts at Dungeness and Plum Orchard, identifying which buildings in each area could be preserved and which could not. Oddly, they suggested preservation of the exterior of the Recreation House. Unfortunately, senior Park Service officials rejected the plan because of its inadequate attention to archaeological resources. Nevertheless, with one exception the agency continued in every plan revision to offer the same prescriptions for separate and individual districts and selective preservation of buildings. The aberrant suggestion of saving anything other than the foundation of the Recreation House was never repeated.[66]

National Park Service intransigence continued to generate negative public relations. In February 1980 the Georgia Trust for Historic Preservation invited the Park Service to explain its preservation efforts on Cum-

berland Island. In attendance were representatives of the Georgia preservation office, the National Trust for Historic Preservation, and the Camden
County Historical Commission. John Garner of the regional office made
the presentation. He later wrote to the regional director that these groups
apparently had collaborated to present a document that strongly criticized
Park Service cultural resource management. However, Garner's presentation seemed to mollify them, and the meeting ended with promises to com-
municate and cooperate.[67]

The National Register nomination process proved to be a time-
consuming one, even without all these problems. Because of the conflict
over aspects of the plans for Dungeness and Plum Orchard, the only district
to be listed on the register during the seashore's first decade was High
Point–Half Moon Bluff. Denver Service Center historian Edwin Bearss initially prepared data for the nomination in October 1974. Over the next several years, Park Service officials had to determine ownership of several private tracts in the Half Moon Bluff area and acquire them as well as settle
the issue of retained rights in the Settlement. In addition, they had to negotiate with the Candlers in order to inspect and list the private High Point
estate.[68]

In January 1978 the agency submitted its proposal to the Advisory Council on Historic Preservation and the Georgia state historic preservation
officer. Elizabeth Lyon, the acting officer, promptly refused to certify it because of her disagreement with the Park Service's nomination of separate
districts. National Park Service historians elected to ignore the state office
and pushed ahead anyway. The High Point–Half Moon Bluff district was
added to the register in December of that year (fig. 6.4).[69]

Yet even successful establishment on the National Register did not end
conflict over Cumberland's historic resources. Six months after the accept-
ance of the High Point–Half Moon Bluff district, Carol Ruckdeschel notified the regional office that the nomination contained numerous mistakes.
After listing them one by one, she stated that only "three modified structures out of 20 to 30 original ones" were left of the nineteenth-century north
end hotel complex. She added that the Candlers had eleven new buildings
that "further erase the atmosphere of the hotel period." She also challenged
the common interpretation of the Settlement as the home of freed slaves.[70]

Some Park Service officials reacted negatively to her evaluation. They believed that she had ulterior motives of trying to limit the size of the historic
district and blunt plans to bring visitors to it. A briefing statement for Di-

Fig. 6.4. The First African Baptist Church in the Settlement in the 1980s

rector Whalen blandly suggested that she professed to be concerned about this intrusion in a planned wilderness area, but that her position, if adopted, would "coincidentally" assure her of more privacy. Nevertheless the agency adopted most of her corrections and submitted a second nomination in 1979. Final approval of the amended documentation came in January 1980, more than five years after Bearss's original report.[71]

By the end of 1979, National Park Service historic preservation on Cumberland Island and the criticism it drew had crystallized. The agency had formulated a set of policies to nominate sites for the National Register, to prioritize renovation and adaptation of salvageable buildings, and to identify and protect archaeological sites. Various offices of the Park Service had spent slightly over $2 million on repairs and stabilization. One district had been listed on the National Register, effectively ending the state's argument for an islandwide historic district. Nominations for four others were in various stages of preparation. Seashore personnel or contractors had worked on more than sixty-four structures and adapted eleven buildings for use by seashore personnel.[72]

However, a fundamental distrust of the agency had settled in among many members of the local public, the island residents, and some historic preservation experts in other agencies. This distrust fed on itself until nearly every step taken by the Park Service was questioned. Public relations became one of the most important jobs for cultural resource managers. During the next twenty years, much of the criticism of the National Park Service and a lot of its public relations work would focus on the Plum Orchard mansion.

Evaluating the Seashore's First Seven Years

Late in 1979 a Park Service official wrote a nine-page summary of the history of the national seashore from 1972 through 1979. After dispensing with the Carnegie period and the issues surrounding the unit's establishment, the author summarized the accomplishments of the Park Service. The agency had preserved and stabilized historic structures but learned it had little power to stop destruction on retained estates. It had opened the seashore to visitors and experimented with various transportation options. It had conducted critical natural resource research and started some monitoring programs. Planning began with a 1973 environmental statement for the preestablishment master plan and continued with public hearings in 1975, 1976, and 1977 and release of a general management plan and wilderness recommendation in August 1977. Land acquisition continued, and programs for historic preservation, natural resource protection, and interpretation were well under way. The author concluded: "In the seven years since Cumberland Island was established considerable progress has been made. An outstanding seashore area has been preserved and made available to the public. . . . The work is not complete, but it has been well begun."[73]

However, this terse summary put a positive spin on what had been a conflict-plagued period. A bitter argument with state historic preservation interests over the island's cultural resources continued. Early general management and wilderness plans had met bristling rejection from environmentalists and many island residents. Some natural resource actions, such as the attempted elimination of pigs, also drew unexpectedly negative responses.

A year later the Park Service released a "final" environmental impact statement for the general management plan and wilderness recommendation. It bore little resemblance to the 1971 master plan or to the agency's understanding of appropriate development for a recreation area. At the same time a Park Service effort to close a road in the future wilderness failed miserably under intense political pressure. The National Park Service had learned that it faced enormous competition to shape the future of both the physical and the legislative island that is Cumberland.

Historic preservationists wanted the entire island declared a single historic district constrained by the National Historic Preservation Act of 1966. Environmentalists wanted the entire island designated wilderness constrained by the Wilderness Act of 1964. Retained-rights holders insisted that their contracts with the National Park Foundation and the Park Service allowed them to live in semi-isolation as they always had, driving the island's roads and beach, modifying structures to suit their needs, and comanaging the national seashore. The conflicts inherent in these legally binding policies focused stressful attention on a badly underfunded public agency. By 1980 nearly all the pernicious issues of the seashore's first three decades had surfaced. In the following twenty years, a few would be solved, but many others were just beginning.

Contested Paradise: The 1980s and Early 1990s

The fifteen years beginning in 1980 saw a maturation of management at Cumberland Island National Seashore. Park Service officials completed a general management plan and several development plans for historic and visitor-use zones. Congress established wilderness on the northern half of the island and added a layer of protection for the natural ecosystem. Agency historic preservation officers successfully nominated three more multiple-resource districts and several individual sites to the National Register of Historic Places. Natural resource management benefited from a substantial amount of research as well as the institution of monitoring programs for geological and biological processes. Visitors enjoyed the island, and the cadre of loyal Cumberland Island devotees grew ever larger. The national seashore became a storied as well as valued element of the national park system.

Yet many of the intransigent problems that arose during the seashore's early years persisted. The Park Service grudgingly accepted a general management plan that continued operations at what was originally intended to be a "trial" level. Enactment of wilderness legislation sharpened the conflict between retained-rights holders and environmental activists. Plum Orchard mansion swallowed Park Service maintenance funds as well as private donations while a long-term solution for its use and care remained elusive. Finally, natural resource threats continued at disturbing levels. Erosion marred sound-side archaeological sites. Horses multiplied in spite of their poor health and continued to impact dunes and marshes. In addition, a second and successful introduction of bobcats to prey on the island's excessive deer and pig populations incensed some residents and visitors. Finally, the agency's relationship with St. Marys continued to be tinged with disagreement and suspicion.

During the decade and a half, the staff at the national seashore increased, the budget for operations rose by 43 percent to $1,125,000 in 1994, and visitation plateaued at roughly 46,000 people per year. However, the interpretation program drastically diminished. Across the entire park system, special resource needs and the spiraling demand for law enforcement siphoned both personnel and funds from interpretive programs. Cumberland had little crime, but fire management drew increasing attention after a frightening 1981 blaze. By 1994 the interpretation program shrank to a few talks or walks coordinated with the arrival and departure of the ferries.[1]

The fundamental management philosophy at Cumberland Island continued to be shaped by conflict: use versus preservation, natural versus historic resources, active recreation versus passive inspiration. The island continued to serve as an arena for fundamental debate over agency values and policy. The irony of an idyllic retreat for nature and solitude embroiled in such passionate conflict continued unabated.

Another Try for a General Management Plan

The National Park Service released yet another draft general management plan and wilderness recommendation to the public on February 25, 1981 (map 7.1). At the time, Cumberland Island was still a quiet backwater for a national park unit. Only 300 visitors—two ferry loads—came to the island on the busiest days. Island transportation between Dungeness and Sea Camp had been suspended two years earlier. Only one developed campground existed, the Sea Camp facility built years earlier by Charles Fraser. Three slightly improved primitive camp areas were located at Hickory Hill, Yankee Paradise, and Brickhill Bluff (Brickhill). No concessions, no beach facilities, and no stables for horseback riding were on the island. At the two debarkation points, the Park Service used existing structures, most of them historical, for limited interpretation and island management. On the mainland the seashore staff greeted visitors, prepared them for their visits to the island, and conducted planning, resource management, and all the other

Map 7.1. The National Park Service's 1981 general development plan for Cumberland Island. (National Park Service, Dec. 1980, *Final Environmental Impact Statement, General Management Plan, [and] Wilderness Recommendation,* CINS Library)

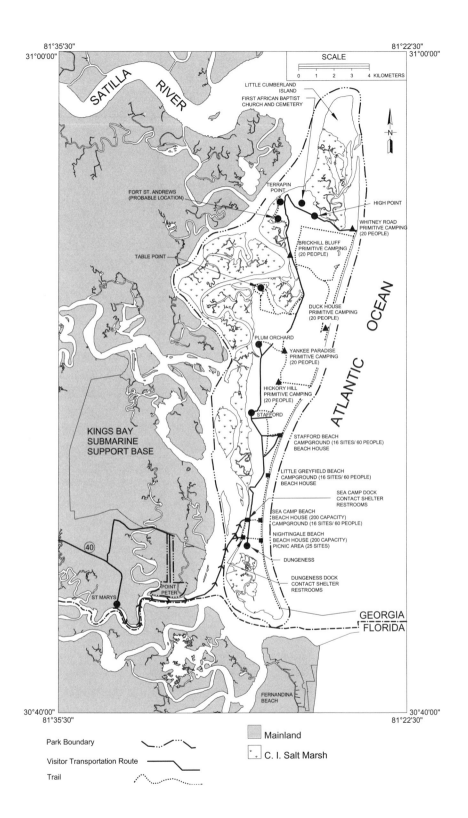

SCALE

0 1 2 3 4 KILOMETERS

SATILLA RIVER

LITTLE CUMBERLAND ISLAND

FIRST AFRICAN BAPTIST CHURCH AND CEMETERY

-N-

FORT ST. ANDREWS (PROBABLE LOCATION)

TERRAPIN POINT

HIGH POINT

WHITNEY ROAD PRIMITIVE CAMPING (20 PEOPLE)

BRICKHILL BLUFF PRIMITIVE CAMPING (20 PEOPLE)

TABLE POINT

DUCK HOUSE PRIMITIVE CAMPING (20 PEOPLE)

ATLANTIC OCEAN

PLUM ORCHARD

YANKEE PARADISE PRIMITIVE CAMPING (20 PEOPLE)

HICKORY HILL PRIMITIVE CAMPING (20 PEOPLE)

STAFFORD

KINGS BAY SUBMARINE SUPPORT BASE

STAFFORD BEACH CAMPGROUND (16 SITES/ 60 PEOPLE) BEACH HOUSE

LITTLE GREYFIELD BEACH CAMPGROUND (16 SITES/ 60 PEOPLE) BEACH HOUSE

SEA CAMP DOCK CONTACT SHELTER RESTROOMS

SEA CAMP BEACH BEACH HOUSE (200 CAPACITY) CAMPGROUND (16 SITES/ 60 PEOPLE)

NIGHTINGALE BEACH BEACH HOUSE (200 CAPACITY) PICNIC AREA (25 SITES)

DUNGENESS

(40)

POINT PETER

ST MARYS

DUNGENESS DOCK CONTACT SHELTER RESTROOMS

GEORGIA
FLORIDA

FERNANDINA BEACH

Park Boundary

Visitor Transportation Route

Trail

Mainland

C. I. Salt Marsh

administrative functions from a hopelessly inadequate dockside building with extremely limited parking.[2]

Compared to all the previous drafts, the new document suggested only moderate changes to this scenario. The most significant were a proposed move of the mainland visitor center and the classification of more than 20,000 acres as wilderness. The new draft plan also called for a reestablishment of transportation at the south end of the island to carry visitors between the two docks, two developed campgrounds, the wilderness boundary near Stafford, and the Dungeness historic district. Very limited additional tours to Plum Orchard and High Point–Half Moon Bluff also would be available until retained rights were extinguished. The plan proposed several new backcountry campsites and a visitor limit of 1,460 per day, less than 15 percent of the figure offered in the 1971 master plan. Regional Director Joe Brown promised a thirty-day period during which interested agencies, organizations, and individuals could comment. If the public responded favorably, the plan would go into effect in early March.[3]

The public did not respond favorably. Most of the more than 4,000 respondents accepted the wilderness proposal with relatively little comment. However, they positively erupted over the development proposals. Abetted by a strident media, especially the *Atlanta Journal and Constitution,* the public directed withering criticism at nearly every aspect of the scaled-down recreation plan. As in the past, the greatest anger focused on the level of visitation and on visitor transportation into the proposed wilderness. In addition, the choice of Point Peter for a mainland center met collateral criticism, much of it whipped up by the town of St. Marys.[4]

Within a few days letters began to arrive. Atlanta resident Patricia Koester wrote a typical view: "I was horrified to read in the Feb. 27 edition of the Atlanta Constitution that the National Park Service plans to open Cumberland to as many as 1,460 visitors daily! This is criminal! . . . Cumberland has been so special to so many of us. A chance to sit peacefully beneath those magnificent oaks, to encounter free animals, especially the horses, at any turn of the trail, oh, so many, many things. How can you dare to threaten all of this?"[5] Robert Coram of the *Atlanta Constitution* seized the story as a personal crusade. He had visited Cumberland many times and even worked for the Park Service briefly some years earlier. Over the next month Coram followed the unfolding drama with articles almost daily denouncing the plan and its adherents. He attended environmentalists' meetings, interviewed agency officials, and both encouraged and reported the growing correspondence antagonistic to the new plans. Regional Director Brown later said, "Coram chose his angle and worked it to the nth degree." The newspaperman, like many others, did not believe that the National Park Service took the public's comments seriously.[6]

The reaction of the public completely surprised National Park Service planners once again. In part this was due to the muted response from environmental organizations. The initial statements of the Georgia Conservancy, Atlanta Audubon Society, Sierra Club, and Wilderness Society were cautiously favorable. Each group was relieved to see the new wilderness proposal, and together they expressed preliminary agreement with the goals if not the details of the development plan. They too were surprised when Coram and his colleague Robert Ingle published articles accusing them of abandoning their principles and "caving in" to the Park Service. A spirited exchange of letters to the editor and caustic newspaper articles followed, the latter primarily from Ingle.[7]

As letters poured into the regional office, Park Service officials tried to defend the plan, citing a need to serve more visitors. They were belatedly supported by local congressman Ronald "Bo" Ginn. In a letter to the *Atlanta Constitution,* he called the existence of private lands the most serious threat to the island. He urged readers to turn their attention to helping the Park Service acquire the remaining tracts before the prices escalated out of reach and the current benign owners were replaced by some who might develop mass tourism. Finally, he warned, "With daily visitation on the island held at the current level of 300 persons daily, I expect that no funding is likely to be requested [by the Park Service] for Cumberland acquisition during any year of the current [Reagan] Administration."[8]

By the time Ginn wrote that letter, however, the National Park Service had already conceded defeat. On March 18, after receiving permission from Washington, Regional Director Brown scrapped the 485-page plan and ordered park planners to start over once again. Two weeks later Brown quit his position as Southeast Region director. Veteran official Robert M. Baker replaced him two months later.[9]

In April 1981 the Park Service decided to divide planning for the general management plan and the wilderness recommendation into two separate procedures. The generally favorable response to the wilderness proposal convinced agency officials that they could rush through a report to Congress, now nearly six years late. In the meantime, seashore general management planners could try once more to satisfy the island's clamorous constituency.[10]

The Wilderness Bill

Free to focus on a wilderness bill for the island, the Georgia Conservancy and other conservation groups refined their idea of a proper proposal and submitted it to the Park Service and Congressman Ginn. In particular, they sought to have the Main Road north of Plum Orchard reclassified from potential wilderness to full wilderness status and the portion of the road from Plum Orchard southward to the wilderness boundary from nonwilderness to potential wilderness. Ginn introduced that bill to the House of Representatives on October 7, 1981, where it met a favorable response. Nevertheless, H.R. 4713 still faced an uncertain fate. Secretary of the Interior James Watt had promised that he would block the creation of any more wilderness areas during his tenure.[11]

On October 16 the House Subcommittee on Public Lands and National Parks held a brief hearing on the bill. Despite Watt's antagonism, Park Service director Russell E. Dickinson recommended that the bill be passed. All the environmental organizations concurred except for the Coastal Georgia Audubon Society. With remarkable prescience its vice president for conservation, Verna McNamara, warned against including the Main Road south of Plum Orchard. She further recommended that Congress take out all the land south of Plum Orchard and west of the road. Her reasons were economic and visitor-oriented. She suggested that the cost of maintaining the mansion by boat only would be prohibitive. Also, taking visitors to Plum Orchard by boat would cost more and take up too much of the day visitors' limited time on the island. She added that disabled and elderly visitors deserved the right to see the mansion, that the road was already in use by retained-rights holders, and that the area her group wanted removed from wilderness designation was insignificant in acreage.[12]

Thornton Morris and William Ferguson spoke on behalf of island residents and submitted a long letter of comment as well. They reported that the residents supported the proposed bill in principle but wanted some clarification added. After the squabble over use of South Cut Road, the residents wanted to make sure that the retained-rights agreements superseded the restrictions of wilderness designation. Furthermore, Morris and Fergu-

son urged the representatives to make those same rights available to any future land sellers. The latter stipulation carried particular weight for the
Candlers, who had recently signed an option with the Park Service for their
estate. Finally, Morris reminded the congressmen that the government had
promised Lucy Ferguson that it would not try to acquire her land, most of
which lay in the proposed wilderness area. They categorized all these
caveats as "valid existing rights" and suggested that the phrase be inserted
in the bill.[13]

The Georgia Department of Natural Resources sent a short but important telegram to the committee. It too supported the bill but not without
certain understandings and stipulations. Commissioner Joe Tanner allowed that designation of the marshes as potential wilderness, even those
where the state held jurisdiction, was acceptable but only if hunting, fishing, and the use of motorized boats could continue. He called these activities "traditional and legitimate rights" of the people and insisted that the
Park Service respect them.[14]

Finally, Hans Neuhauser spoke in favor of the legislation that he essentially wrote. He eloquently described the value of the island for inspiration
and a "wilderness experience" and recapitulated the long ten-year history
leading up to the proposed bill. He requested two changes in the language
of the bill to clarify the intent of the legislation. First, the Park Service's proposed wilderness map still showed the roads as "potential wilderness additions." He asked that the representatives order the agency to correct the map
and manage the roads as wilderness. Second, to counter the recommendations of the Coastal Georgia Audubon Society, he urged the committee to
state clearly that the intertidal lands on both coasts were to be immediately
absorbed into the wilderness whenever the state transferred ownership to
the Park Service. These lands included the portion south of Plum Orchard
and west of the Main Road.[15]

After the House hearing the Cumberland wilderness bill entered the tor-
tuous process of congressional enactment. Initially it appeared that the bill
might become law before the end of 1981. Concurrent with Ginn's introduction of the House bill, Georgia senators Mack Mattingly and Sam Nunn
offered a similar bill in the upper chamber. The Senate passed it on October
21, only five days after the House hearing. That bill, S. 1119, immediately
went to the House Committee on Interior and Insular Affairs. On December 10 the committee recommended passage of the Senate version but
changed the acreage amounts of both the wilderness and potential wilder-

ness additions. The Department of the Interior requested the changes in order to include a 65-acre parcel of private land recently acquired by the Park Service in the area to be designated wilderness and another 268 acres composed of the waters from the island eastward to the national seashore's boundary one quarter mile east of the mean high-tide line.[16]

With these changes the House passed the bill on December 15, and all parties awaited what was expected to be a simple voice vote in the Senate to accept the minor alterations. Anticipating resistance from President Ronald Reagan, the bill's proponents attached it to another bill that proposed changes to the boundaries of Crater Lake National Park, legislation the president was known to favor. Hopes for passage of the bill evaporated, however, when it went before Senator Malcolm Wallop (D.-Wyoming), chairman of the Subcommittee on Public Lands. Despite the absence of any opposition to the new dual-purpose bill, Wallop claimed that he had not been thoroughly briefed and wanted time to review it. Senators Nunn and Mattingly lobbied furiously for its immediate passage, but to no avail. Whether fellow Wyoming native James Watt had any hand in this delay is unknown.[17]

In May 1982 Senators Mattingly and Nunn introduced a new bill, S. 2569, which was identical to the former S. 1119 as amended by the House of Representatives. On June 24 the Senate Subcommittee on Public Lands and Reserved Waters held a hearing at which only legislators and environmentalists spoke about Cumberland wilderness.[18] A second bill concerning U.S. Forest Service lands also was under consideration in the hearing and drew its own set of speakers.[19] All the participants favored the Cumberland legislation. Representatives of the four main environmental groups, the Georgia Conservancy, Sierra Club, Wilderness Society, and Atlanta Audubon Society, explained the checkered history of Park Service planning and the repeated demands by the public for wilderness designation. The subcommittee then referred the bill to its parent Committee on Energy and Natural Resources.[20]

While the congressional procedure lumbered on, another complication arose. The National Park Service let it be known that it would forbid any motorboats from landing on the shores of the designated wilderness zone. Furthermore, although tidal creeks in the wilderness were excluded from the wilderness and, hence, open to motorboat traffic, those boats could not land.[21] Camden County officials, led by Commissioner Jack Sutton, reacted with outrage, as did the local public. Sutton claimed that local fishermen had been able to land on the beach and siene for fish for generations. This claim naturally drew much popular support in spite of the history of antagonism to trespassers from the Carnegies in general and Lucy Ferguson in particular.

Sutton called for a "boat-in" protest off the western shores of Cumberland Island for two days beginning on July 31. He confidently predicted 200 watercraft would gather around his houseboat, which he moored near the Dungeness estate. When the weekend came, fewer than 60 boats showed up, carrying approximately 100 people, picnic lunches, and coolers full of drinks. A country and western band played on the roof of Sutton's houseboat, and all the participants seemed to have a good time. Sutton later expressed disappointment at the turnout and blamed a Jacksonville, Florida, kingfishing tournament for drawing away many supporters.[22]

The event did draw the attention of the Georgia House of Representatives, which passed a resolution urging the National Park Service to "respect the recreational needs and concerns of Georgia residents in its management of Cumberland Island." However, Camden County's protests had no effect on Congress in Washington or on the Park Service. The Senate Energy Committee voted unanimously to recommend passage of the bill, and the Senate did so on August 19. In spite of a last-minute effort by Secretary Watt to kill it, President Reagan signed the bill on September 9, 1982.[23]

Public Law 97-250 gave the National Park Service wilderness designation for 8,840 acres on Cumberland Island and potential wilderness status to another 11,718 acres. The legislative history of the act did address some concerns central to administering the island's many resources. The legislators allowed the Park Service to take small groups of visitors to Plum Orchard and High Point–Half Moon Bluff by vehicle over the Main Road. However, they also made it abundantly clear that they favored stringent adherence to the Wilderness Act of 1964 wherever and however possible. Congress expected to be notified of any Park Service deviance from these principles. To that end the lawmakers charged the Park Service to use water transportation unless it proved completely impractical and unaffordable. The final law also included protection for "valid existing rights" as requested by Thornton Morris and William Ferguson. President Reagan, in his statement on the new law, echoed Congress when he admitted that houses and traffic in a wilderness area were inconsistent, but he noted that the island would develop into a proper wilderness as the retentions ended. Also like Congress,

Reagan offered no specific recommendations for how to manage this apparent legal contradiction.[24]

Wilderness Use

With management of the Cumberland Island wilderness left unclear by Congress, seashore administrators faced a number of challenges. For guidance, they relied on two pieces of legislation: the Wilderness Act of 1964 and the Cumberland Wilderness Act. The 1964 law expressed an idea to strive for. Section 4c of the act spelled out what activities were prohibited:

> Except as specifically provided for in this Act, and subject to existing private rights, there shall be no commercial enterprise and no permanent road within any wilderness area designated by this Act and, except as necessary to meet minimum requirements for the administration of the area for the purposes of this Act (including measures required in emergencies involving the health and safety of persons within the area), there shall be no temporary road, no use of motor vehicles, motorized equipment or motorboats, no landing of aircraft, no other forms of mechanical transport, and no structure or installation within any such area.[25]

Although the Wilderness Act allowed for nonconforming activities associated with retained rights, the 1964 lawmakers never anticipated the number of intrusive uses present at Cumberland Island. Superintendent Morgan described the situation to a university professor who was researching wilderness across the country:

> The legislation [to establish Cumberland's wilderness] established an area of 8840 acres within which there are seven estate reservations with homes, rights of vehicular access through the wilderness, right of ingress and egress to the reserved estates, right to use various roads which bisect the wilderness area, an underground power cable which traverses the wilderness, goes to each resident structure and requires periodic maintenance; there are fourteen estate reservations on either side of the wilderness with the right of passage through it and there is an airstrip (reserved estates) on either side of the wilderness lands. Also, the surrounding water is non-wilderness so motorized craft can access the land at many points. We expect the valid existing rights will continue for 50+ years.[26]

The intensity of interest-group attention increased the difficulty of managing the wilderness area. The same environmental groups that shaped the

Cumberland Island wilderness legislation now insisted on management as close to the principles of the 1964 law as possible. Island residents protected their rights just as fiercely. Backpackers challenged the residents' right to drive any roads or the beach within the wilderness zone. They also complained about low-flying aircraft, including those trying to land at the two airfields abutting the wilderness. The Sierra Club insisted that the Park Service itself should not use vehicles or power tools, that people renting from the residents not be allowed to drive, that the Greyfield Inn not be allowed to give motorized tours in the wilderness, and that maintenance of the Plum Orchard structures be accomplished by boat. Wilderness enthusiasts arriving on the island expected to find a pristine reserve like those in western national parks. Instead, they found houses with no-trespassing signs and cars.

In 1986 a University of Georgia researcher conducted a study of visitor satisfaction and expectations. The results showed that the greatest focus of conflict was between hikers and drivers on the Main Road.[27] Hikers naturally used the wide throughway for an easy trail. Other than the beach it was the only unobstructed path. When a vehicle came, the incensed backpackers would spread across the road so it could not pass. This in turn outraged the drivers, especially the Carnegie and Candler heirs who resented the dismissal of the rights they had earned by preserving the island and creating the seashore. They furiously defended their retained rights. The wife of one of the Greyfield owners asked a reporter: "How many times have any of you been driving on a right-of-way to your property and had people curse at you, give you the 'finger' (try explaining what that means to a frightened 4-year-old sitting beside you in your vehicle), spit or throw things at your vehicle?"[28]

Seashore officials sought advice and legal counsel from the regional office. A few months after the national seashore wilderness was established, Superintendent Morgan wrote to the regional director asking for a definition of wilderness or potential wilderness status for each of the roads on the island and for an explanation of what constituted an emergency that would allow the Park Service to use vehicles in the wilderness. His final comment would become a common administrative plaint: "Management of the designated wilderness area is a hot issue and actions taken are highly visible. I need some guidelines on operations, within the wilderness, that will be defen[sible] and hold complaints to a minimum."[29]

Acting regional director Neal Guse responded that the entire main road and all the side roads were potential wilderness and that the seashore rangers could use the roads not only for emergency purposes but also "for law enforcement and for administrative purposes necessary to meet minimum requirements for the administration of these areas as wilderness." The specific language came from the Senate's remarks on the seashore wilderness bill. He ignored the House version, which was much stricter.[30]

Somehow this letter reached the various conservation groups, and the four principal ones immediately fired an angry letter to Robert Baker demanding that the Park Service respect wilderness law. They rehashed the legislative history and insisted that all the roads except the one to Plum Orchard were full wilderness and that rangers could only use them for emergencies. The environmentalists also promised to monitor the Park Service carefully as it performed its Cumberland Island duties.[31]

Two problems exacerbated the wilderness driving issue. One concerned short-term renters who brought their cars to the island. National seashore managers were never quite certain what to do about retained-rights holders who rented their homes. Most of the deeds specified that the rights to an estate did not include any commercial use of the property. However, at Cumberland those same deeds did not classify renting one's home as a commercial use. Island residents fell back on the legal phrase "and their assigns" to justify extending their right to drive to the renters.

While the National Park Service did not transport vehicles to the island, the Greyfield Inn did. With more than 100 derelict cars still lying around in 1985, this was a sore point for the environmentalists. The problem appeared to be solved in late 1986 when the Greyfield owners voluntarily stopped transporting cars for anyone staying less than a year. Although this gratified the Park Service and environmentalists, it met resistance from the other residents who rented their homes. Mary Bullard was particularly upset, claiming that she would lose renters. The Fergusons mollified her by promising to transport cars for her relatives even if they only stayed for a few days.[32]

A second problem also involved the Greyfield Inn. The Park Service found it difficult to influence a private operation on private land. The 1964 and 1965 court decisions dividing the Carnegie Trust's lands guaranteed that these heirs could drive the Main Road. It did not address other roads, and the other heirs' deeds of sale and retained-rights contracts were often vague. However, because the Carnegies had always used all the public roads, Greyfield continued to do so. When the Park Service reviewed the deeds and questioned this practice, the residents claimed that it was a traditional use. They cited promises made by Stewart Udall, George Hartzog, and land negotiator George Sandberg not to interfere with such uses. In the case of the 1979 South Cut Road controversy, this argument had prevailed.[33]

The Greyfield Inn also conducted tours of Plum Orchard, the Settlement, and other island sites for groups of eight to ten of their guests. This practice quickly attracted the attention of environmental groups because it seemed to flout the 1964 Wilderness Act's stipulation against any "commercial enterprises" in a wilderness area. However, the Fergusons argued that it was part of their legal right to traverse the roads on the island. Cumberland officials requested help from regional solicitor Roger S. Babb. He advised that, in spite of the vagueness of the legislative history and conflicting decisions in similar court cases, the Park Service could ban the Greyfield tours in the wilderness. Although Plum Orchard, the primary attraction, was not itself in the wilderness, it required transport along the Main Road, which was. Babb did admit, however, that the matter probably would have to be settled in court.[34] Apparently the Park Service elected not to pursue this contentious course because every land protection plan through the 1990s listed motorized tours into the designated wilderness as an activity "beyond the management control of the NPS."[35]

One side effect of the retained rights to drive in the wilderness was that the Park Service had to maintain the roads. As in the case of South Cut Road, island residents challenged any agency decision to close or stop maintaining a particular road. When the government acquired the national seashore in 1972, it contained a latticework of rough roads. The agency immediately began to convert some of them to trails. Nearly all of today's trails were at one time roadways. Although most of these closures did not meet coordinated and combined opposition from the residents, virtually all of them irritated somebody. Even Carol Ruckdeschel, the indefatigable lobbyist for stringent wilderness management, complained when Bunkley Road at the north end was closed to vehicle traffic. Another side effect was a decision to build a trail through the wilderness so that hikers could avoid both the Main Road and the beach.[36]

From time to time other wilderness issues arose that further strained the delicate balance that the Park Service tried to maintain between those favoring a strict interpretation of wilderness law and those who insisted that retained rights to drive rendered the wilderness open to other atypical activities. One of those activities was the use of bicycles. During the mid-1980s a spirited debate arose throughout the entire park system over the propriety of these nonmotorized mechanical devices in a wilderness area. The

Wilderness Act's ban on "mechanical transport" seemed to prohibit bicycling. However, mountain bikers argued that the human effort involved in
riding along rough wilderness trails made it acceptable. The debate intensified nationwide after Point Reyes National Seashore in California banned
bikes in its wilderness in 1985. One year later the *Federal Register* carried an
announcement banning them in all wilderness areas but supporting their
use on park roads or specially designated nonwilderness trails. Cumberland
Island officials interpreted this to mean visitor bicycling in the island's
wilderness area could continue because of the elaborate road system used
by both automobiles and bicyclists from Greyfield. However, Superintendent Rolland Swain finally stopped wilderness bicycling in 1995 except for
guests at Greyfield.[37]

National Park Service wilderness management at Cumberland Island
frustrated both environmentalists and retained-rights holders. Both groups
complained that officials made spontaneous decisions for each situation
without consistent policy guidelines. They added that the national seashore
needed a wilderness management plan that would bring consistent enforcement of specified policies. Yet in 1994, with the Cumberland wilderness twelve years old, the seashore still did not have one.

Designing a wilderness plan for Cumberland Island, however, would be
a difficult and somewhat intimidating task. Such poor congressional guidance in the 1982 act and so many diametrically opposed interest groups
made it hard to know where to start. Seashore planners would have to set
firm parameters for beach and road driving, determine the rules for maintaining historic structures in the wilderness, and decide whether to sanction
traditional uses by the island residents or stick to the letter of the law and
the specifics of individual retained-rights agreements. In all likelihood lawsuits would result from any managerial choices. As 1994 drew to a close, the
Park Service grappled with how to begin the long process of designing a
wilderness plan.[38]

The Final General Management Plan

As the wilderness issue took on a life of its own, Park Service officials resumed work on the overall general management plan. On June 29, 1981, the
Atlanta Constitution published a picture of new Park Service regional director Robert M. Baker smiling and holding a box crammed with some of the
4,000 letters protesting the withdrawn general management plan. Baker told

the newspaper, "The revised plan will be a reflection of what the public wants to happen on Cumberland." The catastrophic response to the original plan had chastened veteran agency planners. Baker added, "We can't afford to make a decision in a back room, float it, then go with it." As the National Park Service began again on the general management plan, it took extraordinary steps to include public input, especially from environmental organizations.[39]

Stinging from media ridicule for their tepid response to the earlier plan, the environmental organizations seized the opportunity to get even more involved in shaping the new one. On July 18 the Georgia Conservancy held a Cumberland Island workshop in Atlanta. Approximately 125 people attended, with a predictable majority of environmentalists. The Conservancy tallied the numerical results of the workshop. Fifty-three people supported the wilderness designation, and one said it was too large. Sixty people wanted either boat tours to Plum Orchard or no tours at all while four wanted limited vehicle tours. The record on St. Marys as a visitor center was 25 yeas and 1 nay. Most other elements of the management plan received similar input. This time the Georgia Conservancy wanted to make sure the media would not brand it "soft" on Park Service development.[40]

In September the Park Service held a two-day workshop on Cumberland Island. Regional Director Baker hosted a group consisting primarily of environmentalists and island residents. He announced, "This weekend, I am the audience and you are the experts." Representatives of the Georgia Conservancy and the Sierra Club spoke in favor of leaving the island at the current level of development. Thornton Morris added: "You've got the best data you can ever get about keeping it [island visitation limit] at 300 people per day. More than 4,000 people wrote you saying to leave the island alone. I'm told it was the second-greatest outpouring of letters the Park Service has ever received."[41]

One Park Service employee reminded reporters that a 300-person limit actually meant 200 because most visitors only went to the island if they could return on the later 150-passenger ferry. Nevertheless, at the end of the workshop, Baker promised that the new plan would be "scaled down considerably."

Two weeks later Robert Ingle of the *Atlanta Constitution* gloated over the changes. He commended the new regional director and explained that Baker "had not come up through the NPS and thus isn't stuck with the good-old-boy network way of doing things." The reporter also approved of new superintendent William Harris, whom he called "a dean of superintendents, which means he won't have to prove anything to anybody or build monuments to himself." He blamed the former general management plan on departed superintendent McCrary, the Georgia Conservancy, and federal planners. He then insulted Denver Service Center personnel and complained some more about the dead plan.[42] By this time the National Park Service recognized that its primary constituency was Atlanta, not Camden County. Much work would need to be done to repair the public relations debacle of the rejected management plan.

Local governments did not take their relegation quietly. Jack Sutton claimed that Camden County commissioners were not invited to the September workshop. A colleague agreed and added, "The local folks around Camden get the feeling that they [the Park Service] are more interested with what the folks in Chicago, New York and Southeast Asia think." Sutton also reiterated that Camden County had given up a lot of taxable land to the federal government. Echoing many earlier Camden complaints, he called the federal government a "private property–consuming monster" and summed up their actions: "It sees. It likes. It takes." In the case of the Park Service, he grumbled, "The least they could do is allow us to have some input on how the island is run." Subsequently, the Park Service claimed that it had sent an invitation to the county board of commissioners.[43]

The Park Service mailed 6,000 copies of a summary revised general management plan on November 19, 1981 (map 7.2). The pamphlet described the previous planning efforts, the proposals of the previous plan, and the new recommendations. It included most of what the environmentalists and residents wanted. Visitation would stay at 300 per day, and the visitor facility would remain in St. Marys. In response to heavy criticism, the agency promised to improve its telephone reservations system. Many potential park visitors tried to call, only to receive busy signals. Later when they finally got through, no spaces were left in the campgrounds or on the ferries. The Park Service planned to limit camping to one developed site at Sea Camp, one "transitional" campground with a bathroom but no fire pits or tables near Stafford, and five primitive camping areas in the wilderness. Two of the latter, located at Duck House and Lake Whitney, would be new.

Map 7.2. The revised 1981 general development plan for Cumberland Island National Seashore. (National Park Service, Nov. 19, 1981, *Summary of the Revised General Management Plan for Cumberland Island, Georgia,* Denver Service Center, NPS 1829, copy in SERO Planning files)

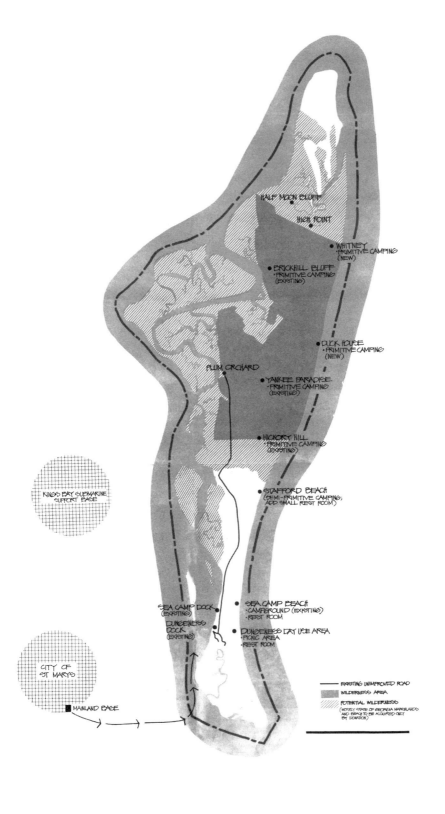

HALF MOON BLUFF

HIGH POINT

WHITNEY
•PRIMITIVE CAMPING
(NEW)

BRICKHILL BLUFF
•PRIMITIVE CAMPING
(EXISTING)

DUCK HOUSE
•PRIMITIVE CAMPING
(NEW)

PLUM ORCHARD

YANKEE PARADISE
•PRIMITIVE CAMPING
(EXISTING)

HICKORY HILL
•PRIMITIVE CAMPING
(EXISTING)

STAFFORD BEACH
(SEMI-PRIMITIVE CAMPING;
ADD SMALL REST ROOM)

KINGS BAY SUBMARINE
SUPPORT BASE

SEA CAMP DOCK
(EXISTING)

DUNGENESS
DOCK
(EXISTING)

SEA CAMP BEACH
•CAMPGROUND (EXISTING)
•REST ROOM

DUNGENESS DAY USE AREA
•PICNIC AREA
•REST ROOM

CITY OF
ST MARYS

MAINLAND BASE

———— EXISTING UNIMPROVED ROAD

■ WILDERNESS AREA

▨ POTENTIAL WILDERNESS
(MOSTLY STATE OF GEORGIA MARSHLANDS
AND SPACE TO BE ACQUIRED (NOT
BY DONATION)

With the lower visitation limit, the agency abandoned a day-use development at Nightingale Beach, scaled down plans for the two docks, and canceled island transportation except for occasional tours in sixteen-passenger vehicles to Plum Orchard. It also promised to study seriously the option of boating tourists to the mansion.[44]

The new general management plan met instant approval. One day after its official release, one of Lucy Ferguson's grandchildren complimented it by calling it "the ultimate" plan. The Georgia Department of Natural Resources also enthusiastically endorsed the new document.[45]

However, even this popular plan received some criticism. Camden County residents were upset that almost no facilities existed outside the wilderness to dock their boats. The Park Service planned to install mooring buoys. A boat could still come to the Sea Camp or Dungeness Dock to disembark passengers, but then the captain would have to tie up to the buoy and swim or wade ashore. Commissioner Sutton, already fighting the planned exclusion of motorboats from the wilderness, was particularly incensed. He asked the Park Service to design a dock for private watercraft and called the ferry operation "a damn hassle." In answer to criticism from environmentalists, he added: "I don't understand why the Sierra Club would fight this. We just want to dock and go to the beach. They can have the north end of the island as wilderness."[46]

Meanwhile, the principal environmental organizations expressed pleasure with the changes incorporated in the new planning goals but were cautious. A representative of the Wilderness Society said the new plan appeared good but "there are a lot of things that aren't covered at all." William Mankin of the Sierra Club praised the Park Service's new management direction but offered a number of suggestions. He proposed that the Park Service should use horses for patrols north of Sea Camp and boats for tours of Plum Orchard, move the deer-hunting camp out of the wilderness to Plum Orchard, and delete the proposed campgrounds at Lake Whitney and the Duck House areas because of their ecological fragility. As for wilderness access by the elderly and infirm, he blandly stated, "People who can't walk ten miles can walk 50 yards into the wilderness if they wish." Hans Neuhauser echoed the Sierra Club's recommendations and added requests that the Park Service develop a fire management plan and conduct a full public review of any modification of the docks to accommodate private boats.[47]

In March 1982 Superintendent Bill Harris and Assistant Superintendent Wallace Hibbard met with Neuhauser, G. Robert Kerr, and Carol Ruckdeschel to review the status of general management planning. The Park Service agreed to drop the Lake Whitney and Duck House campgrounds. In addition, the officials promised to reassess the National Register listing of structures at High Point and Half Moon Bluff and to manage retained rights in the wilderness "more closely." To that end they had requested copies of the final deeds between the residents and the National Park Foundation or the Park Service.

Two months later the agency released an "informal draft" of its detailed plan. Neuhauser still found problems with the document, claiming that it lacked detail and explanations of agency policy in matters where the legislative history provided conflicting guidance. He pointed out the issue of Park Service vehicle use in the wilderness as a case in point. However, the document was such a vast improvement over previous ones that he and the environmental community elected to accept it.[48]

One more significant task remained. The Park Service had to submit the general management plan to the Georgia State Historic Preservation Office and the Advisory Council on Historic Preservation. Both those agencies then had to sign a "Memorandum of Agreement" approving the plan and its implications for cultural resource management. Regional Director Baker forwarded the plan to those agencies in early November 1982. After a series of interagency battles described below, the historic preservationists approved it in January 1984. The Park Service issued a "record of decision" adopting the plan the following month. The 1984 general management plan reflected the voluminous input from environmental organizations, Atlanta residents, and retained-rights holders. The only differences from the summary revision of November 1981 were the deletion of two environmentally sensitive campgrounds and reclassification of all of the Main Road as wilderness. The voices of tourism interests and the desires of Camden County were unceremoniously swept away.[49]

Mainland Operations

The enactment of wilderness legislation and adoption of a popular general management plan freed the Park Service to concentrate on defining and implementing the new directives. One of the first tasks the Park Service faced was developing a permanent visitor center and ferry embarkation point in St. Marys. Late in 1983 Congress passed legislation eliminating the Point Peter site from the seashore's land acquisition schedule.[50] The law ended a ten-year quest by the Park Service to establish a mainland base in a parklike setting. For eight of those years the focus had been on the marshy peninsula east of St. Marys.

Originally, seashore officials had planned to acquire a 100-acre area that would contain the visitor center and ferry dock plus a maintenance facility, employee housing, picnic tables, and a self-guided nature trail (map 7.3). There, visitors would encounter an integral part of the native ecology and environmental education of the national seashore. Thus the Park Service could satisfy those visitors who might not have the time or opportunity to visit the island. As an added benefit, Point Peter was closer to Cumberland. National Park Service officials had convinced Congress to authorize acquisition of the area in the 1978 act that raised the funding ceiling for land acquisition.[51]

From the moment it became public, the Point Peter proposal met implacable enmity from Frederick G. and Jean Lucas Storey, the couple who owned the property. According to the *Savannah Morning News,* when the Park Service aired its plan to obtain Point Peter, it caught the Storeys completely off guard. Frederick Storey bluntly responded, "We have, through our lawyers, this week put the park service on notice that we do not intend to sell our property under any circumstance to anyone at any price for any purpose." As Park Service officials asked each other why no one had contacted the Storeys before they released the plan, the couple began a long and furious campaign of letter writing and legal actions to block condemnation by the federal government.[52]

Initially the town of St. Marys accepted a decision by seashore officials to move their operation when visitation reached a level of 600 per day for the summer months. However, it reversed its position for two reasons. First, public antagonism to any increase in visitation suggested that the number might not go above 1,000 per day. City fathers believed that St. Marys could handle that number.

The second reason was a 1979 plan for the city's waterfront, conducted under contract with the Coastal Area Planning and Development Committee. Consultants from Roberts and Eichler submitted a remarkably ambitious proposal to turn the downtown area into a major tourist attraction based on the city's long history and its advantageous proximity to Interstate 95 carrying tourists to Florida, the Kings Bay submarine base, and Cumberland Island National Seashore. The Park Service presence was the key to the level of development. Uncertain whether the seashore's offices would

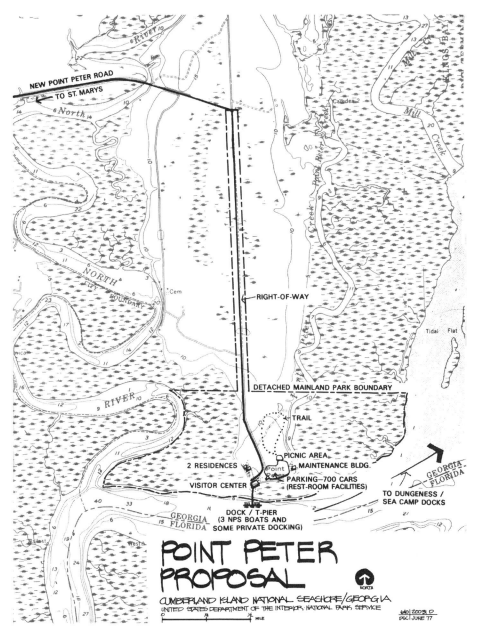

Map 7.3. The National Park Service development proposal for a Point Peter headquarters and main embarkation center. (National Park Service, Dec. 1980, *Final Environmental Impact Statement, General Management Plan, [and] Wilderness Recommendation*, CINS Library)

stay or not, Roberts and Eichler drew up two plans. If the Park Service stayed in St. Marys, the town would re-create an eighteenth-century seaport, develop extensive new commercial operations, and design a trolley system to bring tourists from remote parking lots. If the Park Service moved to Point Peter, the town would have to drop plans for a seaport museum and trolley system and vastly reduce commercial development. Camden County was the poorest county in the United States containing a national seashore or lakeshore. Both the town and the county were anxious to bring tourism to support development in other economic sectors.[53]

When the Park Service chose to remain in St. Marys, the city and agency planners worked together to develop a waterfront plan to suit the needs of both. In November 1985 the seashore adopted a development concept plan under the assumption that St. Marys officials liked it. The plan called for Park Service acquisition of three properties, two of which contained structures on the National Register. The primary property was a block-long stretch of the shoreline containing the rented building then used by the seashore plus a scenic but decaying seafood operation known locally as Miller's Dock. It consisted of a small pier and commercial building. Diagonally across the street from that property was another block with a historic building known as the Bachlott house and an oak-studded field. The third block lay just west of the Bachlott block (map 7.4). The Park Service proposed to raze its current headquarters and Miller's Dock and build a large, new visitor center and ferry landing on the site. The Bachlott house would be renovated to house administrative offices while the rest of that block and the adjacent one would serve as parking for seashore visitors. Both parking blocks would remain unpaved and would accommodate a total of 190 cars. This would relieve the city of visitor parking along the streets, which displaced local people conducting business downtown.[54]

Before the Park Service could secure funds to act, however, problems surfaced. In its own separate 1985 plan, St. Marys decided to limit parking on the street to two hours. This created a crisis for the Park Service because it did not yet own the proposed parking blocks, and visitors had nowhere to leave their cars for the day.[55]

At the same time, several developers began to investigate the town's waterfront. Their attention focused primarily on Miller's Dock. After preliminary inspections and negotiating, the Sea Winds Development Corporation of Atlanta offered the dock's owners $385,000 for the complex. The

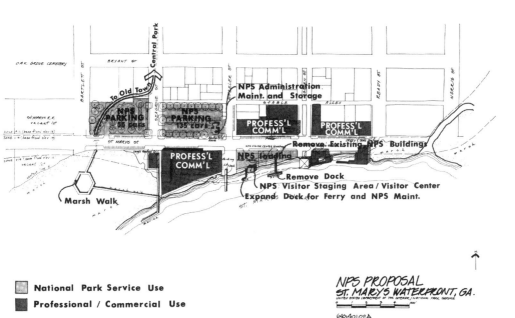

National Park Service Use

Professional / Commercial Use

NPS PROPOSAL
ST. MARYS WATERFRONT, GA.
UNITED STATES DEPARTMENT OF THE INTERIOR / NATIONAL PARK SERVICE

Map 7.4. The National Park Service development plan for St. Marys. (National Park Service, July 1985, *Draft Development Concept Plan and Environmental Assessment, Cumberland Island, St. Marys Waterfront, Georgia,* CINS Library)

Park Service appraised the property and offered $100,000 less. The Millers rejected the government's offer. At that point Superintendent Ken Morgan informed all parties that the agency might condemn the property. This information had already scared off one potential developer two years earlier.[56] In late 1988 the Park Service learned that Sea Winds had purchased Miller's Dock anyway and planned to build a 371-foot pier into the river plus a restaurant and gift shop in the old building. Subsequently the same corporation bought the Bachlott house as well.[57]

The town of St. Marys took the position that a major commercial facility on taxable property was better than an expanded government operation on land removed from the tax rolls. In addition, it began to romanticize the picturesque and familiar old fishing complex. Local historic preservation groups opposed the idea of razing the structure. From 1989 through most of 1991, the Park Service quietly investigated a condemnation action. During that same period city and county officials, historic groups, and local businesses loudly supported Sea Winds and resurrected the menacing image of a grasping, dictatorial federal government.[58] Ken Morgan tried to downplay the local image of federal government bullying. He told the *Camden County Tribune:* "It's not a case where the property owners are thrown out on the street. They're willing to sell; we just can't make up the difference in price. At least at Cumberland Island, the people who went to condemnation proceedings came out better than those who sold the property outright."[59]

The Park Service sought other ways to expand operations but met more resistance from St. Marys. In June 1990 it submitted a proposal to redesign the area east of its current building and convert the town's waterfront pavilion to an embarkation point. Under this arrangement St. Marys would donate the pavilion. A large crowd of local people enthusiastically supported the city councilmen as they unanimously turned down the proposal. U.S. Congressman Lindsay Thomas cautioned the locals not to be too antagonistic as he tried to work out a compromise: "I want to [be] sure that the park service also maintain[s] its departure point for Cumberland right here in St. Marys. I think we all know that Fernandina would be happy to have that." Eventually, the financial condition of Sea Winds ended the stalemate. The corporation declared bankruptcy in 1991, and the Park Service negotiated to buy both Miller's Dock and the Bachlott property.[60]

Just as it appeared that the seashore's controversies in St. Marys might have ended, another issue arose. After purchasing Miller's Dock, agency architects found it too dilapidated to renovate. In a repeat of the Dungeness Recreation House conflict, the Park Service let it sit until part of the dock collapsed. The agency then notified the Advisory Council and state historic preservation offices of its intention to tear the complex down. Although the St. Marys Historical Preservation Commission eventually agreed with this action, many Camden County people cited this as more evidence of National Park Service disdain for their historical resources.

In early 1994 St. Marys and the Park Service briefly considered a memorandum of agreement with the city for federal development at the waterfront. The Park Service would agree to keep the Miller's Dock site as a "passive recreation area" until it built a new structure and to give high priority to renovating the Bachlott building. Thereafter it would either use the building itself or lease it as a historic property. The city, in turn, offered a structure known as the Coastal Bank Building on a nearby street for a reduced price with the understanding that the Park Service would turn it into a museum for both the seashore and the town. Although the two government parties never signed the agreement, the Park Service followed many of its prescriptions. Despite the acrimony of the past, both groups promised a harmonious working relationship in the future.[61]

Natural Resources

Management of natural resources at Cumberland Island advanced significantly during the 1980s and early 1990s. A number of important scientific studies took place, many of them sponsored by the cooperative park studies program at the Institute of Ecology, University of Georgia. The Park Service began monitoring programs for horses, sea turtles, deer, and pigs. Scientists also collected baseline data on the vegetation, dunes, water quality, and fire history. New technological tools, such as a geographic information system at the regional office, enabled researchers to begin compiling a long-term spatial profile of the island's many natural processes and human modifications.[62]

Unfortunately, natural resource management got off to a controversial start in the 1980s. Late in the evening on July 15, 1981, a lightning-caused fire broke out north of South Cut Road in vegetation so thick that one "could literally walk through 20 or 30 feet on top of palmetto without touching the ground."[63] At 4:30 the next morning a turtle researcher discovered the blaze. Superintendent Paul McCrary was immediately notified, and seashore rangers began arriving by 6:15. They initially decided to use Charles Fraser's two bulldozed roads as firebreaks and cleared more brush to start backfires. However, by late afternoon the fire jumped the North Cut Road and moved toward the Candler property. Firefighters then decided to change their strategy to protecting buildings.

In the afternoon the staff received warning of thunderstorms with high winds. McCrary called the Georgia Forestry Department, the U.S. Forest Service, and the Park Service's regional office seeking aircraft to assist in the battle, but these groups were either bogged down fighting other fires or reluctant to fly with potential gale winds on the way. Over the next five days, the Park Service brought in two pieces of heavy equipment and a helicopter to fight the fire. Crews from the Forest Service, Georgia Forestry, and Everglades National Park arrived, as did an eleven-man fire-fighting crew from the regional office. They widened South Cut Road and started backfires while watering homes at the north end to keep them from burning. By July 22 crews of more than forty firefighters, several tanker planes, and a light rain brought the fire under control.[64]

On July 27, after a few rainless days, the fire flared up again. Firefighters burned 100 acres themselves to deprive the renewed fire of fuel. Crews continued to battle the flames until August 2 when rain damped down the blaze. Wary of another reignition, fire crews surveyed the area from the air until August 24 when a tropical storm arrived to finally end it. By then the "South Cut Fire" had burned more than 1,700 acres (map 7.5). Due to the firefighters' focus on saving buildings from the first full day of the blaze, no structures were lost.[65]

Before the fire was out, firefighting agencies gathered to analyze the Park Service's preparedness and response. A clash of missions between the Park Service on one side and Georgia Forestry and the U.S. Forest Service on the other quickly became apparent. The central focus of the debate was the lack of heavy equipment on the island to fight fires. Not only did the seashore not have any, it did not request any from the other two agencies during the early days of the fire. The Park Service explained that such equipment was unacceptable because it damaged both natural and archaeological resources. A 310-acre fire at Table Point four years earlier had not damaged its rich prehistoric remains precisely because the Park Service did not bring in large road graders of the type the other agencies routinely used.

The debate then focused on management priorities. Fire specialists from the Park Service argued that their agency took a long-term view in management and that this same area reportedly had burned a quarter century earlier with no apparent damage to the vegetation. The island would recover. Georgia officials, however, would not accept the danger to personal property and to state resources. One official projected that the fire would invite an insect infestation that would then spread to the mainland. Both sides claimed that the island residents supported their position.[66]

In the meantime, local and state officials publicly denounced the Park Service. Senator Mattingly, in particular, called Secretary of the Interior Watt and demanded an investigation. The claims of Georgia Forestry that the Park Service was unprepared and apparently unwilling to fight a fire properly continued to resound in the press and at various interagency meetings.

One immediate casualty of the episode was Superintendent McCrary. His stubborn refusal to accept large, potentially destructive equipment made him a particular target for Georgia Forestry and Senator Mattingly. Regional Director Robert Baker told the *Atlanta Journal* that "Paul understands that his personality is part of the issue" and removed him in the last days of July. After two months with an acting superintendent, the regional

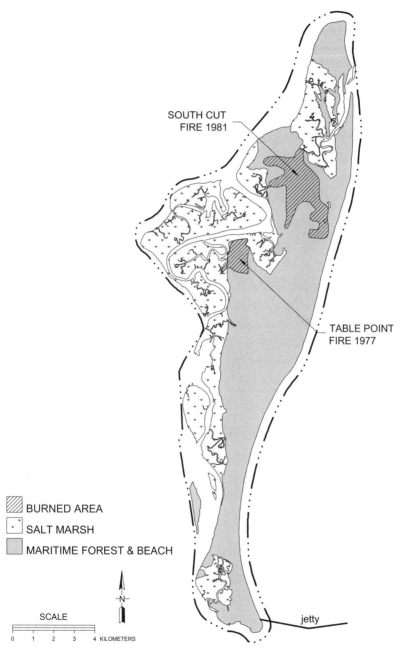

SOUTH CUT
FIRE 1981

TABLE POINT
FIRE 1977

BURNED AREA

SALT MARSH

MARITIME FOREST & BEACH

N

SCALE

0   1   2   3   4 KILOMETERS

jetty

Map 7.5. The Table Point and South Cut fires

office opted to appoint an experienced crisis-management team to lead the seashore. William Harris, former superintendent of Cape Hatteras National Seashore, took over in mid-September. Another veteran official, Wallace Hibbard, became assistant superintendent. The Park Service hoped these two old pros could lead it out of the quagmire of resource protection and planning crises that plagued the seashore.[67]

The final outcome of the South Cut Fire was an agreement between the National Park Service and Georgia Forestry. The federal and state agencies would keep in constant communication and instantly cooperate in the event of another fire on Cumberland Island. Furthermore, the Park Service moved onto the island the very fire equipment it philosophically opposed. The agency pursued a modest program of prescribed burning until the political outfall from the massive 1988 fire at Yellowstone National Park brought total fire suppression to the entire system. Ironically, studies by fire ecologists and archaeologist John Ehrenhard showed no significant damage from the South Cut Fire.[68]

Monitoring Natural Processes on Cumberland

Coastal erosion continued to be one of the most worrisome issues that the seashore faced. In early summer 1980 the U.S. Navy released an environmental impact statement for its proposed nuclear submarine base at Kings Bay. Both the National Park Service and U.S. Fish and Wildlife were invited to evaluate it. On July 9 Superintendent McCrary wrote to Park Service officials in Washington that the navy gave only cursory attention to environmental effects like water pollution, danger to threatened species, and destruction of wetlands. Shortly thereafter, Assistant Secretary of the Interior James Rathesberger notified the navy that his office opposed the Kings Bay plan.

Navy planners subsequently amended the environmental impact statement to satisfy Interior officials but also suggested that they might widen the ship channel from 400 to 500 feet. Rathesberger wrote to complain that this would require a new environmental impact study. The navy then flatly turned down the call for a new study, claimed that the original one was adequate, and informed Rathesberger that it would monitor the impacts after construction.[69]

The Department of the Interior was unable to stop this particular action, but a few years later it did escape a much more intrusive proposal for Cum-

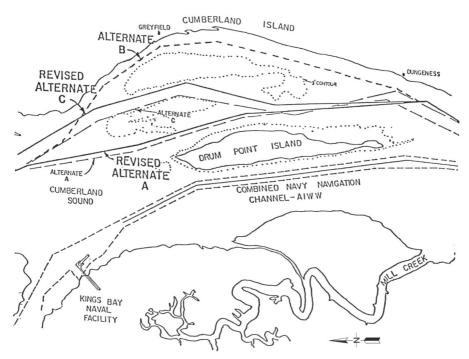

Map 7.6. The Atlantic Intracoastal Waterway Channel and proposed changes to its course. (Michael J. Harris, July 1986, "Fish and Wildlife Coordination Act Report on Reroute of the Atlantic Intracoastal Waterway, Cumberland Sound, Georgia," U.S. Fish and Wildlife, Southeast Region, Atlanta, CINS Library, 24)

berland Sound. In 1985 the navy announced that in order to adapt the Kings Bay base for its new, large Trident submarines, it would have to separate its military ships from commercial vessels. Navy planners claimed that traffic on the Intracoastal Waterway was too heavy to allow for safe and unrestricted passage by Trident submarines. They proposed that the commercial waterway be moved eastward toward Cumberland Island. Seashore officials believed that the worsening erosion of archaeological sites and the Plum Orchard estate was caused by wakes from passing ships, and therefore movement of the waterway closer to Cumberland Island would exacerbate the problem.[70]

The navy offered three alternative routes for the new waterway that would lie 300, 1,200, or 1,500 feet closer to Cumberland (map 7.6). However, all three routes were east of Drum Point Island, which served as a buffer against ship wakes for the area from Greyfield to Dungeness. Once again the navy offered its assessment of the potential environmental costs, and once again the Park Service and U.S. Fish and Wildlife called it inadequate and poorly researched. Georgia Department of Natural Resources and respected coastal engineer Robert Dolan also found the proposal disturbing. Soon environmental organizations took up the campaign. They found the project objectionable not only because it passed through the boundaries of the national seashore but also because it intruded into a zone designated by the Coastal Barrier Resources Act of 1982 as a protected coastline. The navy answered that this project was a "military activity essential to national security" and thus exempt from the Coastal Barrier Resources Act.[71]

As the agency charged with evaluating and, if approved, executing the channel shift, the U.S. Army Corps of Engineers undertook a review of the project. After examining the navy's proposal, the corps narrowed it to two alternatives: one midway between Drum Point and Cumberland and the other only 100 to 200 feet off the shore of the bigger island. The corps also suggested the possibility of depositing dredge spoils on Drum Point Island. After a review, the Fish and Wildlife Service recommended the route closest to Cumberland Island as the lesser of the evils because it would have less impact on benthic communities.[72]

Meanwhile, the environmental groups' campaign against the project mounted. Hans Neuhauser told the Corps of Engineers that the legislation establishing the national seashore required that a project such as this one must be "mutually acceptable to the Secretary of the Interior and the Secretary of the Army." In his mind this meant that the project required full compliance with NEPA. He opposed dumping spoils on Drum Point Island, and if a new channel would have to be chosen, he would reject the route immediately adjacent to Cumberland Island. The Georgia Chapter of the Sierra Club agreed and announced in a press release that the corps's environmental assessment "has failed to evaluate so many aspects of the project that it has been, in effect, a wasted exercise." Soon the Georgia Conservancy announced not only that it opposed the new waterway route but that the Department of Defense should transfer Drum Point Island and Raccoon Keys to the national seashore. Finally, the Conservancy engaged the Southern Environmental Law Center to challenge the corps's quick approval of the project. This action plus continued bad publicity soon had both the navy and the corps on the defensive. In early March 1987 the navy announced it would scrap plans to establish a new channel because of "a lack of hard requirement and the controversial nature of the project."[73]

Despite the defeat of this serious threat, erosion along the western shore

Fig. 7.1. Riprap at the Sea Camp Visitor Center protects it and its archaeological relics from coastal erosion.

of Cumberland continued to expose and then sweep away archaeological resources (fig. 7.1). Once the huge Trident submarines began to ply the waterway, they became the focus of suspicion from Park Service scientists and environmental groups. During the debate over rerouting the Intracoastal Waterway, the navy agreed to fund a program for monitoring Cumberland Island. Research and monitoring began in 1988 and resulted in a number of useful reports. In July 1993 the *Florida Times-Union* reported that the navy-funded study of erosion showed that the submarines were not the cause. Other experts remained unconvinced, and research continued to seek the reasons for the accelerated erosion that they measured.[74]

One additional concern about the Kings Bay submarine base was the impact of deepening the ship channel on the aquifer that supplied Cumberland Island wells. Development along the mainland coast had already reduced the supply of groundwater to the national seashore. In 1887 the Carnegies dug a deep well to the Florida Aquifer that flowed at an estimated 800,000 gallons per day and had an estimated head of fifty-one feet above sea level. By the 1990s water barely reached the surface. Many of the wells for campground use were shallow ones that accessed the Pliocene-Miocene aquifer. Scientists worried that deepening the ship channel would lead eventually to saltwater intrusion into this aquifer.[75]

Ecological Resources

The richness of Cumberland Island's biotic assemblage received another mantle of recognition during the mid-1980s when it was designated an "international biosphere reserve." Scientists and diplomats gathered at a 1970 conference of the United Nations Educational, Scientific, and Cultural Organization (UNESCO) had conceived the Man and the Biosphere Program (MAB) to coordinate research and management of representative terrestrial and coastal/marine ecosystems. These biosphere reserves function as examples and scientific monitoring sites for surrounding regions. Each unit must meet criteria based on its ecology, history, land use, and management. A reserve fulfills three complementary functions: preservation in the core area, sustainable development, and education. Each reserve contains one or more core units, such as a national park, that are securely protected and monitored, an adjoining buffer area of compatible use, and a transition area of other uses like agriculture that cooperate with the core unit(s). The United States unilaterally established units in 1974 while UNESCO scientists developed the conceptual framework. In 1996 the United States boasted 47 reserves encompassing 99 legally protected sites. In addition, eighty-four other countries protected 290 reserves.[76]

In 1981 William Gregg, Park Service ecologist and cochair of the U.S. chapter of MAB, coordinated a research project by Yale graduate student Judith Gale to identify potential sites for a Carolinian–South Atlantic Biosphere Reserve. After several years of study, the team recommended three units: the Outer Banks and continental shelf at Cape Lookout, North Carolina; the Santee Delta–Cape Romain area of South Carolina; and the Georgia Sea Islands. The latter unit focused on Sapelo Island. The Georgia Department of Natural Resources, however, balked at the idea of cooperative management with other states, so the U.S. committee shifted its Georgia core unit to the federal property at Cumberland Island. In April 1986 UNESCO added the Carolinian–South Atlantic reserve to the system.[77]

International biosphere reserve status brought no change to Cumberland Island or its management, however. Incompatible land use in the South Carolina and North Carolina portions of the reserve and a bizarre rise in the public's xenophobic paranoia about internationalism stymied any action. Cumberland Island's inclusion in a biosphere reserve was quietly forgotten, except as a vague recognition of its superior resources. In response to a National Parks and Conservation Association questionnaire sent in 1997, Cumberland Island ranger Newton Sikes answered ten questions about the MAB status the same way: "To our knowledge tourism [or any other aspect of the seashore] has not been affected in any way due to the park being a biosphere reserve. There has been no activity at all regarding biosphere matters. No publicity. Nothing."[78]

In spite of inactivity in the biosphere program, research continued at Cumberland Island. Susan Bratton of the cooperative studies unit at the University of Georgia conducted or directed much of it. Botanists looked at vegetation dynamics, response to fire, and exotic species. Among the latter were tung trees, which had begun to spread from their original two agricultural plots. Wildlife biologists focused on sea turtles, turkeys, armadillos, manatees, deer, pigs, horses, wood storks, and a variety of seabirds. Of particular concern were the effects of feral animals and deer on dune vegetation and stability. Monitoring programs in conjunction with the navy erosion studies or as independent projects were established for sea turtles and most of the higher mammals.[79]

Three animals drew not only intense scientific study but also considerable public attention. Despite up to five annual hunts per year, the population of white-tailed deer continued to overbrowse the vegetation on Cumberland Island. Elsewhere in the park system, resource managers sought to reintroduce predators that would correct ecological imbalances. Soon Cumberland's managers also began to investigate this possibility. In a 1984 study Edward B. Harris proposed the reintroduction of bobcats to the island to restore ecological balance. According to Harris's research they had been extirpated in 1907. The reintroduction attempted in 1972-73 had failed. Harris noted that the chief prey of bobcats were rabbits, rodents, and the occasional old or injured deer. A few years later Robert J. Warren of the University of Georgia took charge of the program to reintroduce the bobcats.[80]

Before any reintroduction could take place, however, NEPA required that the Park Service issue an environmental assessment for public comment. In August 1988 the agency released the assessment and received seventeen written comments. The Georgia Department of Natural Resources and all the major environmental groups supported the plan. However, opposition came from hunters who were afraid the competition would drastically reduce deer numbers, from some island residents and others who worried about the fate of the wild turkeys, and from the St. Marys City Council, which regarded the scheme as "inhumane." The Park Service issued a "finding of no significant impact" a few months later and began introducing fifteen bobcats per year. Subsequent studies showed that the predators had established themselves, reproduced, and were affecting the deer herd primarily by taking fawns. Research on the impact of bobcats on the turkey population is still inconclusive. Although the controversy died down in the press and among animal enthusiasts, some Camden County and island residents still fume over what they believe was an irresponsible and destructive action.[81]

Feral Animals

Unfortunately, the reintroduction of bobcats had no impact on the population of feral pigs, nor did most other actions. Each general management and resource plan advocated complete extermination of the pigs, citing all the familiar reasons. They dug up and ate turtle eggs, destroyed endangered plant species, took resources away from native animals, and caused a nuisance by rooting around on the private estates. By early 1982 the national seashore had conducted thirty-five trapping operations and removed 1,346 pigs in seven years. Rangers began shooting them thereafter, which, added to the hunts, led to a significant reduction in their numbers.

In 1984, in an ironic twist, Cumberland's resource management chief ordered rangers to stop killing pigs because they had been so successful. The pigs had become so shy that Robert Warren could not find enough to conduct research on feral hog ecology. In 1988, four years after the cessation of shooting, Warren's study reinforced the common beliefs of the seashore staff. The pigs were healthy, ate mostly plant material, and were descended from domestic stock. Despite their addition to the list of animals taken by hunters, their numbers had grown. Warren recommended that rangers return to shooting them on sight to control the population. Even with that, he did not think it was possible to eradicate them.[82]

In 1992 the Park Service escalated its war on feral pigs by bringing professional trappers with dogs to the island. For five days the trappers used "bay dogs" to corner pigs so they could be removed from the island. One man brought a pit bull trained to attack the swine. Island residents and environmentalists alike reacted with alarm. Carol Ruckdeschel, Bill Mankin of the Sierra Club, and others worried especially about attacks on visitors and residents by packs of dogs. According to Superintendent Rolland Swain, some complained that the hog catchers were "good old boys who wanted a chance to hunt hogs and have fun in the park." The Park Service also received letters from backpackers who were frightened when they met the trappers and their dogs in the wilderness area. Soon the Camden County Humane Society complained about the methods that trappers used to load the pigs for transport to the mainland. Negative publicity did not take long to appear in local and Atlanta newspapers.[83]

The Park Service responded to the complaints by reiterating the reasons for pig removal and allowing the trappers to continue without using dogs. Concerned letters continued to arrive and included one from a senior professor of biology who claimed that damage and disruption by the dogs was itself unacceptable. Some letters did support the Park Service and included one from Thornton Morris. In 1993 the Park Service added a "pigs only" public hunt to the four annual hunts for deer and pigs. The following year rangers renewed the practice of shooting them whenever possible. In spite of all this effort, the swine population continued to rise and expand its range southward from the wilderness.[84]

The difficulty of managing the pig population under public scrutiny was insignificant compared to that of horse management. Concern about the environmental impact of the feral horses continued to grow as research results arrived. Censuses taken by University of Georgia researchers and seashore personnel estimated 114 horses in 1981, 154 two years later, and 220 by 1990 despite the brief appearance of eastern equine encephalitis. A study of the effect of horses on salt marshes showed a reduction of the vegetation of up to 98 percent in heavily grazed areas (fig. 7.2). The author suggested that the carrying capacity for horses on Cumberland Island was between 45 and 73 individual horses. Research on the health and mortality rate of the island horses determined that 30 percent of the foals died in an average year. A study at Assateague Island National Seashore and the adjacent Chincoteague National Wildlife Refuge showed a similar mortality rate at the national park unit but a much lower one at the refuge where a local fire department removed horses each year to sell.[85]

In 1985 Robert Warren and Susan Bratton of the University of Georgia proposed that the Cumberland horse population needed to be reduced and then controlled. They suggested a brand new drug that would prevent conception for two years. Bratton noted that this approach could use "a

Fig. 7.2. Horses are one of the major attractions for visitors, but they are responsible for destruction of dunes and marshes throughout the island.

biodegradable bullet" to administer the drug which would reduce both the possibility of injury to the animals and protests from animal rights groups. The following year one of Warren's graduate students, Robin Goodloe, began formal research on the efficacy of contraception for feral horses. Over the next five years, Goodloe and other researchers reached two conclusions from their work. First, genetic testing showed that the wild horses on Cumberland were descended primarily from stock released during the Carnegie period. Little genetic evidence of the "marsh tackies" of romantic lore appeared in the tests. The second conclusion was that the new drug did not work. Even a population of treated mares kept in Florida continued to foal. Subsequent research showed that "booster shots" of the drug were necessary every four weeks for at least the first few months in order to make the drug effective.[86]

As Goodloe's project neared its conclusion, the National Park Service grappled with the feral horse issue not only at Cumberland but also at Assateague Island, Cape Lookout, and Cape Hatteras National Seashores. New immunocontraceptive drugs tested at Assateague proved more effective.

Goodloe's annual census showed that the horse population on Cumberland continued to increase partly due to releases of new horses by island residents to "improve the stock." Another outbreak of equine encephalitis in 1991 killed at least one and probably five horses, but it had little impact on the population increase.

Finally in 1992 the Park Service decided to develop a horse management program. New resource management specialist Jennifer Bjork had experience from the other seashores with horses and headed the planning team. In spite of the agency's formal scientific position that horses should be entirely removed, the seashore planners chose to reduce the herd to a small number, perhaps forty, and distribute them on the south end of the island so that at least 66 percent of the visitors would see them. In order to maintain such a herd, Warren and others recommended that the Park Service heavily manipulate the water sources and forage at the south end of the island, especially around the Dungeness estate. The Cumberland Island staff then turned to developing an official horse management plan for the NEPA process. Finally the Park Service appeared ready to take planned and coordinated action on one of the most troubling natural resource issues.[87]

Cultural Resources

Like natural resource protection, cultural resource management advanced during the 1980s and early 1990s. Park Service historic preservation officials completed a multiple-resource nomination to the National Register for three more historic districts, two individual historic sites, and two archaeological districts. Maintenance continued on the many preservable structures around the island, and several more were adapted for use by seashore personnel. Important historical research took place, and development concept plans were drawn up for Plum Orchard and Dungeness. In addition, the Park Service completed new plans for managing the "historic landscapes" of Dungeness and the Stafford plantation. The agency finally tackled some overdue management issues such as establishment of a proper facility for storing furnishings and other objects from the island's history. Finally, cooperation with the Georgia State Historic Preservation Office improved by the end of the 1980s.[88]

The accomplishments in cultural resource management did not always come easily, however. The Georgia office rejected several management plans before the Park Service could satisfy its standards. Charles Hauser once again threatened important historic resources on the reserved estate that he rented. The debate over management of historic resources in the wilderness intensified. Finally, the care and use of Plum Orchard proved a constant source of frustration. Ultimately, island residents and other interested parties formed an organization to search for a solution. Friction between this group and the Park Service grew when the residents claimed that the agency did not care about the mansion.

Once again most of the trouble came during planning processes. In September 1981 the Park Service met with the Advisory Council on Historic Preservation and the National Council of State Historic Preservation Officers in Washington, D.C. They amended an earlier cooperative agreement to improve interagency communications during historic resource planning and nominations to the National Register. In spite of this leadership, Cumberland continued to have problems securing approval for its historic preservation programs. As part of a draft of the general management plan in 1982, the seashore staff submitted a cultural resource plan which ran into immediate trouble. Elizabeth Lyon of the state office wrote that while the plan appeared to provide for adequate protection of historic structures, it did not have a timetable for preservation actions. She added that the Park Service was still remiss in keeping the state office informed. Two months later the Advisory Council said the plan was "vague in describing proposed actions and leaves open possibilities for adverse effect."[89]

The agency also encountered difficulties in preparing a multiple-resource nomination to the National Register of Historic Places. Southeast Region officials sent the nomination to Washington in early 1984. This time the Park Service's own chief historian, Edwin Bearss, rejected part of the document. He accepted the sections on the Dungeness district, Main Road, Duck House, and Rayfield archaeological area. However, he returned the portions on the Plum Orchard and Stafford historic districts and the Table Point archaeological district. After revisions, both the cultural resources management plan and the National Register nomination finally were approved in 1984. Cumberland Island boasted six discrete historic or archaeological districts plus the isolated Duck House and the Main Road (see map 6.1). The seashore's list of classified structures totaled sixty-four. Over the next decade new surveys and corrections to the list increased the number to eighty-two structures despite destruction of the Duck House by an illegal camper's fire (fig. 7.3). Those surveys also more than doubled the number of identified archaeological sites.[90]

Aside from the planning and register nominations, the Park Service

Fig. 7.3. The Duck House was listed on the National Register in 1984, but an unidentified illegal camper burned it down a few years later.

faced a difficult task in maintaining the cultural resources on the island. From 1980 through 1994 most of the structures received some maintenance. At Dungeness seashore workers, Youth Conservation Corps volunteers, U.S. Navy Seabees, and contractors hired for specific jobs restored and re-modeled the carriage house, dairy manager's house, carpenter shop, ice house, and one of the dormitories for adaptive use. They also repainted and reroofed other buildings at Dungeness and Plum Orchard. The Park Service rebuilt all the government docks and restored several seawalls as well. After a line-item appropriation of $800,000 for historic maintenance in fiscal year 1981, Plum Orchard received extensive rehabilitation. Almost every year thereafter the mansion received further attention. In addition, land-scape clearing, especially at Dungeness, removed encroaching vegetation from the many historic buildings.[91]

Despite this program, annual inspections showed that the Park Service was falling behind. A briefing statement in January 1988 reported:

> Some of the cultural resources at Cumberland Island have received atten-tion in the last few years, but there are still many unmet needs in this area. Our ongoing monitoring program has revealed that National Register–listed archaeological sites on the west side of the island are being eroded by wave action to the point that substantial cultural material is beginning to be lost. Other sites suffer from wind erosion and damage caused by an exotic animal species, the armadillo. The curatorial collections on the island include about 15,000 uncataloged objects ranging from furnishings to architectural ele-

ments to prehistoric artifacts, scattered through a variety of substandard storage areas. Cataloging these objects and providing safe storage for them remains a pressing need. Most of the 74 [at that time] historic structures on the island need some sort of work to slow their progressing decay, though some work has been done on Plum Orchard, the old Carnegie mansion. . . . No NPS monies are currently available to restore the interior of Plum Orchard so a local friends group is trying to raise money to restore the house room by room.[92]

The retained-rights estates on the island occasionally complicated resource protection. In the 1970s the Park Service removed a decaying and dangerous water tower from a retained estate held by Nancy Butler. A similar water tower at Stafford began to lean in 1982, and retained-right holder Franklin Foster requested that the Park Service remove it. Enmeshed in expensive repairs to Plum Orchard, the Park Service told Foster that maintenance of structures on reserved estates was the responsibility of the landholder. The Carnegie heir responded that the Park Service had done some work on a tabby wall on his estate, and he did not understand why it would not act now. The ensuing debate soon involved the regional solicitor, who informed Cumberland Island officials that unlike other agreements with the National Park Foundation, Foster's did not include a statement that he was responsible for the maintenance. In the end, the seashore staff not only had to remove the water tower but also had to fix the roof and replace 40 percent of the stucco on the Stafford mansion and instigate a continuing maintenance program for all the historic structures on the Stafford estate.[93]

A more troubling question of historic preservation arose when Charles Hauser decided to renovate an old stable on the estate he rented. In 1983 he began to modify the building into a group of small rental apartments. During the course of the work, he bulldozed part of his estate to clear a vista for the new units. Although seashore personnel had questioned the regional office about the legitimacy of these actions, the issue blew up when agency archaeologist John Ehrenhard came upon the scene while on his way to another site. According to Ehrenhard, Hauser had pushed trees more than 150 years old plus other brush and a substantial amount of soil into the marsh. Ehrenhard entered the property and found that the cleared area included part of a known Indian midden as well as a foundation dating from the early plantation period.

A flurry of correspondence followed as Hauser filed a complaint about Ehrenhard's trespass, and seashore officials appealed to the regional solicitor for help. In this case, Hauser did not get his way. The retained-rights agreement with Thomas Johnston clearly spelled out the prohibition against the sort of commercial activity Hauser intended with his apartments. Furthermore, the Park Service determined that it could enter an estate and stop actions that were blatantly destructive to archaeological resources. Superintendent Morgan later wrote that Hauser "has been put on official notice that future action would put the life estate in jeopardy."[94]

Of all the buildings on Cumberland Island, the most difficult and expensive to maintain was the Plum Orchard mansion. By 1983 the 21,724-square-foot house suffered from leaks in the roof, old and potentially dangerous electrical wiring, and rapid deterioration of the cement plaster on the outer walls. The interior also needed rehabilitation, especially the wallpaper. The public tours of the house had been running for four years, and they too contributed to the problems. Rangers reported incidents of theft and petty vandalism. The cost of maintaining the big house could easily consume the bulk of the seashore's historic preservation budget for a number of years.[95]

On March 11, 1983, a group gathered on Cumberland Island to search for a long-term solution. The Park Service had secured a $355,000 appropriation for the most necessary work that summer. However, everyone recognized that even this large amount was a mere Band-Aid. Several Carnegie heirs, other island residents, and historic preservation advocates joined Park Service officials from the seashore and the regional office. Conspicuously missing were environmentalists, although Jane Yarn had been invited. The star of the meeting was William Penn Mott, former director of California's state park system and future director of the National Park Service. The California system included the gigantic William Randolph Hearst estate with its magnificent mansion at San Simeon. Hearst Castle had threatened to bankrupt the golden state's park system until it was made into a lavish tourist site. According to Mott it netted $500,000 profit per year for the California parks. Clearly he could give some advice on what to do to save Plum Orchard.[96]

After some introductory discussion Mott took the floor and delivered a lengthy lecture on the island's visitor carrying capacity, wilderness, historic resources, and the way things were done at Hearst Castle. He also introduced a number of interesting ideas to secure money for Plum Orchard's maintenance. First, he suggested that a foundation be established to get around the bothersome bureaucracy that the Park Service had to follow.

Such a group could raise money and contract for repairs without going through the tedious bidding process. He also suggested that the foundation could hold a gala on the mansion grounds to attract donors. This had worked quite well at Hearst Castle. Finally, Mott suggested that the price to visit the house be raised substantially in order to keep the house in highest-quality condition. Only then would people pay to see it. As an afterthought he added that the Park Service needed to fight environmentalists who wanted to restrict visitor numbers both within and outside the wilderness.[97]

The discussion that followed resurrected various ideas for using the mansion, including a lease to a corporation, a research facility, and a museum. Participants also raised a number of related questions. Access by boat or horse-drawn buggy was feasible for visitors but not for maintenance. Others, including Janet "GoGo" Ferguson of Greyfield, suggested that the limit on visitors needed to be raised, countering the island residents' long insistence on keeping island visitation to an absolute minimum. Fred Storey, owner of the Point Peter land, thought that "the 300 limit could be varied and easily overcome." However, Thornton Morris warned that increasing the visitor capacity to ensure more tourists at Plum Orchard would not be "politically expedient." One island resident, Cindy McLauchlan, stressed that interpretation at the mansion must include all of the history of the area. To her, the island was "not a Carnegie shrine! . . . [There is] so much more here than just Carnegie."[98]

Everyone left the meeting with an enthusiastic sense of hope. The following September many of the same people gathered to form the Cumberland Island Historic Foundation. Soon many historic preservation officials joined along with more island residents and some environmentalists. The foundation began to seek money for repairs to the interior of the house while working with the seashore officials to compose a development concept plan (DCP) for the mansion.

The Park Service released its draft DCP for Plum Orchard in November 1984. Because the plan required a full NEPA process, the public had a chance to review the ideas for Plum Orchard. The Park Service proposed three alternatives. The agency preferred a plan to lease the mansion to a corporation or organization. Under this arrangement the lessee would assume the burden of maintenance costs while the government retained the right to bring visitors through the ground-floor rooms. Optimistic seashore planners even projected that the payments would help defray other historic preservation costs on the island. A second alternative called for the Park

Service to rehabilitate the house room by room as funds became available and keep the first two floors open for visitor tours. In the final option seashore officials would mothball the mansion until a more propitious time for leasing.[99]

Coming on the heels of the contentious general management plan, the reaction to the Plum Orchard plan was subdued but nevertheless negative. Some who wrote opposed the preferred alternative because it would bring more people and cars into the wilderness and because visitors to Plum Orchard would count against the 300-per-day limit on island visitation. After this response, the Park Service reconsidered the options and soon cobbled together a new plan. It called for the agency to delay leasing the mansion until it became financially feasible, to perform minimum maintenance but no historic restoration unless unexpected funds became available, and to continue the very limited visitor tours. Seashore officials also decided to refurbish part of the house for a ranger residence in order to protect the property. In May 1995 the regional office issued a "finding of no significant impact" officially adopting the development concept plan.[100]

Putting the house on low-maintenance hold shifted the burden of financing its rehabilitation to the Cumberland Island Historic Foundation. Many in the organization expressed concern over what appeared to be a dereliction of duty by the government. In response to their criticism, the Park Service repeatedly tried to explain how the mansion fit into the program of historic preservation in the region. Robert Baker told the foundation at one meeting: "There are some globally significant resources in the Southeast that are in danger of being destroyed. This group needs to understand on an overall basis what [the] NPS is trying to do—that is to preserve the cultural and natural fabric of America. I don't think there is enough money on the face of the earth to do everything everyone would like to do in terms of preservation of natural and cultural resources for everyone's interest."[101] What Baker left unsaid was that the Park Service classified the mansion as a building of only "local significance" while other parks in the Southeast Region contained scores of structures of national or even international importance.[102]

The historical foundation began a two-pronged campaign to save the house. Acting upon Mott's advice, the group held a "Plum Orchard Gala" at the mansion in 1988 that raised more than $7,500. A second gala followed in 1989 and was called a success although it drew fewer people than the first one. Over the decade from 1984 to 1994, the Cumberland Island Historic

Foundation raised more than $700,000 for the house and various out-buildings. At the same time, the Cumberland historical group refused to accept the Park Service's inaction. Various members approached major corporations like Coca-Cola seeking both donations and a potential lessee for the estate. In 1994 GoGo Ferguson led an effort to create a Carnegie-Cook Center for the Arts. This organization planned to secure funding from corporations and foundations to lease and refurbish Plum Orchard as a retreat for "education and research in the broad field of American fine and performing arts." Supported activities would include seminars, workshops, classes, and symposia as well as full support for artists in residence for specified periods of time.[103]

The National Park Service, especially Superintendent Rolland Swain, enthusiastically greeted this proposal. The Carnegie-Cook group readily accepted the requirement that seashore visitors would still tour the first floor. In addition, the arts group agreed to restore the nearly disintegrated Plum Orchard carriage house for additional rooms. In turn it wanted permission to house 30 staff members and long-term guests plus bring over up to 300 additional people four times per year for colloquia and other meetings. Swain seemed amenable to excusing these operations from the 300-person visitor capacity. Finally, the Park Service seemed to have a solution to the emotion-charged Plum Orchard problem.[104]

As 1995 began, the future for Cumberland Island National Seashore looked better than at any time in the past. The Park Service was established in St. Marys with several pieces of property and plans for plenty of office and parking space. New horse and fire management plans were in the review process while seashore officials were ready to fashion one for wilderness management. Seashore personnel and cooperating university scientists had established robust research and monitoring programs for coastal erosion, turtles, horses, bobcats, and many other elements of the natural resources. Park Service leaders in Atlanta and Washington, D.C., promised funds to eliminate pigs and control the horse population. The possibility of leasing the Plum Orchard mansion gave reason to hope that money might be freed to maintain the many other historic structures and archaeological sites. Perhaps most surprising, Andrew Rockefeller and the Fergusons at Greyfield expressed tentative interest in selling portions of their private lands to the Park Service.

Hope for the
New Century

Going into the last five years of the twentieth century, National Park Service officials still sought closure for four types of problems. Natural resource managers needed to expand their research and monitoring of coastal erosion as well as resolve the question of feral animals. Historic preservation suffered not only from a woeful shortage of funds but also disagreement over the priority of cultural resources. Private landholders still controlled several critical island-straddling private plots while the limits of retained rights remained vague. Finally, Cumberland Island National Seashore desperately needed a wilderness plan. The level of conflict between various interest groups grew more emotional and vicious with each passing year. The agency's reactive and somewhat erratic management faced almost certain lawsuits over the wilderness in the near future. Designing a wilderness plan in the midst of such acrimony was a forbidding prospect. Superintendent Rolland Swain understated the complexity of wilderness planning when he noted that "it is almost certain to be difficult and contentious."[1]

Unfortunately, events during 1995 and 1996 crushed the hard-won optimism of the early 1990s. In each management area the National Park Service saw its plans and proposals unravel in the face of renewed public criticism and the worst interest-group conflict in the history of the seashore. Yet as the new century dawned, the agency reevaluated the Cumberland legislation and retained-rights pacts and decided to plow ahead with wilderness planning. At the same time, many seashore officials reluctantly admitted that only a court of law or Congress could settle the myriad problems of Cumberland Island National Seashore.

Natural Resources

Park Service and cooperating independent scientists continue to research and monitor natural resources and to plan for their management in the new century. The agency released another version of a fire plan in 1996, which again failed to satisfy the prescription for the wilderness. Turtle research has expanded, as have efforts to protect the nests from marauding pigs. Seashore officials now even relocate nests that are in vulnerable areas. Other species of fauna continue to draw attention, while the agency sporadically pursues eradication of tung trees and tamarack. Nevertheless, natural resource management is focused predominantly on three issues: coastal erosion, pigs, and horses. U.S. Navy monitoring and research culminated with a report in December 1997 stating that erosion on the western shore of Cumberland Island was not the fault of its submarines. Instead, the contract scientists blamed the Cumberland Island jetty, Intracoastal Waterway dredging, and the deepened St. Marys channel. Although this study did not satisfy some independent coastal geomorphologists, the navy summarily terminated its investigation of the destructive erosion on Cumberland Island.[2]

Pigs continue to destroy sea turtle nests, and this has proven especially difficult to solve. Rangers or turtle program volunteers patrol the long Cumberland beach each night during the summer nesting season for loggerheads. Upon locating a fresh nest, a resource specialist covers it with a wire grate that allows the baby turtles to escape but blocks access by raccoons to the deeper portions of the nest. Unfortunately, pigs are intelligent enough to learn how to dig open the grated nests. Seashore resource management specialist Jennifer Bjork has estimated that from 1995 to 2001, hogs consumed 7,800 turtle eggs, a disheartening figure for an endangered species. In October 2000 the agency released an environmental assessment for feral hog management alternatives. The preferred option calls for the agency to alternate trapping and hunting (fig. 8.1). This sequence may offset the pigs' ability to learn to avoid one method. The large traps must be moved over trails widened for the trucks that carry them. Even then, removal of trapped pigs to the mainland will take place only if they are found to be disease free. Testing for porcine disease threatens to become a tedious and expensive process. As the pigs' number dwindles, the Park Service plans to allow hunters with dogs to eliminate the last and smartest of them. Public response has been muted and generally supportive. However, when the actual destruction of pigs begins, it will surely bring renewed criticism from animal lovers.[3]

Horse management continues to be the most irksome of natural resource issues for seashore officials. In early 1995 the Park Service held a

Fig. 8.1. Live trapping and transportation off the island is one difficult and expensive way to control the hog population on Cumberland Island.

meeting of twenty-six agency, state, and university specialists to consider horse management at both Cumberland Island and Cape Lookout. For Cumberland, the group was unified in its solution: remove the horses, or at least reduce their population. With this renewed support, as well as that from the environmental organizations, the agency developed an environmental assessment of various horse management alternatives. These options included complete removal, reduction of the herd to 60 horses kept at the south end of the island, use of a onetime herd reduction and immunocontraception to stabilize an islandwide population at 120, or the no-action alternative required by NEPA.[4]

On April 8, 1996, the seashore staff released the assessment that identified the 120-horse option as its preferred one. Three hearings were held in Brunswick and Kingsland, Georgia, and Fernandina Beach, Florida. The Park Service conspicuously neglected to schedule one for Atlanta. Jennifer Bjork presented the data on the increasing horse population and their destruction of marshes and dunes. The response of the public was predictable. Locals united with island residents to oppose all the forms of herd reduction. They refused to believe either the Park Service or the scientific data. Some insisted that the horses had been there for centuries with no effect on the environment. A few even claimed that the horses were native. Island residents, still annoyed about the reintroduction of bobcats, insisted that the feline was also exotic to the island. If the Park Service could introduce an exotic, they reasoned, why persecute the horses? Speakers claimed that horses were part of the history, part of the beauty of the island, and the key to visitor satisfaction with the national seashore. Furthermore, they cited a *Florida Times-Union* article that showed people who adopted feral horses and mules rounded up in the West by the Bureau of Land Management often sent them to slaughterhouses. Soon people from Atlanta joined the clamor. Letters and petitions from more island residents, children's groups, and animal rights activists poured into various Park Service offices.[5]

Not all the respondents to the environmental assessment favored uncontrolled horses on the island. Environmental groups and ecologists continued to call for their removal. Robert Coram of the *Atlanta Constitution* blamed overpopulation for the condition of the horses, which he called "spavined, disease-ridden, glue-factory material with a foal mortality rate of 29 percent." Coram claimed that the Park Service was backing away from the option to remove horses completely and worried that it would "bow to irrational, and in this case destructive, emotion rather than doing its job."[6]

The horse management issue dragged into autumn of 1996 as National Park Service officials wavered in the face of yet another bristling public response to its planning. At this point local congressman Jack Kingston decided to act. He toured the island with one of the residents and suddenly claimed that the horse numbers in fact had decreased and that he could find little evidence of damage caused by them. Without discussing the issue with the Park Service, he added a rider to the fiscal year 1997 budget bill banning all horse management at Cumberland Island National Seashore. The rider suggested that the Park Service use the money allocated for the horse program to further sea turtle protection.[7]

Seashore officials, as government employees, were forced to remain publicly quiet on this extraordinary step, but environmentalists erupted. Judy Jennings of the Georgia Sierra Club told the *Florida Times-Union,* "What does he know about managing a horse herd on Cumberland Island? The Park Service is relying on expert managers to do their job." Kingston held a forum in Kingsland on October 20, 1996, to defend his action. According to the *Times-Union,* environmentalists pressed him to say whether he had noted any damage from horses on the island, but he refused to answer. The Florida paper added that in response to a question about why he did not contact the Park Service or any environmental groups, he replied, "If it was such an important issue to them, why didn't the environmental groups or the National Park Service contact me?" At present, the Park Service continues to take no action with horses other than the annual census.[8]

Cultural Resources

Each superintendent of a national park unit seeks to leave a legacy of his leadership. For Rolland Swain, solving the dilemma of Plum Orchard's maintenance was paramount. During 1995 and 1996 he pursued the bureaucratic steps to complete a memorandum of agreement (MOA) giving the Carnegie-Cook Center for the Arts a fifty-year lease on the mansion. At the same time, opposition grew among environmental groups and, surprisingly, some Camden County residents. Conservationists warned that people living in or visiting the center would inevitably spill into the adjacent wilderness and damage its resources. In addition they cited concerns such as increased automobile traffic on the Main Road, the probability of a new wave of construction at Plum Orchard and on private lands, and the detrimental effect that a fifty-year lease would have on later planning and wilderness legislation.[9]

In the meantime, Camden County state representative Charlie Smith Jr., whose father once challenged the Park Service ban on motorboat access to the island, was incensed by what he regarded as a "sweetheart deal" between the Park Service and the arts center. He complained that after years of pushing for a higher visitor ceiling on the island, locals were furious over the agency's plan to allow "an elite group of persons interested in the arts and restoration causes" to evade the 300-person limit. Smith called for an investigation of the Park Service on Cumberland Island. He told the *Southeast Georgian*, "I can't imagine how the Park Service could ignore the public's right to visit Cumberland Island in favor of a private institution."[10]

Despite a January 24, 1995, meeting between the Cumberland Island Historic Foundation, the Carnegie-Cook Center, and the Park Service, who favored the MOA, and the environmental groups, who opposed it, the two factions continued to polarize. Even the National Parks and Conservation Association, a seventy-five-year-old organization pledged to support preservation of all national park system resources, questioned the proposed agreement. Don Barger, the organization's southeast regional representative, wrote to Swain: "Rolland, I am not an attorney, but it seems clear to me that the public's interests are not being adequately protected by this agreement. I am amazed that the Interior solicitors would even consider the numerous hooks and loopholes I believe are in this document. In fact, it seems as though the public's interests are being *increasingly sacrificed* as this process moves forward." He followed with a line-by-line review of the proposed MOA, questioning nearly every aspect of the plan. He assured Swain that while he was gravely concerned, he was also "willing to be convinced I'm wrong."[11]

Barger's moderate tone quickly drowned in a debate that increasingly involved grander philosophical beliefs and strong personalities. The president of Earthwatch, a research and public resources institute, lauded GoGo Ferguson and called restoration of Plum Orchard her "rendezvous with destiny." Norman Owen of the Georgia Sierra Club doubted that 20 people in the mansion would share bedrooms while 10 others stayed at the rebuilt carriage house. Even though the Carnegie-Cook faction dropped its plan to build new structures and agreed that the 300 people participating in the quarterly colloquia would stay on the mainland, Owen did not believe that the center could resist expansion. Environmentalists saw the agreement as a window of opportunity for development in the middle of the wilderness. They resolved to stop it at all costs.[12]

In February 1996, despite a voluminous letter-writing campaign by environmentalists, the Park Service issued a "finding of no significant impact" approving the lease of the mansion to the Carnegie-Cook group, now renamed the Plum Orchard Center for the Arts. This in turn spurred a desperation move by a new organization that had appeared during the Plum Orchard MOA debate. Atlanta attorney Hal Wright founded the Defenders of Wild Cumberland to oppose what he saw as a serious threat to the wilderness status and natural resources of the island. On April 17, 1996, Wright filed a lawsuit to block the memorandum of agreement for Plum Orchard. This froze the plan and set Park Service solicitors to work analyzing the entire scheme. Apparently they did not like what they saw because in mid-June Regional Director Robert Baker told GoGo Ferguson and Nancy Parrish of the Center for the Arts that the Park Service would "discontinue" the MOA proposal. He cited the need to initiate a new lease procedure under guidelines specified in the National Historic Preservation Act and NEPA. Baker added that the future of Plum Orchard should be considered as part

Fig. 8.2. The Plum Orchard mansion is the largest and costliest structure on the island. Surrounded on three sides by wilderness, the magnificent house is the focal point of controversy over Park Service management. (National Park Service photograph by Elbert Cox during the 1957 survey)

of the wilderness management planning due to commence very soon. Bitter Carnegie heirs and historic preservationists had to start all over looking for a way to save the historic house (fig. 8.2).[13]

The Museum Collection

While preservation of Plum Orchard and other historic structures on Cumberland Island continues to be dogged by controversy, the Park Service did find a solution to the conservation of the historic furnishings left in those buildings. Yet even this positive step has raised questions about the worth of the island's cultural resources and the spending priorities of a financially strapped Park Service. The term *historic furnishings* includes not only furniture and household goods but clothing, linens, paper materials, photographs, and personal possessions of the former residents. The museum collection at Cumberland Island includes two other categories of items: natural resource specimens and archaeological remains. When the Carnegies turned over the land and buildings to the Park Service, the total of the three cate-

Fig. 8.3. Perhaps the most significant archaeological find on the island was this Native American canoe. It is now preserved at the Smithsonian Institution in Washington, D.C.

gories ran to more than 12,000 items. The agency's Southeast Archaeological Center in Tallahassee, Florida, took most of the archaeological resources for preservation and study (fig. 8.3). The Cumberland Island staff has the responsibility to preserve, catalog, and properly store the remaining items, which range in size from individual straight pins to horse-drawn wagons.[14]

The majority of the valuable historic items came with the donation of Plum Orchard mansion by the Johnston family. The house contained several rare Tiffany lamps, Carnegie family china, crystal, and fine linens. However, a dispute arose in November 1978 when the Johnston group removed a number of items for division among themselves, including the china, crystal, silverware, and pieces of furniture. Approximately 1,400 of the items had been cataloged for the national seashore's museum collection. Superintendent McCrary contacted Arthur Allen, the Park Service's chief of museum services at Harpers Ferry, West Virginia. He in turn wrote to the regional office in Atlanta and pointed out that many of those items not only were Park Service property but as part of the mansion were included on the National Register.[15]

An investigation followed that further muddied the picture. The gift deed from the Johnston branch of the family to the National Park Foundation had no reference to the contents of the Plum Orchard house. Nevertheless, the foundation reported that it had correspondence indicating the Johnstons had reserved the china, glassware, and silver. Foundation officers said nothing about furniture or other objects. Additionally, for some months the foundation even speculated about whether it had conveyed the furnishings to the Park Service or still owned them. A disgusted Superintendent McCrary later wrote, "The only thing we didn't resolve was the care and feeding of the Carnegie heirs who, evidently, still had claim to several items but had showed no interest in them for almost six years [since the seashore's establishment]." Later Margaret Wright returned her 239 items, enabling the Park Service to display some of the china and glassware for tours of the mansion.[16]

For more than twenty-five years, the Park Service stored the remaining furnishings in a variety of island buildings. Periodic inspections showed that the facilities were inadequate and that many museum resources were exposed to unacceptable levels of heat and humidity. Eventually the seashore purchased a "Bally building," a climate-controlled metal shed which was placed inside one of the Dungeness buildings. It provided adequate storage for the most endangered items such as clothing and photographs. Criticism from Carnegie heirs, historic preservationists, and museum experts eventually combined with the opportunity to acquire the Coastal Bank Building in St. Marys. From 1996 through 1999 the agency spent more than $1 million on acquiring, redesigning, and adapting the structure for storage and display of museum objects.

Several problems arose to complicate the process and exaggerate the cost. Builders installed a floor for the second story that regional office specialists claimed might emit harmful gases into the collection. Some months later, the regional office replaced it with a vastly more expensive one. The first elevator in the building also had to be replaced. Even after all that expense, seashore officials submitted a proposal for $40,000 in 2001 to modify the building's facade in order to meet the standards of the St. Marys historic district.[17]

As curators prepared to move the museum materials from their many island caches, they were able to evaluate the quality of the collection. The natural history materials were first-class, the products of nearly thirty years of painstaking collection by rangers and visiting scientists. The historic objects, however, did not match the quality of the natural resources. In 1998 the collection underwent a comprehensive survey to identify the significance and condition of every object. The curatorial specialist evaluated them on a scale from 1 for very significant to 7 for no significance. Excluding the natural resources and the paper items, most of which were photographs, the curator rated more than 76 percent of the collection from 4 to 7. Only 2.5 percent rated the highest level of significance, and those items consisted primarily of the remnant china and fine linens. These results corroborated the belief among most seashore officials that the Johnstons removed the bulk of the valuable items and left behind objects in poor condition or of low value.

The agency's full-time museum curator carefully preserves an amazing array of common and virtually worthless items in the finest museum storage materials available. These include modern straight pins, burned-out lightbulbs, shoe polish, ordinary clear glass ashtrays, a late 1960s Maxwell House coffee can, dog shampoo from the same period, twisted aluminum and plastic lawn chairs, jam jars, and a container of prophylactics. Many items of some original value, such as beds, couches, and chairs, are in terrible condition, with rips in the fabric and broken frames. Even the museum curator questions the propriety of spending so much money on storing these items when significant structures like the Dungeness Tabby House need further maintenance.[18]

Lands and Retained Rights

Land acquisition proceeded in the late 1990s in fits and starts. The Park Service suffered several setbacks with the Rockefeller lands. Andrew Rockefeller elected not to sell his land. His sister Georgia Rose chose to sell but refused to negotiate with the Park Service. Instead, Rose sold her 82-acre cross-island tract abutting Sea Camp to Atlanta developer Chris Allen in 1997. The sale was conducted in secrecy, and for months the Park Service had no idea who the new landholder was. Many speculated that John F. Kennedy Jr., who had been married at the Settlement's African-American church earlier in the year, was the purchaser. Once Allen admitted his ownership, many environmentalists and seashore officials feared he would subdivide or build rental cottages. However, in 2001 Allen claimed that he would only build a personal residence.[19]

The National Park Service regarded the nearly 1,200-acre plot of land in segment 2N offered by the Greyfield Corporation as much more important. This land lay just north of Stafford in the proposed wilderness zone. The agency convinced the Nature Conservancy to secure options and help defray the cost of the purchase. The Conservancy negotiated a five-phase purchase of 1,148 acres for approximately $20 million. Release of funds for the purchase had to be approved by Congress. Money theoretically existed from the 1965 Land and Water Conservation Fund Act to enable such an acquisition. That act, as amended, orders that revenue generated by offshore oil and gas leases, motorboat-fuel taxes, the sale of surplus property, and recreation user fees on public lands be used to acquire and develop property for recreation at both the federal and state levels. However, President Reagan began a practice of diverting the funds for other purposes, and over the years only a tiny fraction of the revenue generated by the various commercial activities has been used to support recreation.[20]

Congressman Kingston followed this trend by objecting to the conditions of the proposed sale. He called for an evaluation of the proposal by the General Accounting Office (GAO). When the GAO concluded that the projected deal was not unusual, Kingston submitted a new plan aimed at avoiding any cost to the government.

The Candlers at High Point had come to regret selling their land to the Park Service. During the Greyfield negotiations they informed Kingston that they would purchase the tract and donate it to Cumberland Island National Seashore if the government in return would reestablish Candler ownership of a plot of land at High Point of equal value. In September 1997 Kingston announced his support for this idea. The Park Service rejected this option based on agency opposition to returning potential wilderness lands to private ownership. In early 1998, despite the expenditure of $4.8 million by the Nature Conservancy for 344 of the Greyfield acres, Kingston made it clear that he would not support any other land acquisition plan. Senator Max Cleland of Georgia supported the Park Service but could not convince Kingston to change his mind. Faced with his intransigence, the Park Service, through the Nature Conservancy, opened negotiations with the Candler heirs while environmental groups sought ways to block the deal.[21]

Driving on the Beach

At the same time, environmental organizations looked anew at other incongruous wilderness activities. Because of its perceived threat to sea

Figure 8.4. Driving on the beach by retained-rights holders and their guests and renters is another focus of conflict at Cumberland Island.

turtles, shorebirds, and sunbathers, an obvious target was the island residents' cherished right to drive on the beach. In February 1996 Cumberland Island began receiving very similar letters from around the South complaining about cars on the beach (fig. 8.4). In each reply seashore officials explained the retained rights of residents and the state's jurisdiction below the high-tide line and asked for patience until the estate agreements ended.[22] However, organizations like the Defenders of Wild Cumberland would not be placated. Soon the seashore also received letters from island residents insisting that their rights be respected.[23]

In this particular case the Park Service was able to pass the responsibility for solving a management debate on to the Georgia Department of Natural Resources, which faced the issue of driving on its beaches on other islands. In 1992 the state legislature passed a regulation that allowed beach driving only with official permits. On Cumberland the DNR intended to issue permits immediately, but a lawsuit filed by Hal Wright forced it to carry out a full public planning procedure. In June 1998 the state agency held public meetings in Atlanta, Savannah, and Kingsland to determine who should receive permits. After the skirmish with Wright and other environmentalists, the DNR proceeded cautiously, hoping to avoid the planning debacles of the National Park Service at Cumberland Island. Among those who spoke were environmental activists, residents of Cumberland and other Georgia islands, and the National Park Service. The latter supported issuance of permits to island residents with private property or retained rights, their immediate families, and their resident employees. However, the agency opposed permits for people renting homes on the retained estates, guests of island residents, and any commercial tours, including the ones operating from the Greyfield Inn.[24]

In December 1998 the DNR issued its final regulations for beach driving. Only island residents, their immediate family members, researchers, and management officials would receive permits. Cumberland Island residents anticipated this decision by voluntarily agreeing to use only existing access roads to the beach, stay on wet sand below the high-tide line, and avoid driving at night during the turtle-nesting season. Nevertheless, Hal Wright brought suit against the DNR when it issued the first permits to island residents. The National Park Service testified for the state office, and the suit was dismissed. At present the state office has issued more than 300 permits for Cumberland Island's beach. However, this has proven to be a minor interruption in Wright's campaign to turn Cumberland into a true wilderness.[25]

Wilderness and Other Plans

The failure of the Plum Orchard Center for the Arts project disheartened Superintendent Rolland Swain. On September 1, 1996, he left to assume leadership at Big South Fork National Recreation Area on the Cumberland River in Tennessee and Kentucky. After several temporary superintendents, the Park Service assigned Denis Davis to the permanent position in December. Davis arrived from Glen Canyon National Recreation Area on the Colorado River with strong credentials in park planning. He hoped to use collaborative input from all the interest groups to formulate a wilderness plan and solve the dilemma over use of the Plum Orchard mansion.

However, Davis underestimated the degree of philosophical differences and the personal hatred that had developed between the factions. Island residents had formed another association, the Cumberland Island Preservation Society, to advocate increased protection for historic structures and defend their rights against aggressive environmental organizations like the Defenders of Wild Cumberland. The residents' new organization is also aggressive, at one point filing suit against Hal Wright for harassment.[26]

After a series of public meetings hosted by the Park Service, the Cumberland Island Preservation Society sponsored a forum on wilderness planning in October 1997. Residents, environmentalists, historic preservation specialists, Park Service officials, and a professional facilitator from New Jersey attended. The results were promising. Participants avoided recriminations and discussed all the ramifications of wilderness planning for the island. They identified a number of issues to be addressed by wilderness planners, including a definition of Cumberland's wilderness and its boundaries, retained rights including vehicle use, visitor activities and the limit on their numbers, identification and preservation of historical and archaeological resources, and management of horses, hogs, exotic species, and fire.[27]

The preservation society hosted two more meetings the following spring. The first met to consider the Plum Orchard dilemma. Historic preservation supporters dominated this forum and struggled to find a way to fund the mansion's rehabilitation and carry out maintenance surrounded by wilderness on nearly all sides. The second forum again addressed wilderness planning and the related issues that had been raised during the October meeting. This time several preservation society members suggested that the Main Road be removed from wilderness designation. In this way building supplies and visitors could be carried to both Plum Orchard and the Settlement by vehicle. This option would be far less costly than access only by boat. The process of removing a road from the wilderness is called "cherry-stemming." In addition, residents called for removal of all features on the National Register of Historic Places from the wilderness as well. They suggested that Congress could add a portion of the island south of Dungeness to compensate for the loss of wilderness acreage on the road. Much of the proposed addition consists of marsh and dredge spoils deposited by the Army Corps of Engineers. The Park Service and environmentalists rejected these ideas.[28]

Subsequently, GoGo Ferguson of the Cumberland Island Preservation Society issued a transcript of recommendations from the second wilderness forum. Included in the summary was a statement that the Park Service "will address all available avenues to enhancing designated and proposed wilderness while permitting unrestricted access to the island's cultural resources, including local discretion, permitted exceptions, cherry-stemming the road, and redefining the wilderness boundary, to add to, not diminish wilderness on Cumberland Island." A week later Denis Davis wrote to Ferguson: "I had some notes that corresponded to what you showed in your transcript, but they were notes of the general discussion, not the group's recommendation. In fact, several of the recommendations in your transcript are clearly in violation of the Wilderness Act and I, and many others, would have objected if that would have been part of the group's recommendation."[29]

This disagreement became moot on June 23, 1998, when Congressman Kingston announced he would introduce a bill to be called the Cumberland Island Preservation Act. The bill had four sections. First, it provided funds for the restoration of Plum Orchard and other historic structures. The congressman cited a lack of funds as the reason for the losses of the Dungeness Recreation House, the Plum Orchard carriage house, and one of the houses in the Settlement. Sierra Club volunteers had razed the latter at the direction of the Park Service. A second provision ordered the High Point–Greyfield land swap. The final two sections cherry-stemmed the Main Road and added the south end land and marsh to the wilderness. The bill matched exactly the proposals of the island residents.[30]

The reaction to Kingston's bill was predictably tumultuous. Island residents, naturally, were elated. The bill caught the seashore officials completely off guard again, and they opposed it. Environmental organizations were furious at what they saw as a betrayal of the collaborative process worked out during the island meetings. Senator Cleland also opposed the bill, as he had opposed the High Point–Greyfield deal separately. During the year since Kingston first proposed the idea, the Nature Conservancy had completed two more phases of land acquisition but still had only options on the remaining 575 acres. During the ensuing months the two legislators negotiated to find a solution. Hoping to break the deadlock, Park Service director Robert Stanton and Assistant Secretary of the Interior Donald J. Barry visited the island and agreed to divert funds from other programs and parks for cultural resource preservation.[31]

In late November, Senator Cleland and Congressman Kingston announced their mutual stand on the Cumberland Island issues. Their letter committed the Park Service to allocate $1 million to rehabilitate Plum Orchard mansion, $500,000 for other cultural resources, and $50,000 for new interpretive exhibits. Cleland and Kingston also promised to increase the seashore's annual base funding for historic maintenance by $300,000. In addition to these welcome increases for cultural resources, the legislators promised that Congress would release $11.9 million for land acquisition. These funds, coupled with a $6 million donation by the Nature Conservancy, would complete acquisition of the Greyfield North tract. However, the lawmakers placed conditions on the land funds. The Park Service and all its many interest groups had to settle the issue of visitor access to Plum Orchard and the Settlement.[32]

During the next two months these senior officials worked with all the parties to reach an agreement. On February 17, 1999, fourteen organizations representing the various advocacies, the Carnegie and Candler heirs, and representatives of the senator and the congressman signed a "Cumberland Island Agreement." This document included provisions that the Park Service would:

1. Provide another $1.4 million for Plum Orchard and $150,000 for stabilization of other island structures in addition to funds already promised for management of historical resources.

2. Establish a Cumberland Island Subcommittee of the national park system advisory board.

3. Renew the request for proposals to lease Plum Orchard.

4. Provide scheduled visitor tours to the mansion and the Settlement by vehicle in the short term but by boat as soon as possible.

5. Build a trail parallel to the Main Road for hikers.

6. Keep the visitor capacity at 300 per day.

7. Carry out the land purchases with government and Nature Conservancy funds.

8. Expand the wilderness planning effort to include discrete plans for natural resources, cultural resources, interpretation, and commercial services (concessions).[33]

This sweeping document would have been unusual at any national park unit. At Cumberland Island it was little short of miraculous. The key factor was linking land acquisition to the public access to historic properties. Within weeks, however, more difficulties appeared. First, the Greyfield people approached the Park Service to change the sales contract they had signed. Second thoughts led them to want to keep more of their land. According to Superintendent Davis, the Park Service refused to renegotiate the contract, but Assistant Secretary Barry ordered them to do so. Ultimately, the Greyfield group kept sixty-five of the potential wilderness acres that had been included originally in the contract.[34]

At the same time, ominous signs appeared regarding the agreement itself. Another environmental organization, Wilderness Watch, announced that it did not approve of the stipulation that the Park Service should take visitors to the historic sites by vehicle through the wilderness. Then, citing similar concerns, the Defenders of Wild Cumberland decided to negate their commitment to the agreement.[35] Harsh words and lawsuits soon reappeared, and Superintendent Davis became a casualty.

According to Davis, Congressman Kingston urged Regional Director Jerry Belson to remove him from the superintendency. Kingston gave two reasons. First, he claimed that the seashore staff had erected a wayside exhibit on horses that insulted the island residents. Second, he accused Davis of demonstrating his contempt for historic resources by putting vinyl siding on his home in the St. Marys historic district. A distraught Davis answered that no horse exhibit existed on the island and that he did not live in the St. Marys historic district. In fact, it turned out that it was a Dennis Davis, not Denis Davis, who had altered the historic home. Then, a few months later, Davis received word that Assistant Secretary Barry wanted him removed immediately. The Park Service leadership resisted, and Davis could have refused to move, but the pressure was intense. He left in December 1999 for the assistant superintendent position at Glacier National Park.[36]

It is noteworthy that the Department of the Interior removed Davis after the National Parks and Conservation Association bestowed on him its annual national award for conservation leadership. Hal Wright summed up the feelings of the environmental community when he wrote to Barry, "There is no longer any doubt whatsoever for whose interests you are attempting to manage the Cumberland Island Wilderness and those interests are certainly not the public's." Yet another superintendent fell to the incessant conflict and political influence that characterized the seashore's entire history.[37]

The Five Plans

As the new century began, Park Service officials at the regional office steadfastly but carefully developed the wilderness and other plans. Regional planner Richard Sussman drew upon expertise from around the national park system to craft a series of drafts for public review. At the same time, the Park Service sought to establish the legal and scientific grounds to support the alternatives it preferred. The regional solicitor's office undertook the most complete and detailed review of retained-rights agreements ever compiled to determine who had rights to the different uses and activities and who did not. In January 2000 one of the Foster retained rights ended. They requested an extension, but the Park Service refused. Ownership passed to the seashore, and turtle management volunteers and other researchers now seasonally occupy the Foster Beach House.[38]

When the five draft plans were released on December 15, 2000, they had something to annoy everyone. The Park Service's preferred alternatives for the wilderness included vehicles for ranger patrols, for one sea turtle–monitoring trip per day, and on rare occasions for maintenance of roads, bridges, and historic structures. The agency proposed two new wilderness campgrounds, a loading ramp at the Plum Orchard Dock, a new dock at the north end, a trail parallel to the Main Road, and visitor tours by concession boat. Planners also asked island residents to "voluntarily" give up driving on four roads east of the Main Road, including South Cut. The draft natural resource plan proposed elimination of the hogs, by the use of dogs if necessary, and reduction of the horses to a "representative" herd across the island. The cultural resource management plan called for the Park Service to remove Plum Orchard's carriage house and boat house plus the ruins of the Recreation House and three other structures at Dungeness. The agency also proposed to find a lessee for Plum Orchard, move the wilderness campground away from the Brickhill archaeological site, and turn Lucy Graves's home, The Grange, into a visitor facility when her lease ends in 2010. Finally, the Park Service planned to encourage kayak and canoe tours to the wilderness.[39]

While the public reviewed the drafts, the Park Service continued its data-gathering effort. In early July 2001 Andrew Carnegie III died. Subsequently, rumors circulated that his trustee Gertrude Schwartz was also dead.[40] This information raised questions about their retained right, rented by Ben Jenkins. The agency also prepared to contract for an official study of the island's visitor carrying capacity. Two studies conducted in the mid-1990s claimed that island visitors did not perceive much recreation conflict or crowding. At the same time, Camden County and its political representatives constantly lobbied for a visitation limit of at least 600.[41]

Between December 2000 and July 2001, the Park Service held eight hearings and received more than 3,500 letters and electronic mailings. Among the respondents were two early players in the Cumberland Island drama. Former secretary of the interior and Carnegie attorney Stewart Udall addressed the question of retained rights on the island:

> It's my understanding that some controversy has arisen about Island residents and their guests driving in the Wilderness and on the beach. This issue was addressed in detail in the individual retained rights agreements when the Seashore was formed. These retained rights must be honored as written, but they should not be extended in any way, as this would be harmful to the Wilderness and the values it was designated to protect. In our work to acquire the private lands to establish the Seashore, I do not recall any commitments made by the federal government to the landowners other than those specifically included in the deeds of conveyance. Needless to say, any commercial tours in vehicles would be contrary to the 1964 Wilderness Act and should not be allowed.

Thereafter he chastised the Park Service for its plans to conduct vehicle tours to Plum Orchard and the Settlement, even if only on a temporary basis. He wrote: "The National Park Service is obligated to carry out its administrative duties in the Cumberland Island Wilderness as it would in any other Wilderness. It should fulfill its stewardship role by setting an example for others to follow."[42]

Another interested reviewer was George Sandberg, the man who negotiated for the National Park Foundation in the early 1970s. In a letter to Thornton Morris, he reiterated that the residents had been promised "convenient access" by vehicle to the beach until a jitney transportation system throughout the island began operations. Sandberg added that George Hartzog had promised these liberal retained rights to offset the low price per acre that the government offered. This argument swayed legislators in 1979 when they forced the Park Service to keep South Cut Road open.[43]

The thousands of other reviewers of the draft plans fell into predictable groups. Those favoring the positions of the main environmental groups decried all traffic in the wilderness. One extremist suggested at a public hearing that the structures in the settlement be burned to cleanse the wilderness. This met an angry response not only from residents and historic preservationists but also from more moderate environmentalists and the National Association for African American Historic Preservation. At the other end of the scale, a number of individuals suggested that the Park Service should either move the historic structures to sites outside the wilderness or eliminate the wilderness designation completely.[44]

As the Park Service fielded these responses, some of its thoughts about the wilderness plan began to change. By July 2001 the agency banned virtually all use of its own vehicles in the wilderness except for emergencies or if justified as the minimum action necessary to complete a task. Ranger patrols now take place on foot or on horseback. Historic preservation activities are implemented with "minimum tools." The latter term denotes use of hand implements unless the job is impossible without a machine. Then the job's importance itself must undergo careful review. Coupled with the regional solicitor's findings, this compliance with the letter and spirit of the Wilderness Act has led the Park Service to review the Greyfield Inn's motorized tours except along the Main Road where it holds a legal right to drive. This ultimately may force the inn's guests to walk from the road to other locations in or near the wilderness such as the Plum Orchard mansion or the Settlement.[45]

Ultimately, the five plans probably will not stand as proposed. The intensity of public interest, distrust of the Park Service, group self-interest, and philosophical differences are too deeply entrenched. Sierra Club representative Bill Harlan writes: "40,000 people come from across the country and across the globe to Cumberland because it is wild. . . . They visit Cumberland because they want to experience wilderness—not to see the mansion of Andrew Carnegie's brother's fifth son George, [a] mansion built less than a hundred years ago and of limited regional historical significance. The international significance of the Cumberland wilderness outweighs any historical value at Plum."[46]

Georgia historic preservationist Gregory Paxton offers a different prescription: "There is more to Cumberland Island than wilderness. An indelible 5,000-year history of human habitation is written on the Island's landscape, and the evidence is everywhere, from the Native American burial grounds and shell middens to the crumbling chimney pots and tabby ruins, from the circa-1870 freed slave settlements to the large estates with numerous outbuildings. These tangible traces of Georgia's history need to be protected along with the areas that have grown wild around them."[47]

Finally, Cumberland Island resident and Candler heir William C. Warren writes: "I believe a wilderness has, virtually, no roads, no houses, no airstrips, no docks, no permanent human habitation, no fences, etc., etc., and this cannot and will not happen in my lifetime or yours. As long as there are retained rights to be honored, there can be no true 'wilderness' or 'potential wilderness' as I define it. Time is on the 'wilderness' advocates['] side—it will be 80 years and they will have it all—until then they'll have to put up with those that made all this possible."[48]

The only thing certain about this ongoing planning process is that it will lead to wilderness management, natural resource management, cultural resource management, concession services, and long-range interpretive plans that will be quite bruised and modified when they go into effect. In fact, their very survival will surely depend on the courts or Congress.

Conclusion

It is obvious that the conflict over Cumberland Island will continue. Nevertheless, the past affords a perspective on the conflicts that have shaped the island's management. Cumberland Island National Seashore has existed for thirty-two tumultuous years. During that time a succession of superintendents, rangers, resource managers, and regional office specialists have struggled to accomplish the three missions of the National Park Service. Protection of natural resources has perhaps fared the best. Threatened and endangered species are protected, and some, like the loggerhead sea turtle, are making a satisfying comeback. Natural vegetation has reclaimed formerly disturbed areas such as Charles Fraser's airfield and North Cut Road. Although the island suffers from erosion, especially along its southwestern shore, the agency and its allies have staved off serious threats to reroute the Intracoastal Waterway and dump spoils at Raccoon Keys and Drum Point Island. The Park Service has made significant progress in dune protection, eradication of exotic tung trees, and reintroduction of a native predator, the bobcat.

Not every natural resource issue has gone well, though. Feral pigs continue to present an almost intractable problem because they seriously harm the programs for protection of native species. Some exotic plants will be nearly impossible to eradicate. The Park Service is unlikely to remove feral horses in the foreseeable future despite their destruction of marsh vegetation and dune stability. Fire management remains trapped in a complete suppression mode despite evidence that fire is a natural feature in most island communities. A rapid rise in the population of armadillos promises more problems for future island managers.

Historic preservation on Cumberland Island has been an amazingly expensive and controversial process. The agency has protected more than 100 historic structures and adapted more than a dozen for modern use. Key features such as the Plum Orchard mansion, the Dungeness ruins, the Dungeness Tabby House, and The Stafford Chimneys have been protected at enormous cost to the Park Service. Interpretive and landscape management plans have been developed to guide further action on these popular resources. Thousands of curatorial items are now stored in one of the finest museums in the entire park system. The island boasts seven separate archaeological or historic areas with dozens of individually protected places.

Nevertheless, criticism and controversy have plagued the preservation process. The Park Service argued with the state historic preservation office and the Advisory Council on Historic Preservation for years, engendering a suspicion that still flares up from time to time. An inability to secure adequate funding doomed several significant but heavily damaged structures, such as the Recreation House, to collapse rather than preservation. This in turn fanned the flames of Carnegie heirs' anger at what they perceived as a capricious dereliction of duty. Their influence over congressmen and senior government officials ultimately secured extraordinary attention and funding for some of the resources they left behind. However, that attention and those funds were diverted from critical projects around the nation into a grand house, a collection of outbuildings and shacks, and a museum full of objects that the Park Service continues to identify as insignificant.

Public use of Cumberland Island has perhaps been the most contentious of the agency's missions. The seashore was originally designated as a recreation unit with plans to satisfy 10,000 visitors per day with trams, horseback riding, and more than a dozen interpretive sites. Within a decade environmental groups, residents, and the public had battered these plans down to an unscientific 300-person daily limit with no special recreation facilities and little in the way of interpretation. At the same time, the Park Service unsuccessfully sought to manage retained rights of use as it did in other park units. New houses and driving on the roads and beaches were only a part of the agency's concerns. The residents also used their influence to modify plans and actions the agency undertook anywhere on the island.

Initiation of wilderness designation on the northern half of Cumberland Island only exacerbated the management problems. Historic preservation, standard Park Service ranger patrols, ecological research, and fire management have all became exponentially more difficult. Suddenly seashore rangers face the prospects of patrolling on horseback, carrying tools and building materials on foot to the Settlement, and myriad other complications that attend the wilderness status. Perhaps worst of all, the Carnegie-Candler rights and their sense of history collided directly with a powerful wilderness lobby. The latter seeks uncompromising purity of the type only found in the most far-flung western mountains. The furious debates, recriminations, and lawsuits of these two factions have drastically prolonged every planning and decision-making process for the Park Service. One of the most difficult and expensive planning efforts ever mounted by the agency finally resulted in a draft wilderness plan nearly nineteen years after the legislation that created the wilderness. Some 3,500 letters representing all the polarized camps in this battle promise that this plan, like all the others, will undergo substantial changes before it is finalized.

The history of Cumberland Island is truly four discrete stories, one affecting only the idyllic island itself and three with implications for the entire national park system and public lands management in general. First, the story of Cumberland Island is one of increasing human modification and then a long return toward its primeval character despite a number of close calls from major development schemes. Native Americans left only a slight impact as they and the few Spanish who sought to convert them disappeared from the scene. The English made only minor improvements, and the island primarily served as a source of valuable ship timbers. Plantation development commenced with the arrival of American sovereignty, rapidly escalated during the Greene-Miller period, and reached its apogee immediately before the Civil War. At that time natural vegetation had been reduced to isolated patches between broad fields of sea island cotton and other crops.

After the war human manipulation declined despite the construction of five mansions, dozens of outbuildings, and a hotel complex on the island. Native vegetation reclaimed many of the cotton fields, and animal life proliferated in the new areas. The legacy of earlier development remained, however, in the form of several extinctions, feral animals, and exotic plants. As the Carnegie family fortune shrank, its rich infrastructure too became a legacy of the past. Establishment of Cumberland Island National Seashore and wilderness designation for the northern half of it further entrenched the process of natural reclamation.

Yet many schemes arose during this postbellum period that could have dwarfed even the modifications of the cotton era. Developers tried to establish large subdivisions on the north end of the island in the late 1890s and on former Carnegie lands during the 1960s. NASA briefly considered turning the island into a space base. Titanium miners nearly gained rights to gut some 7,000 acres in the center of the island. Carnegie heirs also considered conversion of the island to a cattle ranch or a pine plantation. Finally, the Park Service itself nearly turned the quiet retreat into a massive recreation complex with thousands of visitors crammed onto three strips of beach. Over the decades since 1916, the Cumberland Island that the visitor sees today has survived due to a combination of Lucy Carnegie's legal cunning, her heirs' love for its landscape and resources, vigorous public action, and a belatedly protective mantle of Park Service control. The island's biography is indeed a tale of change, adventure, and perilous escapes.

A second story from the saga of Cumberland Island is the tale of conflicting laws and policies. The National Park Service, like all government agencies, is bound by a complex web of regulations affecting its every action. With its multiple missions and resources, these regulations are bound to clash. The intent of the Wilderness Act was to lay aside areas with no roads, no structures, and no use of mechanical devices. Coupled with the Endangered Species Act and other laws, this seems to give preeminence to natural resource protection and ecological processes. The intent of the National Historic Preservation Act and subsequent amendments and executive orders was to study, classify, and protect historic structures and objects. It seems to give preeminence to the cultural resources of the island. Finally, many public laws protect the sanctity of contracts, including those for real estate and reserved rights of use. Most Americans hold a legal contract to be inviolate.

On Cumberland Island all these laws affect the same property. The results are befuddling and infuriating to the various parties affected by and trying to administer these different pieces of legislation. Hence, people with retained rights drive on roads and live in houses within the wilderness area. The Park Service is currently trying to determine if every rights holder has every right that he or she claims. This infrastructure and these activities clearly dilute the wilderness as envisioned in the 1964 act.

Numerous historic resources lie within or nearly surrounded by the wilderness on Cumberland Island. The Park Service must maintain them to the degree financially possible. However, their upkeep requires labor and materials. Resolving to use only animals, boats, or humans to transport these items and hand tools to do the job will exponentially increase the cost of accomplishing the required maintenance. This in turn will leave more resources untreated, exposing them to deterioration that is unacceptable according to historic preservation legislation. What is the agency to do? Which law is to be compromised: the wilderness one or the historic preservation one? A compromise that bends both laws might be possible at other parks or protected areas but not at emotionally charged Cumberland Island. Only a court or Congress will be able to decide the final course for seashore officials to follow legally.

A third story is one of conflict between and influence by a variety of special-interest groups. The National Park Service has often been called the most popular agency in the federal government. Stephen Mather and Horace Albright engineered it that way, and it has become an axiom of Park Service training and culture. Hence most park personnel are quite sensitive to public opinion. No other federal resource agency listens to interest groups and the general public more than the National Park Service. The divisiveness and political clout of various interest groups at Cumberland Island has made this a difficult reality. The only plan the agency submitted that had no public input was the original 1971 master plan. It did not survive the first public meeting. Interest groups and island residents shaped every decision and project the Park Service undertook. Looking back over the seashore's history, the environmental groups, working through the National Environmental Policy Act, probably exerted the greatest cumulative influence on management of Cumberland Island National Seashore. The Georgia Conservancy, especially Hans Neuhauser, and the Georgia Sierra Club led the pushes for fewer visitors, less development, minimization of the area devoted to historic districts, and the wilderness designation. They also contributed their support to natural resource and interpretive programs.

The Carnegie heirs and other island residents also profoundly shaped the seashore and its policies. Of course, they sold the land to the National Park Foundation and then blocked state condemnation efforts to ensure the legislative establishment of the unit. No national seashore would exist to be fought over were it not for their efforts. But in the process they exacted a cluster of rights that have dramatically shaped resource management. Some argue that these rights suborn the seashore's purposes, especially those in the wilderness. Yet historic preservation on the island owes a great deal to the residents, especially at the Settlement and at Plum Orchard. They also succeeded in shaping interpretation of the island to emphasize the Carnegie-Candler period. And they too were instrumental in the campaign to limit visitor capacity and tourism development. Although the residents unsuccessfully opposed reintroduction of the bobcat, they used their influence to stymie Park Service horse management.

Beside these two powerful entities, other interest groups pale by comparison. Nevertheless, the Park Service adapted its plans in accordance with other views. St. Marys succeeded in holding the Park Service headquarters at its waterfront amid a number of competing sites, some of them unquestionably superior. Camden County complaints led the state to withhold donation of the tidal and offshore lands, which considerably complicates resource management. The state's historic preservation office, while it failed to achieve an islandwide historic district designation, provided a critical influence in improving early historic resource planning by the National Park Service. Finally, the general public, especially in Atlanta, periodically arose to destroy certain Park Service plans and programs, coupling with one or more interest groups to bury the obnoxious schemes in floods of outraged letters.

A last story from Cumberland Island is that of national parks as venues for competing visions. The introduction asked what Cumberland Island should be: a recreation playground, a historic landscape, or an ecological wilderness. For most of the twentieth century, the National Park Service wrestled with this three-part mission. Just before the establishment of Cumberland Island, the agency even tried to sort all the system's units into natural, historic, or recreation categories wherein management priorities would be clearly established. It failed miserably because most units had all three types of resources. For Cumberland Island this is especially true.

However, this is only part of the complexity faced by resource managers. Different natural resources have different, occasionally conflicting, needs. The island holds too many cultural resources, and they are too expensive for all of them to be saved. Which ones are most important? At the same time, national seashore recreation can encompass bird watching, backpacking, beach driving, and jet skis. The Park Service must decide which are appropriate.

If these decisions were not already complicated enough, each resource and each use has its advocacy group. From these groups at Cumberland, three main visions of the island dominate. Residents want the island to stay the same as it was in 1970. Furthermore, they want to have the same rights and much of the same influence over its future as they enjoyed with ownership. Their vision is a quiet island where they continue to live and enjoy traditional uses. They seek careful preservation of the buildings, roads, animals, and forests of that time. They often quote an early promise from Stewart Udall that the government would take a snapshot of the island and keep it that way. Visitors may come in small numbers to visit the historic districts at the south end, Plum Orchard, and the Settlement. However, theirs should be a light, transitory presence that in no way affects the residents.

Environmentalists envision a different scenario. Most will allow that historic preservation should occur at the south end of the island but balk at such features in the legal wilderness. They believe the wilderness should return completely to forest and marsh. Roads, retained estates, driving, and the trappings of post-nineteenth-century America should be present only from Stafford southward. They long for the time when a hiker can walk along a trail or the beach, set up camp in the silent forest, and occasionally see an archaeological site or a decaying historic building. In answer to those who complain that this restricts older and disabled visitors, they point out that the south end of the island is available for them.

Camden County residents and advocates of specific forms of recreation espouse a third vision. Many of these people originally wanted a bridge and were quite comfortable with the 1971 master plan. Today they want a higher visitor limit, lodgings and restaurants on the island, and the right to land motorboats and bring jet skis, bicycles, automobiles, and horses to any part of the national seashore. Some merely want these changes to benefit trade in St. Marys. Others want to enjoy their vision of recreation in a public park over which they feel considerable proprietorship. After all, they argue, Camden County lost a good deal of tax revenue when the federal government bought the island.

These visions, of course, represent the public's interest in the three missions of the National Park Service. Clashes between interest groups, confusion over management priorities, public pressure, and an apparently contradictory body of legislation and policy occur at many units of the national park system. Yet on this small 16,400-acre island (the national seashore includes marsh and water and is officially 36,506 acres), the tug-of-war over protection of endangered species, enforcement of wilderness law, proper care for cultural resources, compliance with legal contracts, adherence to Park Service policy, and sensitivity to the public reaches its most intense and virulent level. The history of Cumberland Island National Seashore is riddled with controversy and punctuated by failed attempts to seek compromise and amity. The future no doubt will bring more conflict and almost certainly congressional or court action to settle various disputes. Perhaps when outside forces have solved these issues, all the combatants and the Park Service can collectively enjoy the serenity and beauty of this special island.

APPENDIX A

National Park Service Officials

Superintendents of Cumberland Island National Seashore

Sam Weems—November 12, 1973–September 29, 1974

Bertram C. Roberts—November 8, 1974–December 31, 1975

Paul F. McCrary—February 15, 1976–September 20, 1981

William O. Harris—September 21, 1981–September 18, 1982

Kenneth O. Morgan—September 19, 1982–October 7, 1989

Paul C. Swartz—October 8–November 30, 1989

Mark R. Hardgrove (Acting)—December 1, 1990–February 23, 1991

Zachary T. Kirkland Jr. (Acting)—February 24–June 1, 1991

Rolland R. Swain—June 2, 1991–September 1, 1996

William Carroll (Acting)—September 1996

Mick Holm (Acting)—October 1996

Mary Collier (Acting)—November–December 7, 1996

Denis Davis—December 8, 1996–February 26, 2000

Arthur Frederick— January 16, 2000–present (Davis and Frederick overlapped)

Directors of the Southeast Region Office

Elbert Cox—December 9, 1951–November 7, 1966

Jackson E. Price—November 6, 1966–September 7, 1968

J. Leonard Volz—September 7, 1968–December 12, 1970

David D. Thompson Jr.—December 13, 1970–October 22, 1977

Joseph Brown—October 23, 1977–April 4, 1981

Neal G. Guse Jr. (Acting)—April 5–May 30, 1981

Robert M. Baker—May 31, 1981–May 1997

Jerry Belsen—May 1997–present

Sources: National Park Service, May 1991, "Historic Listing of National Park Service Officials," U.S. Department of the Interior and the Office of Personnel, National Park Service, Washington, D.C.; Cumberland Island National Seashore 2002, Personnel Files, Superintendent's Office, CINS.

Directors of the National Park Service

Conrad L. Wirth—December 9, 1951–January 7, 1964

George B. Hartzog Jr.—January 8, 1964–December 31, 1972

Ronald H. Walker—January 1, 1973–January 3, 1975

Gary E. Everhart—January 13, 1975–May 27, 1977

William H. Whalen—July 5, 1977–May 13, 1980

Russell E. Dickinson—May 15, 1980–March 3, 1985

William Penn Mott—May 1, 1985–April 15, 1989

James M. Ridenour—April 17, 1989–January 20, 1993

Roger G. Kennedy—June 1, 1993–March 29, 1997

Robert Stanton—August 4, 1997–January 2001

Fran P. Mainella—July 18, 2001–present

APPENDIX B

Cumberland Island National Seashore
Operating Base Budgets

1974	$ 314,800	1989	$868,000
1975	$410,400	1990	$899,000
1976	$378,200	1991	$974,000
1977	$ 561,800	1992	$1,068,000
1978	$506,200	1993	$ 1,081,000
1979	$ 547,300	1994	$ 1,125,000
1980	$784,800	1995	$ 1,154,000
1981	$784,900	1996	$ 1,156,000
1982	$ 831,400	1997	$ 1,189,000
1983	$ 818,300	1998	$1,346,000
1984	$856,500	1999	$ 1,350,100
1985	$ 851,700	2000	$1,692,000
1986	$825,000	2001	$1,743,000
1987	$ 835,400	2002	$1,965,000
1988	$ 855,900		

Source: National Park Service, Office of the Budget, Washington, D.C.

NOTES

Abbreviations and Short Citations

CINS   Cumberland Island National Seashore, St. Marys, Ga.

CPSU   Cooperative Park Studies Unit

DSC   Denver Service Center of the National Park Service, Denver

Ga. Archives   Georgia Archives, Atlanta

NA   National Archives, College Park, Md.

NPS   National Park Service

SERO   Southeast Regional Office of the National Park Service, Atlanta

Thornton Morris Papers   Thornton Morris Papers, Morris Law Firm, Atlanta

UGa   University of Georgia, Athens

Introduction

1. The concept of "biography of place" is a useful means by which historical geographers focus on place as a complex entity formed by diverse processes that dynamically changes through time. An early example of this term's use can be found in Marwyn S. Samuels, 1979, "The Biography of Landscape," in *Ordinary Places,* ed. Donald W. Meinig (New York: Oxford Univ. Press), 51–88. Later works include William Wyckoff, 1988, *The Developer's Frontier: The Making of the Western New York Landscape* (New Haven: Yale Univ. Press), and Gay M. Gomez, 1998, *A Wetland Biography: Seasons on Louisiana's Chenier Plain* (Austin: Univ. of Texas Press).

2. A voluminous literature exists on the national park system and its management. Classic overviews include Ronald A. Foresta, 1984, *America's National Parks and Their Keepers* (Washington, D.C.: Resources for the Future); Alfred Runte, 1987, *National Parks: The American Experience,* 2d ed. (Lincoln: Univ. of Nebraska Press); Richard West Sellars, 1997, *Preserving Nature in the National Parks: A History* (New Haven: Yale Univ. Press); and Conrad L. Wirth, 1980, *Parks, Politics, and People* (Norman: Univ. of Oklahoma Press). More specific works on managing preexisting land uses, private holdings, and park threats include John C. Freemuth, 1991, *Islands under Siege: National Parks and the Politics of External Threats* (Lawrence: Univ. Press of Kansas); and Joseph Sax, 1980a, *Mountains*

without Handrails: Reflections on the National Parks (Ann Arbor: Univ. of Michigan Press). Administrative histories for nearly 100 individual national park units are listed at www.cr.nps.gov/history/books/admstdys/Stds.htm. The only other one for a national seashore is Barry Mackintosh, 1982, "Assateague Island National Seashore: An Administrative History," Washington, D.C.: NPS.

1. A Richness of Resources: Cumberland Island to 1880

1. "An Act to Establish the Cumberland Island National Seashore in the State of Georgia, and for Other Purposes," 86 Stat. 1066, Oct. 23, 1972.

2. "An Act to Establish a National Park Service, and for Other Purposes," 39 Stat. 535, Aug. 25, 1916.

3. NPS, 1954, *Cumberland Island Area, Georgia,* a report prepared by the NPS Seashore Recreation Area Survey, Technical Information Center Library, DSC; Hilburn O. Hillestad et al., 1975, *The Ecology of Cumberland Island National Seashore, Camden County, Georgia,* Georgia Marine Sciences Center, University System of Georgia, Skidway Island, Ga., Technical Report Series, no. 75.5, pp. 23–64 plus vegetation map in the back pocket.

4. See especially Tonya D. Clayton et al., 1992, *Living with the Georgia Shore* (Durham, N.C.: Duke Univ. Press), 13–36; William H. McLemore et al., 1981, *Geology as Applied to Land-Use Management on Cumberland Island, Georgia,* Georgia Geological Survey, prepared as part of contract no. CX5000-8-1563, reprinted in 1988 by the Cooperative Park Studies Unit, UGa, pp. 2-1 to 6-7.

5. Clayton et al. 1992, 16–19.

6. A. Sydney Johnson et al., 1974, *An Ecological Survey of the Coastal Region of Georgia,* NPS Monograph Series, no. 3, 13–23.

7. Ibid.; Clayton et al. 1992, 13–36.

8. Hillestad et al. 1975, 45–58; Daniel J. Hippe, Feb. 22, 1999, "Cumberland Island National Seashore, GA, Project Plan for Level 1 Water-Quality Inventory and Monitoring," CINS Resource Management Files under "Water Quality"; McLemore et al., pp. 4-1 to 5-17.

9. Clayton et al. 1992, 37–39; McLemore et al. 1981, 3-1 to 5-17.

10. Ibid.

11. Ibid.

12. Kathleen A. Deagan, 1994, "Cultures in Transition: Fusion and Assimilation among the Eastern Timucua," in *Tacachale: Essays on the Indians of Florida and Southeast Georgia during the Historical Period,* ed. Jerald T. Milanich and Samuel Proctor (Gainesville: Univ. Press of Florida), 89–100; James J. Miller, 1998, *An Environmental History of Northeast Florida* (Gainesville: Univ. Press of Florida), 42–49.

13. Hillestad et al. 1975, 63–104.

14. Ibid.

15. Ibid., 104–14.

16. Ibid., 116–47.

17. Ibid., 63–104.

18. J. M. Adovasio and Ronald C. Carlisle, 1986, "The First Americans, Pennsylvania Pioneers: Meadowcroft Rock-Shelter as a Long-Lost Chapter to American History Books," *Natural History* 95, 2, Dec., 20–27; Brian Fagan, 2001, "The First Americans," *American Archaeology* 5, 3, Fall, 28–32.

19. Miller 1998, 49–57.

20. Stephen A. Deutschle and Robert C. Wilson, 1975, "Known Prehistoric and Historic Resources of Cumberland Island National Seashore," NPS, Southeast Archaeological Center, Tallahassee, Fla., 2–8; John E. Ehrenhard, 1976, *Cumberland Island National Seashore: Assessment of Archaeological and Historical Resources,* NPS, Southeast Archaeological Center, Tallahassee, Fla., 3–6.

21. Deagan 1994, 100–101; Ehrenhard 1976, 6; Jerald T. Milanich, 1972, "Tacatacuru and the San Pedro de Mocamo Mission," *Florida Historical Quarterly* 50, 3, Jan., 283–91.

22. Ehrenhard 1976, 46–50; John E. Ehrenhard interviewed by Joyce Seward, Atlanta, Aug. 14, 1995, transcript in the CINS Archives, 11; Milanich 1972, 283–91.

23. Lary M. Dilsaver, William Wyckoff, and William L. Preston, 2000, "Fifteen Events That Have Shaped California's Human Landscape," *California Geographer* 40, 1–76.

24. Miller 1998, 127.

25. Louis Torres, 1977, *Historic Resource Study, Cumberland Island National Seashore, and Historic Data Section of the Dungeness Area,* Technical Information Center Library, DSC, 7–60; W. Stitt Robinson, 1979, *The Southern Colonial Frontier, 1607–1763* (Albuquerque: Univ. of New Mexico Press), 185–201.

26. William H. Goetzmann and Glyndwr Williams, 1992, *The Atlas of North American Exploration* (New York: Prentice Hall), 22–23, 32–35, 46–47; Clark Spencer Larsen, 2000, *Skeletons in Our Closet: Revealing Our Past through Bioarchaeology* (Princeton, N.J.: Princeton Univ. Press), 121–78; Ann F. Ramenofsky, 1987, *Vectors of Death: The Archaeology of European Contact* (Albuquerque: Univ. of New Mexico Press).

27. Ehrenhard 1976, 6–7; Torres 1977, 7–9.

28. Ehrenhard 1976, 10; Torres 1977, 11–16.

29. Michael V. Gannon, 1965, *The Cross in the Sand: The Early Catholic Church in Florida, 1513–1870* (Gainesville: Univ. Press of Florida), 64.

30. Deutschle 1975, 26–31; CINS, 1996, *Draft Environmental Assessment: Alternatives for Managing the Feral Horse Herd on Cumberland Island National Seashore,* CINS Central Files, N2219.

31. Torres 1977, 18–21.

32. Ibid., 22–32.

33. Mary R. Bullard, 1992, "In Search of Cumberland Island's Dungeness: Its Origins and English Antecedents," *Georgia Historical Quarterly* 76, 1, Spring, 67–86.

34. Robinson 1979, 199–200; Torres 1977, 37–55.

35. Ben W. Fortson Jr. and Pat Bryant, n.d., "English Crown Grants for Islands

in Georgia, 1755–1775," MS, CINS Lands Files under "Family History"; Torres 1977, 55–60.

36. Mary R. Bullard, 1993, "Uneasy Legacy: The Lynch-Greene Partition on Cumberland Island, 1798–1802," *Georgia Historical Quarterly* 77, 4, Winter, 757–88.

37. Ehrenhard 1976, 88–102.

38. Bullard 1993, 768–71.

39. Torres 1977, 66–75.

40. Ibid., 76–79; Bullard 1993, 773–78; Mary R. Bullard, 1995, *Robert Stafford of Cumberland Island: Growth of a Planter* (Athens: Univ. of Georgia Press), 1–41.

41. Bullard 1993, 757–88.

42. Ibid., 785.

43. Torres 1977, 79–90; David G. Henderson, 1977c, "Architectural Data Section [and] Historic Structure Report, Tabby House, Dungeness Historic District, Cumberland Island National Seashore, Camden County, Georgia," issued by the DSC.

44. Lauren Lubin Zeichner, 1987, "The Historic Landscape of Dungeness, Cumberland Island National Seashore, Georgia," Georgia School of Environmental Design, UGa, CPSU Report no. 35; Torres 1977, 113–14, 121–23.

45. Torres 1977, 96–97; Karl Jenkins, 1996, "General Henry Lee, Also Known as Light-Horse Harry: An Enigma," draft for an information circular, CINS Library, Subject Files under "Lee."

46. Torres 1977, 128–35.

47. Bullard 1995.

48. Jonathan D. Sauer, 1993, *Historical Geography of Crop Plants: A Selected Roster* (Boca Raton, Fla.: Lewis Publishers), 101–2; Mary R. Bullard, 2003, *Cumberland Island: A History* (Athens: Univ. of Georgia Press), 3.

49. Bullard 1995, 108, 213–71; Peggy Stanley Froeschauer, 1989, "The Interpretation and Management of an Agricultural Landscape—Stafford Plantation, Cumberland Island National Seashore, Georgia" (M.A. thesis, UGa), printed by the School of Environmental Design, UGa, as CPSU Technical Report no. 59, 27–48; Torres 1977, 153–54.

50. John E. Ehrenhard and Mary R. Bullard, 1981, "The Chimneys, Stafford Plantation, Cumberland Island National Seashore, Georgia: Archaeological Investigations of a Slave Cabin," NPS, Southeast Archaeological Center, Tallahassee, Fla.

51. Froeschauer 1989, 104.

52. Ibid., 96–100.

53. Ehrenhard and Bullard, 1981, vii; Bullard 1995, 249–53.

54. Bullard 1995, 246–71; Torres 1977, 153–56.

55. Torres 1977, 155–57.

2. The Era of Rich Estates, 1881–1965

1. Frederick A. Ober, 1880, "Dungeness, General Greene's Sea Island Plantation," *Lippincott's Monthly Magazine,* Aug., 241–47; for summaries of the

Carnegie acquisition of and residence on Cumberland Island, see Nancy C. Rockefeller, 1993, "The Carnegies of Cumberland Island," MS, Ga. Archives (a copy is also located at the superintendent's office, CINS); Torres 1977, 159–60; Milton Meltzer, 1997, *The Many Lives of Andrew Carnegie* (New York: Franklin Watts), 87.

2. Rockefeller, 1993, 13–40.

3. Ober 1880; Rockefeller 1993, 14.

4. Torres 1977, 159–60.

5. Meltzer 1997, 90–91.

6. Rockefeller 1993, 15–16.

7. NPS, 1983, "Final National Register Nomination for Cumberland Island National Seashore," item A, p. 1, CINS Library; Torres 1977, 271–340.

8. Torres 1977, 271–340.

9. Ibid.; Rockefeller 1993, 78–86.

10. Rockefeller 1993, 78–86; Torres 1977, 167–91.

11. Some writers have suggested that Lucy C. Carnegie built these homes for her children. However, historian and heir Mary Bullard wrote to the NPS that each child received a monetary gift. Some used these gift funds to build houses (CINS Central Files, L. E. Brown Papers under "Cumberland Correspondence").

12. Torres 1977, 170–71.

13. Information on the Carnegie buildings came from these sources: Ehrenhard 1976; David G. Henderson, 1977a, "Architectural Data [and] Historic Structure Report, Plum Orchard Mansion, Plum Orchard Historic District, Cumberland Island National Seashore, Camden County, Georgia," issued by the DSC; NPS 1983; Rockefeller 1993; Torres 1977; Zeichner 1987.

14. "Dungeness Skeletal Remains Form Island Skyline," *Jacksonville Times-Union*, Mar. 13, 1979.

15. Rockefeller 1993, 96–100.

16. Henderson 1977a, 4–5; Rockefeller 1993, 98–100.

17. NPS 1983; Rockefeller 1993, 105.

18. Rockefeller 1993, 41–60.

19. Ibid., 123–27.

20. C. B. Conyers to Andrew Carnegie II, Feb. 11, 1927, Ga. Archives, acc. no. 69-501, box 33, file 8-3-008; also see the will of Lucy C. Carnegie, ibid., box 25, file 7-1-004.

21. Compiled from the records of the Carnegie estate, ibid.

22. Patterson, Crawford, Miller, and Arensberg, Esqs., Dec. 19, 1925, "Second and Final Account of Andrew Carnegie II, Mrs. Margaret C. Ricketson, and T. Morrison Carnegie, Surviving Trustees under the Will of Lucy C. Carnegie, Deceased," ibid., box 29, file 8-1-015.

23. Torres 1977, 215–18.

24. Andrew Carnegie II to heirs and trustees of Lucy C. Carnegie, n.d. (between 1921 and 1927), Ga. Archives, acc. no. 69-501, box 25, file 7-1-004.

25. S. J. Hall to T. M. Carnegie II, Dec. 22, 1927, ibid., box 37, file 8-7-008.

26. D. A. Burnett to Forest Managers, Inc., Feb. 17, 1934, ibid.; J. Pat Kelly to

R. D. Ferguson, Aug. 1, 1949, ibid., box 42, file 9-1-003D; Margaret J. Wright to Mr. Wallis, June 17, 1941, ibid., box 45, file 9-3-024.

27. Guyton DeLoach to N. T. Collett, Apr. 1, 1953, ibid., box 48, file 9-5-014; E. C. Mann to Leo P. Larkin, June 16, 1954, ibid., box 49, file 9-6-021.

28. Carnegie Estate Manager to Leo P. Larkin, Oct. 16, 1950, ibid., box 42, file 9-1-005E; H. H. Sloss to Leo P. Larkin, Feb. 2, 1954, ibid., box 43, file 9-1-011B.

29. T. M. Carnegie II to Norbert L. Cochran, July 12, 1937, ibid., box 39, file 8-8-015.

30. Charles L. Gowan to R. W. Ferguson with attached "Cattle Plan," Apr. 29, 1952, CINS Lands Files under "Family History."

31. "Cattle Plan Memorandum Number 3," July 30, 1952, ibid.

32. Robert D. Ferguson to Robert W. Ferguson, July 30, 1952, ibid.; Robert M. McKey, 1958, "Appraisal: Cumberland Island Properties and Little Cumberland Island, Camden County, Georgia," National Records Center, Denver, acc. no. 079-97-0009, box 3.

33. Susan P. Bratton and Scott G. Miller, 1992, "Historic Field Systems and the Structure of the Maritime Oak Forest," UGa, CPSU Report no. 15; Hillestad et al. 1975.

34. All of the following correspondence comes from the Carnegie Estate Papers, Ga. Archives, acc. no. 69-501: Cumberland Island Caretaker to R. W. Ferguson, Sept. 24, 1948, box 45, file 9-4-012; J. G. Jenkins to E. M. Nix, Oct. 24, 1953, box 49, file 9-6-022; Coleman C. Johnston to Norman Collett, Feb. 11, 1953, box 45, file 9-3-026; H. M. McKay to William Page, Apr. 22, 1920, box 28, file 7-5-004; E. M. Nix to H. H. Sloss, Oct. 14, 1953, box 49, file 9-6-022; R. A. Young to William Page, May 17, 22, 1918, box 28, file 7-5-018.

35. Tung nut production reached 33,000 pounds in 1945, but subsequent competition from China badly depressed U.S. prices (Rolf Buckley to F. W. McLaren, Feb. 16, 1946, ibid., box 39, file 8-8-022).

36. All of the following correspondence comes from the Carnegie Estate Papers, ibid.: Rolf Buckley to F. W. McLaren with attached transcript of a speech, Apr. 8, 1943, box 39, file 8-8-022; Leo P. Larkin to Norman Collett, Mar. 5, 1951, box 42, file 9-1-006; B. F. Williamson to T. M. Carnegie II, Feb. 25, 1929, Nov. 5, 1933, box 39, file 8-8-022.

37. Norman Collett to Robert D. Ferguson, Sept. 18, 1950, ibid., box 42, file 9-1-005C; Edward R. Guy Jr. to John H. Stanley, Oct. 1960 (exact date not shown), CINS Lands Files under "L. E. Brown Papers."

38. This report is among the Carnegie Estate Papers removed from the Tabby House and stored at the Ga. Archives, acc. no. 69-501, box 42, file 9-1-006. The first page and, hence, the exact date and provenance are missing. However, comments within it suggest that it was probably an inspection report by new estate manager Norman Collett in 1951 or 1952.

39. J. Pat Kelly to Robert D. Ferguson, Jan. 25, 1949, ibid., file 9-1-003A.

40. J. Pat Kelly to Robert D. Ferguson, May 10, Aug. 1, 1949, ibid.

41. N. S. Hernandez and G. T. Davis to Estate of Lucy C. Carnegie, Apr. 3, 1952, ibid., file 9-1-008.

42. "Fire Levels Carnegie's Island Castle," *Atlanta Constitution,* June 26, 1959.

43. Hillestad et al. 1975, 20; Leo P. Larkin to H. H. Sloss, Mar. 31, 1954, Ga. Archives, acc. no. 69-501, box 43, file 9-1-011A; "Scotch Origins Evident as Carnegies Negotiate Lease," *Brunswick News,* Apr. 9, 1957.

44. "Scotch Origins," Apr. 9, 1957.

45. Hillestad et al. 1975, 20–21; McKey 1958, 16; "Scotch Origins," Apr. 9, 1957.

46. "Glidden Mining Pact Approved," *Brunswick News,* Apr. 10, 1957; Hillestad et al. 1975, 20–21; "Scotch Origins," Apr. 9, 1957.

47. McKey 1958, 16.

48. "Glidden Mining Pact Approved," Apr. 10, 1957.

49. Ibid.; Nancy R. Copp interviewed by Joyce Seward, Cumberland Island, Ga., Mar. 25, 1996, transcript in the CINS Archives.

50. "Glidden Mining Pact Approved," Apr. 10, 1957; "Cumberland Island Lease Approved by Supreme Court," *Camden County Tribune,* Sept. 13, 1957; Hillestad et al. 1975, 21.

51. Hillestad et al. 1975, 21; J. B. Mertie Jr., 1958, *Zirconium and Hafnium in the Southeastern Atlantic States,* U.S. Geological Survey, Bulletin 1082-A.

52. Lt. P. N. L. Ballinger to William Page, Sept. 1, 1917, and William Page to Lt. Ballinger, n.d., Ga. Archives, acc. no. 69-501, box 28, file 7-5-019; correspondence between Thomas M. Carnegie II and A. A. Ainsworth in 1923 and 1924 and A. A. Ainsworth to Thomas M. Carnegie II, Oct. 14, 1925, ibid., box 37, file 8-7-001; Elmer Dyal to F. W. McLaren, Feb. 16, 1937, ibid., file 8-7-009.

53. Esther Angwin to H. H. Sloss, 1954 (no specific date), ibid., box 43, file 9-2-001; Edward R. Gray Jr. to John H. Stanley, Oct. 20, 1960, CINS Central Files, "L. E. Brown Papers"; Camden County Land Records, Civil Action 391, Lis Pendens Docket 1, 1950–1977, 13–14, Woodbine, Ga.

54. "Cape Cumberland? The 'Final Frontier' of the Wilderness Island Could Have Been Claimed for Space Exploration," *Camden County Tribune,* Jan. 28, 1982.

55. S. H. Brown to F. W. McLaren, June 17, 1930, Ga. Archives, acc. no. 69-501, box 37, file 8-7-005.

56. Rockefeller 1993, 41–176 (quote at 141).

57. Ibid., 41–176.

58. Ibid.; Torres 1977, 215–19; Lucy C. S. Foster, quoted in Rockefeller 1993, 130–31.

59. James S. Rockefeller, quoted in Rockefeller 1993, 126.

60. Louis Torres (1977, 192) refers to this as the Occidental House. However, Mary Miller shows an advertisement for the "Oriental House" in her booklet, 1990, *Cumberland Island: The Unsung North End* (Darien, Ga.: Darien News), 28–29.

61. The data in this section come from: CINS, Dec. 18, 1979, "Amended Documentation on the High Point–Half Moon Bluff Historic District, Cumberland Island National Seashore"; Ehrenhard 1976, 33; Miller 1990; Torres 1977, 192–15.

62. Miller 1990, 28–53.

63. CINS, Dec. 18, 1979; Miller 1990; Torres 1977, 192–215.

64. *High Point, Cumberland Island, Auction,* 1891 (no specific date), brochure located in the Carnegie Estate Papers, Ga. Archives, acc. no. 69-501, folder 11-1-010.

65. Torres 1977, 192–215.

66. Ibid., 208.

67. "Oliver Ricketson II to the four members of the first generation and the fourteen members of the second generation," 1944 (no specific date), CINS Lands Files under "Family History."

68. Ibid.

69. A framed copy of this map can be viewed in the CINS Archives. The Camden County Land Records Office in Woodbine, Ga., also has a copy on file.

70. Cumberland Island Company, Inc., Feb. 9, 1963, "Press Release," CINS Central Files under "L. E. Brown Papers"; "Cumberland Island Divided into Five Equal Shares," *Camden County Tribune,* July 3, 1964.

71. See Camden County Superior Court case no. 1667, July 6, 1964, and the Supplemental Decree issued June 1, 1965, CINS Lands Files under "Family History"; see also "Cumberland Island Divided," July 3, 1964.

72. Ibid.; Rockefeller 1993; newspaper clipping dated July 12, 1962, Bryan Lang Historical Library, Woodbine, Ga., Subject Files under "Cumberland Island."

73. Camden County Superior Court case no. 1667 and Supplemental Decree, July 6, 1964, and June 1, 1965.

3. Creating Cumberland Island National Seashore

1. "An Act to Establish a National Park Service, and for Other Purposes," 39 Stat. 535, Aug. 25, 1916; Barry Mackintosh, 1991, *Shaping the System,* NPS, Government Printing Office.

2. Good surveys of early national park system history include: Horace M. Albright and Marian Albright Schenck, 1999, *Creating the National Park Service: The Missing Years* (Norman: Univ. of Oklahoma Press); Horace M. Albright and Robert Cahn, 1985, *The Birth of the National Park Service: The Founding Years, 1913–1933* (Salt Lake City: Howe Brothers); Ronald F. Lee, 1974, *The Family Tree of the National Park System* (Philadelphia: Eastern National Park and Monument Association); Mackintosh 1991; Runte 1987; Harlan D. Unrau and G. Frank Williss, 1983, *Administrative History: Expansion of the National Park Service in the 1930s,* NPS, DSC.

3. Lee 1974; Mackintosh 1991.

4. The superintendent of Sequoia National Park provided an excellent expression of this attitude of early park managers in John White, 1936, "Atmosphere in the National Parks," reprinted in Lary M. Dilsaver, ed., 1994, *America's National Park System: The Critical Documents* (Lanham, Md.: Rowman and Littlefield), 142–48.

5. Thomas R. Cox, 1988, *The Park Builders: A History of State Parks in the Pacific Northwest* (Seattle: Univ. of Washington Press), 3–13.

6. Unrau and Williss, 1983, 106–28; "An Act to Authorize a Study of the Park,

Parkway, and Recreational Area Programs in the United States, and for Other Purposes," 49 Stat. 1894, June 23, 1936; NPS, 1941, *A Study of the Park and Recreation Problem in the United States,* U.S. Department of the Interior, Government Printing Office.

7. Klaus J. Meyer-Arendt, 1987, "Resort Evolution along the Gulf of Mexico Littoral: Historical, Morphological, and Environmental Aspects" (Ph.D. diss., Louisiana State Univ.), 5–34; A. E. Demaray to Secretary of the Interior with attached report on the seashore survey, Jan. 2, 1935, NPS Archives, Harpers Ferry, W.Va., L75, "Seashore Studies" box 1; Unrau and Williss 1983, 155–60; Wirth 1980, 192.

8. "An Act to Provide for the Establishment of the Cape Hatteras National Seashore in the State of North Carolina, and for Other Purposes," 50 Stat. 669, Aug. 17, 1937.

9. Lary M. Dilsaver and William C. Tweed, 1990, *Challenge of the Big Trees: A Resource History of Sequoia and Kings Canyon National Parks* (Three Rivers, Calif.: Sequoia Natural History Association), 230–32; Louis Torres, 1985, *Historic Resource Study of Cape Hatteras National Seashore,* DSC, 155–56; Unrau and Williss 1983, 156–59.

10. Unrau and Williss 1983, 159.

11. William Everhart to author, Dec. 12, 1994.

12. NPS, 1955, "Our Vanishing Shoreline," CINS Library, 23.

13. Ibid., 29.

14. Howard Chapman interviewed by author, San Rafael, Calif., Dec. 29, 1997.

15. NPS, 1965a, "Chronology of Principal Contacts Made between Service Representatives and Carnegie Heirs or Their Representatives regarding Cumberland Island," CINS Central Files, L58.

16. William Everhart, Dec. 12, 1994.

17. The Andrew Mellon Foundation resulted from the merger and reorganization of the Avalon and Old Dominion Foundations.

18. Advisory Board on National Parks, Historic Sites, Buildings, and Monuments, 1956, "Minutes of the Meeting of March 30, 1956," NPS, Park History Office Files, Washington, D.C., under "Advisory Board."

19. "Cumberland State Park to Be Asked by Odom," *Camden County Tribune,* Jan. 28, 1955.

20. "Odom Reports to People of Camden," ibid., Feb. 4, 1955.

21. Georgia H.R. 166 is described in Cumberland Island Study Committee, 1956, "Report to the 1956 Session of the Georgia Assembly," Georgia House of Representatives, CINS Library.

22. Ibid., 3, 4.

23. Conrad Wirth to Wesley D'Ewart, Mar. 26, 1956, CINS Central Files under "L. E. Brown Papers."

24. Conrad Wirth to Mrs. Floyd [Margaret] Wright, Mar. 30, 1956, ibid.

25. McKey 1958.

26. Little Cumberland Island Association, 1965, "Membership Guide,"

Coastal Georgia Historical Society, St. Simons Island, Subject Files under "Cumberland Island."

27. S. 2010, 86th Cong., 1st sess., 1959; *Congressional Record—Senate,* vol. 5, 86th Cong., 1st sess., May 20, 1959, 8557–59; S. 2460, 86th Cong., 1st sess., 1959.

28. H.R. 8519, H.R. 8445, H.R. 8449, and H.R. 7407, 86th Cong., 1st sess., 1959.

29. *Cumberland Island Company Newsletter* 1, 1, Conrad Wirth Papers, box 17, folder 1, RG 79, NA.

30. NPS, 1959, "Minutes of a Meeting between Officers of the Cumberland Island Company, Inc., and Representatives of the National Park Service in Room 3218, Interior Building, Washington, D.C., Friday, Jan. 8, 1960, at 1:30 P.M.," ibid., box 9, "General Correspondence, 1960."

31. Ibid.

32. S. 2636, 86th Cong., 1st sess., 1959; NPS 1959.

33. Joseph Graves to John H. Stanley, Sept. 3, 1963, CINS Central Files, L58; NPS, 1965b, "Secretary Udall to Visit Georgia's Golden Isles, Nov. 4–5, 1965," press release in NPS Archives, Harpers Ferry, W.Va., L75, "Seashore Studies," box 1.

34. "Udall Says Seashore Park Would Be Tremendous Asset to State, Region," Nov. 4, 1965, newspaper clipping in Conrad Wirth Papers, box 17, folder 1, RG 79, NA.

35. Ibid.

36. "Cumberland Island Fabulous Site for a National Park," *Atlanta Journal and Constitution,* Dec. 5, 1965.

37. Rockefeller 1993.

38. John McPhee, 1990, *Encounters with the Archdruid* (New York: Noonday Press), 88–96.

39. Charles Fraser to Cumberland Island Landowners, Oct. 1965 (no specific date), CINS Central Files, L58.

40. Ibid.

41. Ibid.

42. John W. Bright to NPS Assistant Director, Cooperative Activities, Mar. 21, 1969, ibid., under "Proposed CUIS."

43. Vincent Ellis to SERO Director, Jan. 10, 1969, and Chief Park Historian Frank Pridemore to Superintendent, Kennesaw Mountain, Feb. 17, 1969, ibid., L58.

44. Putnam McDowell to Lucy Ferguson, Oct. 22, 1968, Thornton Morris Papers, file 131A.

45. Nancy C. Rockefeller to George Frazer [Charles Fraser], Mar. 10, 1969, ibid., file 131A2.

46. William C. Warren III, Feb. 24, 1969, "Cumberland Property Wrangle," *Atlanta Journal.*

47. Vincent Ellis to SERO Director, May 14, 1969, CINS Central Files under "Proposed CUIS."

48. Thornton Morris to Cumberland Island Conservation Association Executive Committee, Apr. 18, 1969, Thornton Morris Papers, file 131 Memos; "Dr.

Masters, Charles Fraser to Address Joint Cumberland Island Meeting," *Southeast Georgian,* Mar. 6, 1969.

49. Thornton Morris to Cumberland Island Conservation Association, Mar. 13, 1969, Thornton Morris Papers, file 131 Memos.

50. Vincent Ellis to SERO Director, Mar. 6, 1969, CINS Central Files, L58; Thornton Morris to Cumberland Island Conservation Association, Feb. 18, 1969, Thornton Morris Papers, file 131 Memos.

51. Georgia Senate Bill 229, 1969; Joseph Graves to Coleman C. Johnston, Mar. 5, 1969, Thornton Morris Papers, file 131A.

52. Georgia Senate Bill 260, 1969; Thornton Morris to Cumberland Island Conservation Association, Mar. 18, 1969, Thornton Morris Papers, file 131 Memos.

53. Thornton Morris interviewed by author, Atlanta, Oct. 6, 1999.

54. Robert R. Jacobsen to Theodore Swem, Mar. 7, 1969, CINS Central Files, L58.

55. "Fraser to Sell Cumberland Property," *Savannah Morning News,* Aug. 8, 1969.

56. Vincent Ellis to SERO Director, Nov. 6, 1968, CINS Central Files, L58.

57. Putnam McDowell to Lucy Ferguson, June 23, 1969, Thornton Morris Papers, file 131 section 1A (3).

58. Joseph Sax, 1980b, "Buying Scenery: Land Acquisition for the National Park Service," *Duke Law Journal* 4, 709–40.

59. Andrew Rockefeller to Thornton Morris, Aug. 8, 1969, Thornton Morris Papers, file 131 Memos.

60. Camden County Commission, Nov. 2, 1971, "Resolution on H.R. 9589, a Bill to Create Cumberland Island National Seashore," CINS Central Files under "Congressional Visit."

61. Williamson S. "Bill" Stuckey interviewed by author, Washington, D.C., Sept. 18, 1998; "US Decides Cumberland Belongs in Its Park System," *Atlanta Constitution,* Apr. 12, 1967.

62. Max Edwards to W. S. Stuckey, Mar. 28, 1967, CINS Central Files, L58.

63. Hans Newhauser interviewed by Joyce Seward, CINS, Nov. 21, 1995, transcript in CINS Archives.

64. "Cumberland as a National Seashore Could Mean Big Business for Camden," *Southeast Georgian,* Mar. 30, 1972.

65. NPS, Oct. 1967, *Environmental Assessment, Proposed Cumberland Island National Seashore,* CINS Library; William B. Keeling et al., 1968, *Economic Impact of the Proposed Cumberland Island National Seashore,* Bureau of Business and Economic Research, UGa, Travel Research Study no. 5.

66. NPS, Oct. 1971, *Master Plan, Proposed Cumberland Island National Seashore,* CINS Library.

67. Ingram H. Richardson to Stewart Udall, Oct. 9, 1967, CINS Central Files, L58; "An Act to Establish Cumberland Island National Seashore in the State of Georgia, and for Other Purposes," 86 Stat. 1066, Oct. 23, 1972.

68. See chap. 4 for a detailed account.

69. Putnam McDowell to Thornton Morris, May 5, 1969, Thornton Morris Papers, file 131A.

70. Thornton Morris to Cumberland Island Conservation Association, Mar. 30, June 10, 1970, ibid., file 131 Memos. E. R. Kingman of the Nature Conservancy also reached a tentative agreement with the Johnston family.

71. "An Act to Establish the National Park Foundation, and for Other Purposes," 81 Stat. 656, Dec. 18, 1967.

72. George Hartzog to Ernest Brooke Jr., Aug. 30, 1968, CINS Central Files, L58.

73. Alfred W. Jones Jr. interviewed by Joyce Seward, Sea Island, Ga., Aug. 16, 1995, transcript in CINS Archives; Thornton Morris to Cumberland Island Conservation Association, Apr. 14, May 25, 1970, Thornton Morris Papers, file 131 Memos; "Jones Lauded for Work," *Savannah Morning News,* Nov. 11, 1972.

74. H.R. 15686, 91st Cong., 2d sess., 1970.

75. Ibid.

76. Ibid.; "Cumberland Island Park Discussed by Planners," newspaper clipping, Mar. 11, 1970, Coastal Georgia Historical Society Archives, St. Simons Island, Subject Files under "Cumberland Island."

77. Theodore Swem to Mr. Joseph, Mar. 4, 1970, NPS, Park History Office Files, Washington, D.C., under "Cumberland Island"; "Cumberland Island Park Discussed," Mar. 11, 1970.

78. Georgia House of Representatives, Jan. 1970, "Report of the Georgia Coastal Islands Study Committee" (established by the Georgia House of Representatives Resolution 82-219), CINS Library; Thornton Morris to Cumberland Island Conservation Association, Jan. 23, 1970, Thornton Morris Papers, file 131 Legislation; "State Will Claim Marshlands in Position Paper by Bolton," *Atlanta Journal,* Mar. 12, 1970; Thornton Morris to Cumberland Island Conservation Association, Mar. 20, 1970, Thornton Morris Papers, file 131 Memos.

79. "Attacking an Island," *Atlanta Journal,* Apr. 7, 1970; "Park Service Consulted on Cumberland Project," *Atlanta Constitution,* Apr. 7, 1970.

80. "Park Chief Denies Okay on Island Work," *Atlanta Constitution,* Apr. 10, 1970.

81. "Island Owners Protesting Activities on Cumberland," newspaper clipping, Apr. 22, 1970, and "Cumberland Rally Strikes Light Note for Conservation," newspaper clipping, June 30, 1970, Coastal Georgia Historical Society, St. Simons Island, Subject Files under "Cumberland Island."

82. "Fraser Claims Willing to Sell Island Acreage," *Atlanta Constitution,* Apr. 25, 1970; McPhee 1990, 149–50; "'New Yorker' Tells Story Of Fraser's Island Venture," newspaper clipping, Mar. 26, 1971, Coastal Georgia Historical Society, St. Simons Island, Subject Files under "Cumberland Island."

83. H.R. 9859, 92d Cong., 1st sess., 1971.

84. Hans Neuhauser interviewed by Joyce Seward, Atlanta, Nov. 21, 1995, transcript in CINS Archives; Charles D. Clement and J. Richardson, 1971, "Recre-

ation on the Georgia Coast—An Ecological Approach," *Georgia Business* 30, 11, May, 1–24; see also Charles D. Clement, Jan. 1971, "The Georgia Coast: Issues and Options for Recreation," prepared for The Conservation Foundation by the Division of Research of the College of Business Administration, UGa, vii.

85. "Victory Predicted for Cumberland Plan," *Savannah Morning News,* Nov. 10, 1971.

86. Camden County Commission Resolution, Nov. 2, 1971, CINS Archives under "Congressional Visit."

87. *Hearings before the Subcommittee on National Parks and Recreation of the Committee on Interior and Insular Affairs, House of Representatives,* 92d Cong., 2d sess., Apr. 20–21, 1972, 74–79, 105–14, 79–95.

88. Ibid., 47–52, 90–91, 103–5, 122–24.

89. Ibid., 24.

90. *Hearing before the Subcommittee on Parks and Recreation of the Committee on Interior and Insular Affairs, U.S. Senate,* 92d Cong., 2d sess., May 11, 1972; "An Act to Provide for the Establishment of Cape Hatteras National Seashore in the State of North Carolina, and for Other Purposes," 50 Stat. 669, Aug. 17, 1937; U.S. Senate, 1972, *Establishing the Cumberland Island National Seashore in the State of Georgia, and for Other Purposes,* 92d Cong., 2d sess., S. Rept. 92-972, 5.

91. "Congress Nears Cumberland Island Decision," *Atlanta Constitution,* May 25, 1972.

92. "Conservationists Back Cumberland Island Bill," *Savannah Morning News,* Oct. 26, 1972.

93. A. W. Jones Jr. interview 1995, 3–4; "Jones Lauded," Nov. 9, 1972.

4. Land Acquisition and Retained Rights

1. NPS, 1997, *The National Parks Index, 1997–1999,* NPS, Office of Public Affairs; Joseph Sax 1980b, 714 and n. 28.

2. Ira Hutchison to All Regional Directors, June 20, 1977, CINS Lands Files under "Land Acquisition 73–77."

3. McKey 1958.

4. "An Act to Establish the National Park Foundation, and for Other Purposes," 81 Stat. 656, Dec. 18, 1967; "Report on Residential Occupancy under Special Use Permits," NPS, CINS Central Files, L30.

5. Charles Fraser interviewed by Joyce Seward, Hilton Head, S.C., Aug. 30, 1995, transcript in CINS Archives, 18.

6. Purchase dates and amounts are drawn from two sources: "Master Deed Listing (by Tract Number), Status of Lands as of May 31, 2000," a NPS computer file available from SERO, Lands Division, and "Summary of Retained Estates, May–July 2000," compiled by the same office. Copies of these can be found in the CINS Lands Files. In addition, CINS Lands Files also contain copies of each individual deed and retained-right agreement under "Deeds to the National Park Foundation."

7. "Summary of Retained Estates," 2000, Gertrude Schwartz file.

8. Ibid., Cumberland Island Properties, Inc., file.

9. Ibid., Nancy Butler, Thomas Johnston, Margaret Richards, Lucy Graves, and Marius Johnston files.

10. "Deeds to the National Park Foundation," tract 01-102, Table Point Company, Inc., file.

11. "Summary of Retained Estates," 2000, Lucy C. S. Foster file.

12. "Master Deed Listing," 2000.

13. Land Records, Camden County Lands Office, Woodbine, Ga., Sept. 26, 1967.

14. J. Stillman Rockefeller Jr. interviewed by Joyce Seward, Cumberland Island, Ga., Aug. 2, 1995, transcript in CINS Archives, 61; Robert Harrison interviewed by Joyce Seward, Folkstone, Ga., Aug. 2, 1995, transcript in CINS Archives, 7.

15. Roy L. Gordon, July 5, 1973, "Appraisal of Lands of Millicent S. Monks, Cumberland Island, Camden County, Georgia," CINS Lands Records, deed 58, 10–12; H. Philip Troy, Jan. 17, 1975, "Appraisal Report of Cumberland Island Holding Company Retained Use of Estate," CINS Lands Records under "Reservations," 20–21.

16. Troy 1975, 20–22, Addendum E.

17. Gordon 1973, 13–14.

18. Ibid.; Land Records, Camden County Lands Office, Woodbine, Ga., July 1971.

19. Roy L. Gordon, May 11, 1976, "Appraisal of Tract 03-101, Davis Land Company, Inc., Cumberland Island National Seashore," CINS Lands Records, deed 53.

20. Ibid., 28; "Master Deed Listing," 2000; J. Grover Henderson interviewed by Joyce Seward, St. Marys, Ga., Mar. 27, 1996, transcript in CINS Archives, 6–7.

21. CINS, 1974, "Superintendent's Annual Report, 1973," CINS Central Files, A2621.

22. Richard M. Fairbanks to Richard H. Pough, 1973 (no specific date), CINS Central Files, L1425; NPS, May 1961, "Land Management Handbook," DSC, chap. 7.

23. Jimmy Carter to Rogers C. B. Morton, Oct. 12, 1973, CINS Lands Files under "Land Acquisition, 73–75."

24. Jimmy Carter to Sam Nunn, Nov. 26, 1973, ibid.; "Cumberland Island: Wild Preserve or Wild Speculation," *Atlanta Journal and Constitution,* Sept. 16, 1973.

25. Lawrence C. Hadley to Sam Nunn, Nov. 1, 1973, CINS Lands Files under "Land Acquisition, 73–75."

26. Rogers C. B. Morton to Jimmy Carter, Dec. 14, 1973, ibid.

27. "Cumberland Island: Wild Preserve," Sept. 16, 1973.

28. "Carter Accuses Monks of Cumberland Con Job," Associated Press article, newspaper clipping, Dec. 21, 1972, Conrad Wirth Papers, box 17, folder 1, RG 79, NA.

29. "Cumberland Still for Sale despite Carter Order," *Atlanta Constitution,* Dec. 22, 1972; "Master Deed Listing," 2000.

30. Paul McCrary to David Thompson, Sept. 15, 1976, CINS Lands Files un-

der "Land Acquisition General, 72–73"; Putnam McDowell to Thornton Morris, Feb. 25, 1977, Thornton Morris Papers, file 3267.

31. See file for "Tract 02-185, Phineas Sprague," CINS Lands Files.

32. Troy 1975, iv, 10–12.

33. Charles Fraser to Gary E. Everhardt, Feb. 28, 1975, SERO Solicitor's Files under "Cumberland Island"; CINS, 1983, "Environmental Assessment for Road and Power Service to Cumberland Island Holding Company Retained Estate, 1983," SERO, Planning Files under "Cumberland Island General Correspondence."

34. Ken Morgan to Regional Director, SERO, Apr. 28, 1988, S. Larry Phillips to Ken Morgan, Mar. 25, Apr. 19, 1988, CINS Central Files, L1417.

35. "Master Deed Listing," 2000; "Summary of Retained Estates," 2000, Cumberland Island Properties, Inc., file.

36. "Gift Offer Terms Disputed regarding Cumberland Isle," *Atlanta Journal,* Mar. 2, 1973; "Master Deed Listing," 2000.

37. "Cumberland Isle Buying Snags Are Seen," *Florida Times-Union,* May 21, 1974.

38. I compiled these tract histories from the land records maintained at the Camden County Lands Office in Woodbine, Ga.

39. Finis T. Rayburn, Oct. 31, 1973, "Appraisal of Tract 03-129, William W. Rogers et al.," CINS Lands Files, deed 17.

40. See deed 83, tract 03-130, George Law, in the CINS Lands Files for correspondence and court documents.

41. Gordon 1976; L. Boyd Finch to Associate Director, Park System Management, July 9, 1976, CINS Lands Files, deed 53.

42. Lawrence Miller interviewed by Joyce Seward, St. Simons Island, Ga., Mar. 26, 1996, transcript in CINS Archives, 12.

43. "Federal Private Owners Clash on Cumberland Island," *Florida Times-Union,* Oct. 19, 1975.

44. "Master Deed Listing," 2000; Finis T. Rayburn, Oct. 1, 1974, "Appraisal of Tract 02-150, Wyndham M. Manning III et al.," CINS Lands Files, deed 52.

45. "An Act to Authorize Additional Appropriations for the Acquisition of Lands and Interests in Lands within the Sawtooth National Recreation Area in Idaho," 92 Stat. 3467, Nov. 10, 1978.

46. "Master Deed Listing," 2000.

47. Bert Roberts to Charles E. Bennett, Oct. 28, 1975, CINS Central Files, A3815.

48. I compiled these data by comparing the land purchase records for Cumberland Island maintained at the Camden County Lands Office in Woodbine, Ga., with the "Master Deed Listing," 2000, which lists the prices the NPS paid.

49. U.S. District Court, Southern District of Georgia, Civil Action 281-90. Copies of this and related court documents are located in CINS Lands Files, deed 132.

50. See CINS Lands Files for "Deed 115, Bunkley Heirs," "Deed 79, Olsen/Horton," "Deed 111, O. H. Olsen."

51. "Briefing for Director Whalen," Mar. 20, 1979, CINS Central Files under "Planning Briefs—1979"; see also correspondence and court documents in the file for deed 64, CINS Lands Files.

52. See CINS Lands File named "Henderson/Ruckdeschel Exchange"; CINS, 1985, "Facts and Findings" (this is a chronology of Carol Ruckdeschel's land activities involving Louis McKee and J. Grover Henderson compiled by an unknown NPS official), CINS Lands Files, deed 64.

53. Robert Coram, July 5, 1981, "Life and Death on Cumberland Island," *Atlanta Weekly* (published by the *Atlanta Journal and Constitution*); CINS, Apr. 17, 1980, "Case Incident Report 800008," CINS Law Enforcement Files.

54. "Facts and Findings," 1985.

55. See CINS Lands File for deed 64, tract 01-113; "Naturalist Wins Claim to Land on Cumberland," *Florida Times-Union*, Dec. 8, 1984.

56. Tom Piehl interviewed by author, Atlanta, July 11, 2000; Rockefeller 1976, 61–62.

57. "Master Deed Listing," 2000; Rockefeller 1976, 62–64.

58. CINS, Apr. 1980, "Draft Land Acquisition Plan," CINS Library.

59. Randy Snodgrass to Joe Brown, Feb. 15, 1980, Joe Brown to Randy Snodgrass, Feb. 17, 1980, CINS Central Files, L1425.

60. "Master Deed Listing," 2000; "Summary of Retained Estates," 2000; see also CINS Lands File for deed 112, tract 01-104, "High Point, Inc."

61. "Master Deed Listing," 2000.

62. Marion D. Miller to Regional Director, SERO, Aug. 16, 1978, CINS Lands Files under "Land Acquisitions General."

63. Howard H. Calloway to Nathaniel P. Reed, July 1, 1974, ibid., "Tract 02-138, U.S. Army."

64. "An Act to Establish the Cumberland Island National Seashore in the State of Georgia, and for Other Purposes," 86 Stat. 1066, Oct. 23, 1972; "An Act to Establish a National Wilderness Preservation System for the Permanent Good of the Whole People, and for Other Purposes," 78 Stat. 890, Sept. 3, 1964; "Plans for Cumberland Island Prompt Criticism from DNR," *Florida Times-Union*, Feb. 2, 1978; "Master Deed Listing," 2000.

65. "Summary of Retained Estates," 2000.

66. Martin W. Baumgaertner, July 1975, "Reserved Estates on Cumberland Island, National Park Foundation Acquisition," CINS Central Files under "Retained Rights."

5. Planning and Operating in the 1970s

1. "An Act to Establish the Cumberland Island National Seashore in the State of Georgia, and for Other Purposes," 86 Stat. 1066, Oct. 23, 1972.

2. "Cumberland Island National Seashore, 1972–1979," 1979, summary of issues and progress, CINS Library, Subject Files under "General History," 4; "Cumberland Island: Wild Preserve or Wild Speculation?" *Atlanta Journal and Constitution*, Sept. 16, 1973.

3. "Cumberland Island: Wild Preserve," Sept. 16, 1973; David S. Thompson to Michael A. Doyle, May 31, 1973, CINS Central Files, A3815.

4. "Cumberland Island: Wild Preserve," Sept. 16, 1973; "EPA Will Examine Dump, Junkyard on Cumberland," *Atlanta Journal*, Sept. 21, 1973.

5. Paul McCrary to Martin Gillette, Sept. 27, 1973, CINS Central Files, L1425; Kenneth O. Morgan to Don Nixon, Feb. 18, 1983, ibid., N1623.

6. "Cumberland Island National Seashore, 1972–1979," 1979, 6; Hans Neuhauser interviewed by Joyce Seward, Athens, Ga., Nov. 21, 1995, transcript in CINS Archives.

7. Department of the Interior, Nov. 27, 1973, "Cumberland Island National Seashore to Open near Kingsland," SERO Planning Files under "Cumberland Island General Correspondence"; "NPS May Shift Its Headquarters," *Camden County Tribune*, Mar. 20, 1975; Bert C. Roberts to Richard Daley, May 1, 1975, to George L. Hannaford, May 12, 1975, CINS Central Files, A3815.

8. Regional Director, SERO, to Associate Director, Administration, Aug. 1, 1975, NPS, Park History Office Files, Washington, D.C., under "Cumberland Island."

9. CINS, 1976, 1977, and 1979, "Superintendent's Annual Report" for 1975, 1976, and 1979, CINS Central Files, A3821; Zack Kirkland interviewed by author, Cumberland Island, Ga., June 30, 1998.

10. CINS, 1976–81, "Superintendent's Annual Report" for 1976–80; David S. Thompson to F. T. Davis Jr., including monthly visitation figures from June 1975 through Jan. 1977, Mar. 2, 1977, SERO Planning Files under "Cumberland Island General Correspondence."

11. CINS, Mar. 1970, "Draft Interim Interpretive Prospectus," CINS Central Files under "Cumberland Island"; CINS, 1976–79, "Superintendent's Annual Report" for 1976–79.

12. Acting Regional Director, SERO, to Associate Director, Park System Management, Mar. 25, 1975, CINS Central Files, A3815.

13. Ibid.; David S. Thompson to Lucian Whittle, Aug. 6, 1975, ibid.

14. Zack Kirkland to Art Graham, June 21, 1976, L. Glenn McBay to District Engineer, U.S. Army Corps of Engineers, Sept. 10, 1979, Ben Jenkins to Steven Osvald, Oct. 23, 1979, Jan. 18, 1980, Steven Osvald to Ben Jenkins, Jan. 15, 1980, ibid., L1425; Regional Solicitor, Atlanta, to Regional Director, SERO, Aug. 24, 1976, CINS Lands Files, L1425; Regional Solicitor, Atlanta, to Regional Director, SERO, Oct. 10, 1976, ibid. under "Land Acquisition General."

15. Neal G. Guse Jr. to J. Grover Henderson, Sept. 27, 1978, CINS Lands Files under "Henderson"; Homer Hail to Bert C. Roberts, Dec. 31, 1975, ibid., tract 02-101; "State Transfers Cumberland Law Enforcement," *Southeast Georgian*, June 3, 1982; "GMA Says No to Georgia Senate Bill 764," *Camden County Tribune*, June 24, 1982.

16. Thornton Morris to Joe Brown, Dec. 19, 1978, Thornton Morris Papers, file 3869; Daniel J. Tobin Jr. to Ronald "Bo" Ginn, July 25, 1979, SERO Planning Files under "Cumberland Island General Correspondence."

17. Thornton Morris to Joe Brown, Dec. 19, 1978.

18. Director, NPS, to Ronald "Bo" Ginn, May 5, 1980 (draft), Thornton Morris to Ronald "Bo" Ginn, June 19, 1979, SERO Planning Files under "Cumberland Island General Correspondence."

19. L. Boyd Finch to Roger Buerki, May 7, 1980, CINS Lands Files under "Thornton Morris"; see also Hans Neuhauser to Ronald "Bo" Ginn, Apr. 23, June 13, 1980, Randy Snodgrass to William Whalen, Apr. 21, 1980, Joe Brown to Herman E. Talmadge, June 17, 1980, to Thornton Morris, Sept. 22, 1980, SERO Planning Files under "Cumberland Island General Correspondence"; Thornton Morris, Oct. 9, 1980, "Affidavit Showing Facts Affecting Title to the Land of the United States, Department of the Interior, National Park Service, as Described in a Deed Dated Nov. 2, 1970, and Recorded in Deed Book 97, Page 382, in the Office of the Clerk of the Superior Court of Camden County, Georgia, from Table Point Company, Inc., to the National Park Foundation," court files, Woodbine, Ga.

20. NPS, 1971, *Master Plan Proposed Cumberland Island National Seashore,* CINS Library, 31.

21. Lary M. Dilsaver, n.d., *Saving Our Seashores,* work in progress.

22. "An Act to Establish a National Policy for the Environment, to Provide for the Establishment of a Council on Environmental Quality, and for Other Purposes," 83 Stat. 852, Jan. 1, 1970.

23. NPS 1971.

24. Ibid.

25. NPS, 1972, DES 72-49, CINS Library.

26. Conservation Foundation, 1972, *National Parks for the Future* (Washington, D.C.: Conservation Foundation).

27. "An Act to Improve the Administration of the National Park System by the Secretary of the Interior, and to Clarify the Authorities Applicable to the System, and for Other Purposes," 84 Stat. 825, Aug. 18, 1970.

28. Hans Neuhauser interviewed by Joyce Seward, Athens, Ga., Nov. 21, 1995, transcript in CINS Archives; Hans Neuhauser and Sandy Hobbs, 1974, "Cumberland Island: National Park for the Future," *Conservancy,* Fall, 13–15.

29. Ibid.

30. Albert F. Ike and James I. Richardson, Apr. 1974, "Cumberland Island National Seashore Study, Carrying Capacity," Institute of Community and Area Development, UGa, 15.

31. Albert F. Ike and James I. Richardson, Sept. 1975, "Cumberland Island National Seashore Study, Carrying Capacity, Revised," Institute of Community and Area Development, UGa, 14.

32. Ibid.

33. Ibid., 14–15.

34. "Plans for Cumberland Island," May 1, 1972, newspaper clipping in Bryan Lang Historical Library, Woodbine, Ga., Subject File under "Cumberland Island."

35. Putnam B. McDowell to Lucy Ferguson, May 6, 1974, Thornton Morris Papers, file 3267.

36. Putnam B. McDowell to Thornton Morris, Sept. 5, 1975, ibid.

37. Ibid.; William D. Ruckelshaus to Sam Nunn, Mar. 19, 1975, ibid.; James L. Bainbridge to Sam Nunn, May 5, 1975, Charles C. Corbin to Herman E. Talmadge, Dec. 10, 1974, NPS to Henry M. Jackson, Oct. 16, 1974, CINS Central Files, A3815.

38. NPS, 1975, "Transcript of a Public Hearing Held on Feb. 4, 1975, at Woodbine, Georgia," CINS Library, 56–57; Robert Stanton to Sam Nunn, Dec. 6, 1974, CINS Central Files, A3815.

39. Neuhauser 1995, 20–22; NPS 1975, 57.

40. "Excessive Tourism Threat to Island?" *Atlanta Journal*, Feb. 5, 1975.

41. Ibid.; "An Act to Establish a National Wilderness Preservation System for the Permanent Good of the Whole People, and for Other Purposes," 78 Stat. 890, Sept. 3, 1964; Foresta 1984, 68–71; "Minimal Cumberland Island Backed by 100," *Jacksonville Times-Union and Journal*, Mar. 23, 1975; NPS 1975, 20.

42. Curtis Bohlen to Henry M. Jackson, Sept. 2, 1975, National Records Center, Denver, acc. no. 079-97-0009, box 3, must be accessed through DSC.

43. David S. Thompson to Steve Osvald, Dec. 13, 1974, SERO Planning Files under "Cumberland Island Correspondence 72–75."

44. "Fernandina Access Sought to New Cumberland Island Park," *Fernandina Beach News-Leader*, Mar. 6, 1975; E. A. Williams to Bert Roberts, Apr. 18, 1975, CINS Lands Files, L1425; NPS, 1980, "Possible Embarkation Point for Cumberland Island National Seashore, Reconnaissance Report," CINS Library, Subject Files under "Fernandina"; William Penn Mott Jr. to Lindsay Thomas, Dec. 22, 1988, CINS Central Files, A3815; Denis Davis to Anthony Leggio, July 10, 1997, ibid., A7221; Hans Neuhauser 1995, 12–14.

45. CINS, 1976, *Environmental Review on Environmental Assessment for General Management Plan and Wilderness Study, Cumberland Island National Seashore, Georgia,* CINS Library.

46. Ibid.

47. Ibid.; NPS, 1977a, *Draft Environmental Statement, General Management Plan, [and] Wilderness Study, Cumberland Island National Seashore,* DES 77-25, CINS Library; NPS, 1977b, *Draft General Management Plan and Wilderness Study, Cumberland Island National Seashore,* ibid.

48. NPS, 1977c, "Transcript of a Hearing on the General Management Plan and Wilderness Study on the Cumberland Island National Seashore, Sept. 24, 1977, St. Marys, Georgia," ibid.; NPS, 1977d, "Transcript of a Hearing on the General Management Plan and Wilderness Study on the Cumberland Island National Seashore, Sept. 27, 1977, Atlanta, Georgia," ibid.

49. NPS 1977c, 20–21.

50. Ibid., 43, 73.

51. NPS 1977d, 64–65.

52. Ibid., 62.

53. Ibid., 54–63.

54. Charles H. Badger to SERO, Feb. 24, 1978, includes state of Georgia review of the 1977 *Draft General Management Plan and Wilderness Study* and its *Envi-*

ronmental Statement, SERO Cultural Resource Files under "Cumberland Island—Comments on GMP," 1–8.

55. Ibid.

56. "U.S. Parks Chief to Study Cumberland Jitney Fight," *Atlanta Journal*, Aug. 14, 1978; Carol Ruckdeschel to William Whalen, Feb. 12, 1979, CINS Central Files, A3821.

57. Joe Graves to William Whalen, June 1, 1979, CINS Central Files, L1425.

58. Hans Neuhauser, Apr. 3, 1979, "Cumberland Island National Seashore Boundaries for the North End of the Island," ibid., L48; Acting Director, NPS, to Assistant Secretary for Fish and Wildlife and Parks, Aug. 1, 1979, ibid.

59. NPS, Dec. 1980, "Final Environmental Impact Statement, General Management Plan, [and] Wilderness Recommendation," CINS Library, 7–34.

6. Resource Management in the 1970s

1. The results of these preliminary studies appear in the earliest plans for the seashore. See NPS, 1971, *Master Plan;* NPS, 1972, *Draft Environmental Statement, Proposed Cumberland Island National Seashore, Georgia*, DES 72-49, CINS Library.

2. Hillestad et al. 1975.

3. A. Starker Leopold et al., 1963, "Wildlife Management in the National Parks," report of the Advisory Board on Wildlife Management appointed by Secretary of the Interior Stewart Udall, in *Handbook of Administrative Policies for Natural Areas* (Washington, D.C.: Government Printing Office), 88–103; NPS, 1988, "Management Policies of the National Park Service," Washington, D.C.

4. CINS, 1986, "Geological and Ecological Change on Cumberland Island," briefing report in CINS Central Files, N2215; C. W. Shabica et al., 1993, *Inlets of the Southeast Region National Seashore Units: Effects of Inlet Maintenance and Recommended Action*, NPS, Washington, D.C., 28–33.

5. Stephen V. Shabica to Chief Ranger, CINS, Nov. 1, 1978, CINS Central Files, N22; Chris Baumann to Superintendent, CINS, July 22, 1981, CINS Captain's House Files, N22.

6. "Dredge Arrives to Start Work on 34-Foot Channel at Kings Bay," *Camden County Tribune*, Sept. 2, 1955; "Kings Bay: Use It or Release It, Ginn Says," *Florida Times-Union*, Mar. 24, 1973; David S. Thompson to Steven Osvald, Dec. 13, 1974, SERO Planning Files under "Cumberland Island Correspondence, 72–75."

7. Thompson to Osvald, Dec. 13, 1974; Bo Ginn to James L. Bainbridge, Aug. 21, 1975, David S. Thompson to Sam Nunn, Sept. 22, 1975, CINS Central Files, A3815.

8. Anthony J. Rinck to Superintendent, CINS, July 23, 1974, SERO Planning Files under "Cumberland Island Correspondence, 72–75."

9. Department of the Army, June 1976, *Dredging of Turning Basin and Entrance Channel, Military Ocean Terminal, Kings Bay, Cumberland Sound, Camden*

County, Georgia, CINS Central Files, L76; record of a telephone call from Steven Osvald to Meredith Ingham, Nov. 11, 1975, SERO Planning Files under "Cumberland Island Correspondence, 72–75."

10. "Kings Bay Base 'Official,'" *Florida Times-Union,* Jan. 27, 1978; Larry E. Meierotto to Evert Pyatt, Oct. 21, 1977, CINS Central Files, D3219.

11. SERO Director, Planning and Assistance, to U.S. Fish and Wildlife Service, Feb. 26, 1979, CINS Central Files, L76.

12. Hillestad et al. 1975, 27–31.

13. NPS 1971, 18–19; Hillestad et al. 1975, 165–66.

14. "An Act to Establish a National Park Service, and for Other Purposes," 39 Stat. 535, Aug. 25, 1916; Horace M. Albright, 1931a, "A Forestry Policy for the National Parks," Records of Arno Cammerer, entry 18, RG 79, NA; Horace M. Albright, 1931b, "The National Park Service's Policy on Predatory Mammals," *Journal of Mammalogy,* 12, 2, 185–86; George M. Wright, Joseph S. Dixon, and Ben H. Thompson, 1932, *Fauna of the National Parks of the United States: A Preliminary Survey of Faunal Relations in National Parks,* NPS and Government Printing Office; Newton B. Drury, 1943, "The National Park Service in Wartime," *American Forests,* Aug. issue; Leopold et al. 1963.

15. Bert C. Roberts to Mrs. R. W. Ferguson, May 12, 1975, CINS Central Files, N1427.

16. Ibid.; "Cumberland Island Developed . . . or Saved," *Miami Herald,* Dec. 22, 1973.

17. NPS, May 1, 1991, *Historic Listing of National Park Service Officials,* Washington, D.C.; "NPS Names Superintendent for Cumberland Island," *Tomorrow,* Feb. 1, 1975.

18. Bert C. Roberts to Mrs. R. W. Ferguson, May 12, 1975, Thornton Morris to Bert C. Roberts, May 27, Aug. 4, 1975, CINS Lands Files under "L. Ferguson"; Bert C. Roberts to Mrs. R. W. Ferguson, Oct. 31, 1975, CINS Central Files, N1427.

19. Bert C. Roberts to Mrs. R. W. Ferguson, Dec. 5, 1975, CINS Central Files, N1427; Thornton Morris interviewed by author, Cumberland Island, July 20, 2000; Superintendent, CINS, to Regional Director, SERO, Jan. 2, 1976, CINS Lands Files under "L. Ferguson."

20. Superintendent to Regional Director, Jan. 2, 1976, CINS Lands Files under "L. Ferguson."

21. "Moving Pigs Just 'Disgruntling,'" *Florida Times-Union,* Jan. 15, 1978; CINS, 1978a, "Report on Feral Pigs," CINS Library, Subject Files under "Hogs."

22. CINS, 1976, "Superintendent's Annual Report for 1975"; CINS, Oct. 1979, "Wilderness Recommendation," CINS Library, 12; D. L. Stoneburner to Z. Kirkland, Nov. 20, 1978, CINS Central Files, L7617; Superintendent, CINS, to Research Ecologist, SERO, Apr. 13, 1979, ibid., N14.

23. Hillestad et al. 1975, 116–25, 173–74.

24. NPS Director to Chairman, Advisory Board on National Parks, Historic Sites, Buildings, and Monuments, Feb. 17, 1976, CINS Central Files, N1427.

25. Acting Director, SERO, to Superintendent, CINS, Aug. 7, 1975, Acting Superintendent, CINS, to Regional Director, SERO, Aug. 26, 1975, CINS Library, Subject Files under "Least Terns."

26. Paul F. McCrary to T. Destry Jarvis, Nov. 29, 1976, CINS Central Files under "Turtles."

27. Ibid.

28. Chief, SERO Land Acquisition Division, to Regional Director, SERO, Apr. 16, 1979, CINS Captain's House Files, L1425.

29. NPS 1971, 20; Chief Historian, NPS, to Dr. Connally, Jan. 19, 1976, CINS Central Files, H30.

30. "An Act for the Preservation of American Antiquities, and for Other Purposes," 34 Stat. 225, June 8, 1906; "An Act to Provide for the Preservation of Historic American Sites, Buildings, Objects, and Antiquities of National Significance, and for Other Purposes," 49 Stat. 666, Aug. 21, 1935; "An Act to Establish a Program for the Preservation of Additional Historic Properties throughout the Nation, and for Other Purposes," 80 Stat. 915, Oct. 15, 1966.

31. "Protection and Enhancement of the Cultural Environment," Executive Order 11593, May 13, 1971, *Code of Federal Regulations,* Title 3, 1971–1975 Compilation, 559–62.

32. Ibid.; NPS 1988.

33. NPS Director to Regional Director, SERO, Oct. 10, 1968, to All Field Directors, Aug. 16, 1972, CINS Central Files, H30; Edwin Bearss to Supervisory Historian Luzader, Aug. 26, 1974, Regional Director, SERO, to Superintendent, CINS, Sept. 13, 1974, NPS, Park History Office Files, Washington, D.C., under "Cumberland Island."

34. Ehrenhard 1976.

35. Torres 1977; Louis Torres to John F. Luzader, Apr. 20, 1976, CINS Central Files under "L. E. Brown Correspondence."

36. Henderson 1977a; David G. Henderson, 1977b, "Architectural Data Section [and] Historic Structure Report, Recreation/Guest House, Dungeness Historic District, Cumberland Island National Seashore, Camden County, Georgia," DSC Historic Preservation Division; Henderson 1977c; Ehrenhard and Bullard, 1981.

37. Mary R. Bullard, 1982, *An Abandoned Black Settlement on Cumberland Island, Georgia* (De Leon Springs, Fla.: E. O. Painter Printing Company); Mary R. Bullard, 1984, *Pierre Bernardy of Cumberland Island,* NPS; Bullard, 1995.

38. Historical Architect, SERO, to Associate Director, SERO, Dec. 11, 1975, CINS Central Files, H30.

39. Ibid.

40. John C. Garner to Associate Regional Director, SERO, Mar. 3, 1976, ibid.

41. Historian, Planning and Design Division, SERO, to Associate Regional Director, Professional Services, SERO, Dec. 3, 1975, SERO Planning Files under "Cumberland Island General Correspondence."

42. Ibid.

43. Lenard Brown, May 1975, "Trip Report on Cumberland Island," ibid.

44. CINS, 1977, "Superintendent's Annual Report for 1976"; Historical Architect, SERO, to Chief, Planning Division, SERO, Nov. 14, 1975, SERO Planning Files under "Cumberland Island General Correspondence"; Henderson 1977a, 123.

45. John Bryant to Ron Walker, Jan. 21, 1974, NPS, Park History Office Files, Washington, D.C., under "Cumberland Island"; QRC Research Corporation, Aug. 1, 1974, "Feasibility of Establishing a Specialized Conference Center at Plum Orchard, Cumberland Island, Georgia," report to the National Park Foundation, SERO Planning Files under "Cumberland Island General Correspondence."

46. James L. Bainbridge to Regional Director, SERO, June 26, 1975, CINS Central Files, H30.

47. Nancy Copp interviewed by Joyce Seward, CINS, Mar. 25, 1996, transcript in the CINS Archives.

48. Henderson 1977b, 37–46.

49. Ibid., 49–69, 76.

50. Camden County Historical Commission to the NPS, Jan. 28, 1980, CINS Library, Subject Files under "General History"; Historical Architect, SERO, to Associate Regional Director, SERO, Mar. 10, 1976, CINS Central Files, H30.

51. SERO, 1980, "Briefing Statement, Southeast Region Historic Preservation Program," CINS Central Files, H42.

52. Stephen A. Deutschle, 1974, "Preliminary Archaeological Reconnaissance of the Cumberland Island Mainland Development," NPS, Southeast Archaeological Center, Tallahassee, Fla.

53. George Sandberg to Director, SERO, Oct. 26, 1972, CINS Lands Files under "Georgia Reservations"; Donald L. Crusoe to Chief, Southeast Archaeological Center, Sept. 26, 1974, CINS Captain's House Files under "Archaeology—Cumberland Island"; CINS, 1974, "Environmental Assessment Land Exchange between United States and Coleman C. Johnston," SERO Planning Files under "Cumberland Island"; Lewis H. Larson Jr. to Jackson O. Lamb, Dec. 5, 1974, Robert Stanton to Robert Garvey, Mar. 13, 1975, ibid. under "Cumberland Island General Correspondence"; John D. McDermott to Robert Stanton, Mar. 21, 1975, includes Memorandum of Agreement, NPS, Park History Office Files, Washington, D.C., under "Cumberland Island."

54. T. M. C. Johnston to Bert Roberts, Oct. 16, 1975, CINS Captain's House Files, H30; Regional Director, SERO, to Regional Solicitor, Atlanta Region, June 10, 1975, CINS Central Files, H30.

55. Acting Regional Director, Atlanta Region, to Regional Director, SERO, July 15, 1975, NPS, Park History Office Files, Washington, D.C., under "Cumberland Island"; Donald Spillman to Regional Director, SERO, Dec. 21, 1976, CINS Central Files, L1425.

56. Henry W. Pfanz to Associate Director, Professional Services, Dec. 11, 1975, NPS, Park History Office Files, Washington, D.C., under "Cumberland Island."

57. Historian, Planning and Design Division, SERO, to Associate Regional

Director, SERO, Dec. 3, 1975, SERO Planning Files under "Cumberland Island General Correspondence."

58. Ibid.

59. Historic Architect, SERO, to Associate Director, Professional Services, SERO, Mar. 10, 1976, NPS, Park History Office Files, Washington, D.C., under "Cumberland Island"; CINS, Sept. 23, 1976, *Environmental Review on Environmental Assessment for General Management Plan and Wilderness Study, Cumberland Island National Seashore, Georgia,* SERO Regional Solicitor's Files under "Cumberland Island," 2.

60. Acting Regional Director, SERO, to Manager, DSC, Feb. 28, 1977, CINS Central Files under "L. E. Brown Correspondence"; CINS, 1982, "Cultural Resources Management Plan," ibid., H42; Barry Mackintosh, 1986, *The National Historic Preservation Act and the National Park Service: A History,* NPS, Park History Office, Washington, D.C., 38–39.

61. David M. Sherman and Lewis H. Larson Jr. to David D. Thompson, Mar. 28, 1977, CINS Central Files under "L. E. Brown Correspondence."

62. Acting Keeper of the National Register to Acting Federal Representative, NPS, June 17, 1977, ibid., H30.

63. Ibid.; Historian, Planning and Compliance, SERO, to Associate Regional Director, Planning and Assistance, SERO, Aug. 24, 1977, Superintendent, CINS, to Regional Director, SERO, Aug. 4, 1977, ibid.

64. Charles H. Badger to SERO, Feb. 24, 1978, SERO Cultural Resources Files under "Cumberland Island," 3; SERO, Mar. 2–3, 1978, "Record of a Telephone Call from Len Brown to Don Klima," CINS Central Files under "L. E. Brown Correspondence"; Joe Brown to Holly Miller, Mar. 14, 1978, ibid., H4217.

65. Chief, Planning and Compliance Division, SERO, to John Murphy, Apr. 9, 1978, CINS Central Files, H4217.

66. CINS, 1981, "Briefing Statement for Superintendent [on cultural resources and National Register nominations]," CINS Central Files under "Briefing Statements"; CINS, 1984, *General Management Plan,* CINS Library; CINS, 1978b, *Draft Historic Resource Management Plan, Cumberland Island National Seashore,* CINS Central Files, H30.

67. Chief, Cultural Preservation, to Regional Director, SERO, Feb. 26, 1980, includes a report entitled "Unity Meeting—Historic Preservation Program, Cumberland Island National Seashore," CINS Central Files, H30.

68. Edwin Bearss to John Murphy, Oct. 31, 1974, includes "Cumberland Island Project No. 1082," NPS, Park History Office Files, Washington, D.C., under "Cumberland Island."

69. CINS 1981; Elizabeth A. Lyon to James L. Bainbridge, Feb. 17, 1978, CINS Central Files, H4217.

70. Carol Ruckdeschel to Joe Brown, June 13, 1979, Joe Brown to Carol Ruckdeschel, Sept. 17, 1979, ibid., under "L. E. Brown Correspondence."

71. CINS, Mar. 20, 1979, "Briefing Statement for Director Whalen," ibid., under "Planning Briefs"; F. Ross Holland Jr. to Regional Director, SERO, Jan. 14, 1980, ibid., H32.

72. CINS 1981; Acting Regional Director, SERO, to Superintendent, CINS, Feb. 25, 1983, includes revisions and additions to the list of classified historic structures, CINS Central Files, H42.

73. "Cumberland Island National Seashore, 1972–1979," 1979.

7. *Contested Paradise: The 1980s and Early 1990s*

1. James A. Giammo, 2000, "Cumberland Island National Seashore, Georgia, Operating Bases from 1974 to 1999," NPS, Office of the Budget, Washington, D.C.

2. CINS, Dec. 1980, *Final Environmental Impact Statement, General Management Plan, [and] Wilderness Recommendation, Cumberland Island National Seashore,* D-1194B, CINS Library.

3. Ibid.; NPS, 1971, *Master Plan;* SERO, 1981, "Park Service Nearing Approval of Plan for Cumberland Island," press release in SERO Planning Files under "Cumberland Island National Seashore General Correspondence."

4. Robert Coram, Mar. 7, 1981, "Can Georgia's Island Jewel Be Preserved and Developed?" *Atlanta Journal and Constitution;* "Save Cumberland," *Atlanta Constitution,* Mar. 17, 1981; "Conservationists Divided on Park Plan," *Rome News-Tribune,* Mar. 17, 1981.

5. Patricia H. Koester to Superintendent, CINS, Mar. 28, 1981, CINS Central Files, A26-A2623.

6. Coram, Mar. 7, 1981; "Save Cumberland," Mar. 17, 1981; Bob Ingle, Mar. 18, 1981, "Turnabout for Some Conservationists," *Atlanta Constitution;* G. Robert Kerr, Mar. 13, 1981, "Conservancy Explains Action," ibid.; "Conservationists Divided . . . ," Mar. 17, 1981; Brown's comment about Coram quoted in Lucy Y. Herring, 1989, "Cumberland Island National Seashore: The 1981 Controversy over the Proposed General Management Plan," MS, Thornton Morris Papers, file 4719.

7. Kerr, Mar. 13, 1981; Ingle, Mar. 18, 1981; Hans Neuhauser, Mar. 20, 1981, "Wilderness for Cumberland," *Atlanta Constitution;* "Cumberland Saved," ibid., Mar. 19, 1981.

8. Ronald "Bo" Ginn to Hal Gulliver, Apr. 9, 1981, SERO Cultural Resources Files under "Cumberland Island National Seashore General Correspondence."

9. "Cumberland Saved," Mar. 19, 1981; SERO, Mar. 18, 1981, "Statement by Southeast Regional Director Joe Brown on Cumberland Island General Management Plan and Wilderness Study," press release in SERO Cultural Resources Files under "Cumberland Island National Seashore General Correspondence"; NPS, May 1991, *Historic Listing of National Park Service Officials,* Washington, D.C.

10. SERO, Mar. 18, 1981; G. Robert Kerr, Apr. 1981, "Cumberland Plan Beached," *Georgia Conservancy Newsletter,* 10, 4, Apr. 1.

11. Hans Neuhauser interviewed by Joyce Seward, Athens, Ga., Nov. 21, 1995, transcript in CINS Archives; CINS 1980, 9; U.S. House of Representatives, Dec. 10, 1981, *Correcting the Boundary of Crater Lake National Park in the State of Oregon, and for Other Purposes,* H. Rep. 97-383, 97th Cong., 1st sess., 4–5; "Island Wilderness Bill Submitted," *Savannah Morning News,* Oct. 8, 1981; H.R. 4713, 97th Cong., 1st sess., 1981.

12. U.S. House of Representatives, Oct. 16, 1981, *Additions to the National Wilderness Preservation System,* hearing before the Subcommittee on Public Lands and National Parks of the Committee on Interior and Insular Affairs on H.R. 1716, H.R. 3630, and H.R. 4713, 97th Cong., 1st sess., Serial no. 97-9, pt. 4, 85–86.

13. Ibid., 94–100.

14. Ibid., 109.

15. Ibid., 89–93.

16. See the legislative history for Public Law 97-250, "An Act to Correct the Boundary of Crater Lake National Park in the State of Oregon, and for Other Purposes," 96 Stat. 709, Sept. 8, 1982; U.S. House of Representatives, Dec. 10, 1981; Donald Paul Hodel to Morris K. Udall, Dec. 15, 1981, CINS Central Files, L48.

17. "Cumberland Island Bill Put on Hold," *Savannah Morning News,* Dec. 17, 1981; "Turning Point for Cumberland Island," *Chattahoochee Sierran,* 7, 12, June 1982.

18. U.S. Senate, June 24, 1982, *Conveyance of Certain National Forest System Lands and Additions to the National Wilderness Preservation System, Georgia,* hearing before the Subcommittee on Public Lands and Reserved Water of the Committee on Energy and Natural Resources on S. 705 and S. 2569, 97th Cong., 2d sess.

19. S. 705, 97th Cong., 2d sess., 1982.

20. U.S. Senate, June 24, 1982, "Conveyance of Certain . . . ."

21. "Camden Attacks Cumberland Plan to Restrict Boats," *Florida Times-Union,* July 10, 1982.

22. "Cumberland Boat-In to Protest NPS Plan," *Camden County Tribune,* July 22, 1982; "Boat-In Turnout Pleases Sutton," *Southeast Georgian,* Aug. 5, 1982.

23. "House Urges Park Service Cooperation," *Southeast Georgian,* Aug. 12, 1982; U.S. Senate, Aug. 18, 1982, *Cumberland Island National Seashore, Georgia, Wilderness,* Committee on Energy and Natural Resources Report no. 97-531, 97th Cong., 2d sess.; "Reagan Signs Bill for Island," *Florida Times-Union,* Sept. 11, 1982; Neuhauser 1995, 30.

24. "Statement by the President," Sept. 9, 1982, the White House, Office of the Press Secretary, CINS Central Files, A4027.

25. "An Act to Establish a National Wilderness Preservation System for the Permanent Good of the Whole People, and for Other Purposes," 78 Stat. 890, Sept. 3, 1964.

26. K. O. Morgan to Stephen F. McCool, Mar. 22, 1989, CINS Central Files, N1623. Although Morgan refers to twenty-one estate reservations, there are, in fact, only eighteen. Furthermore, motorboats are not permitted to beach on the island but may anchor off the shore after disembarking passengers. Although Cumberland Island National Seashore is a drastically altered landscape, the Eastern Wilderness Act established that cutover or otherwise heavily used lands could enter the system if they were on their way to recovery. See "An Act to Further the Purposes of the Wilderness Act by Designating Certain Acquired Lands for In-

clusion in the National Wilderness Preservation System, to Provide for Study of Certain Additional Lands for Such Inclusion, and for Other Purposes," 88 Stat. 2096, Jan. 3, 1975.

27. Superintendent, CINS, to Public Information Officer, SERO, Oct. 2, 1984, CINS Central Files, A72; K. O. Morgan to Don Nixon, Feb. 18, 1983, ibid., N1623; K. O. Morgan to Cathy Reynolds, May 16, 1986, ibid., A36; James D. Absher, 1988, "Down by the Seashore: Resource Management and Social Conflict on the Georgia Coast," paper delivered at the annual meeting of the Rural Sociology Society, Athens, Ga., Aug. 20–23, transcript in ibid., N4615; Susan P. Bratton, June 1986, *Foot and Vehicle Traffic Patterns: The Main Road, Cumberland Island National Seashore,* Institute of Ecology, UGa, CPSU Report no. 28.

28. Thornton Morris interviewed by author, CINS, Oct. 6, 1999; "Reader Addresses Greyfield Misconceptions," *Camden County Tribune,* Oct. 9, 1996. Island resident and historian Mary Bullard later maintained that the intense scrutiny by environmental activists was caused by the actions of Charles Hauser. After he successfully defied the Park Service and modified historic structures on the reserved estate he rented from T. M. C. Johnston, Hauser decided to build a set of rental apartments. Bullard noted that the environmental groups were outraged and drew the "benignly-blind eye of the National Park Service" to the problem (Mary Bullard to Thornton Morris, Apr. 1, 1987, CINS Central Files, A3821).

29. Superintendent, CINS, to Regional Director, SERO, Feb. 25, 1983, CINS Central Files, L48.

30. Ibid.; Acting Regional Director, SERO, to Superintendent, CINS, Mar. 18, 1983, ibid.

31. G. Robert Kerr et al. to Robert Baker, Apr. 5, 1983, CINS Resource Management Files under "Wilderness."

32. Thornton Morris, Oct. 6, 1999; Karen Langshaw to Mary R. Bullard, Mar. 26, 1987, Superintendent, CINS, to Regional Director, SERO, Aug. 21, 1986, Mary R. Bullard to K. O. Morgan, Apr. 29, 1987, CINS Central Files, A3821; see also the files for the individual estates and retained-rights agreements in the CINS Archives.

33. Thornton Morris, Oct. 6, 1999.

34. Roger Sumner Babb to Robert M. Baker, Aug. 17, 1987, CINS Central Files, C38.

35. See CINS's Land Protection Plans and updates for 1985, 1986, 1989, 1991, and 1994, ibid., L14.

36. Carol Ruckdeschel to Ken Morgan, Apr. 3, 1985, ibid., A3821; CINS, Sept. 18, 1981, "Cumberland Island, Summary Coastal Processes Workshops," SERO Cultural Resources Files under "Cumberland Island."

37. "Bike Use Limited at Point Reyes," *San Jose Mercury News,* July 8, 1985; *Federal Register,* 51, 115, June 6, 1986, 21844; Rolland Swain to Joseph M. Schmidt, Apr. 13, 1995, CINS Central Files, A36.

38. Carol Ruckdeschel and C. R. Shoop, July 1991, "Cumberland Island Management," "fact sheet," CINS Library, Subject Files under "Wilderness Manage-

ment"; Peter Kirby, Hans Neuhauser, and Bill Mankin to Paul Swartz, July 31, 1990, CINS Central Files, N1623.

39. "Polishing a Plan for Georgia's Jewel," *Atlanta Constitution,* June 29, 1981.

40. Hans Neuhauser to Robert Baker, Aug. 3, 1981, includes a summary of the Cumberland workshop hosted by the Georgia Conservancy, SERO Planning Files under "Cumberland Island General Correspondence."

41. "Cumberland Island's 300 Daily Visitors Judged Enough," *Florida Times-Union,* Sept. 15, 1981; SERO, Sept. 18, 1981, "Cumberland Island Summary, Coastal Processes Workshop," SERO Planning Files under "Cumberland Island."

42. "A New Morale on Cumberland Island," *Atlanta Constitution,* Sept. 25, 1981.

43. "Cumberland Management Criticized," *Southeast Georgian,* Sept. 24, 1981; "Camden Longs for Access to Cumberland," *Florida Times-Union,* Oct. 18, 1981.

44. NPS, Nov. 19, 1981, *Summary of the Revised General Management Plan for Cumberland Island National Seashore, Georgia,* copy in SERO Planning Files.

45. "Revised Cumberland Plan Is Called 'The Ultimate,'" *Atlanta Constitution,* Nov. 20, 1981; "Island Revisions Praised," *Florida Times-Union,* Nov. 20, 1981.

46. "Cumberland Island Park Service 'Quiet' on Its Review," *Savannah Morning News,* Nov. 29, 1981.

47. Ibid.; William E. Mankin to Robert M. Baker, Dec. 13, 1981 (includes "Comments on Summary of the Revised General Management Plan for Cumberland Island National Seashore, Georgia"), Hans Neuhauser to Robert M. Baker, Dec. 17, 1981, SERO Cultural Resources Files under "Cumberland Island."

48. Hans Neuhauser to Friends of Cumberland Island, Mar. 16, 1982, CINS Central Files, A4031.

49. Robert M. Baker to Robert R. Garvey Jr., Nov. 4, 1982, Elizabeth Lyon to Don Klima, Jan. 27, 1984, Hans Neuhauser to Neal Guse, Sept. 21, 1983, and SERO, Feb. 21, 1983, "Record of Decision, General Management Plan, Cumberland Island National Seashore, Georgia," SERO Cultural Resources Files under "Cumberland Island."

50. "An Act to Amend the Boundaries of Cumberland Island National Seashore, and for Other Purposes," 97 Stat. 1116, Nov. 29, 1983; Chief, Land Resources Division, NPS, to SERO Director, Jan. 11, 1984, CINS Central Files, L1417.

51. DSC, NPS, Apr. 1977, "A Planning Analysis of Mainland Base Sites for Cumberland Island National Seashore," CINS Library; CINS, Dec. 1980, 14–17; Vollmer Associates, 1975, "Mainland Site Alternatives, Cumberland Island National Seashore," report to the NPS, CINS Superintendent's Office.

52. "Point Peter Owner against Sale," *Savannah Morning News,* Oct. 9, 1976; Frederick G. Storey to Herman E. Talmadge, Oct. 29, 1976, Jean Lucas Storey to Neal G. Guse Jr., Mar. 21, 1980, CINS Lands Files under "Storey."

53. Roberts and Eichler, Inc., 1979, "A Plan for St. Marys Historic Waterfront," a report to the Coastal Area Planning and Development Commission and the St. Marys Waterfront Development Committee, CINS Lands Files.

54. CINS, July 1985, *Draft Development Concept Plan and Environmental Assessment, Cumberland Island, St. Marys Waterfront,* CINS Library; SERO, Dec. 6, 1985, "Plan Approval and Finding of No Significant Impact for Development Concept Plan, St. Marys Waterfront," CINS Central Files, L3215.

55. [First name missing] Hardy, 1985, "Downtown/Waterfront Revitalization for St. Marys: A Development Strategy." This is a portion of a larger plan with an attached note indicating the author's last name, SERO Planning Files under "Cumberland Island–St. Marys DCP," 11.

56. "Akle Outlines His Plans for Waterfront," *Southeast Georgian,* Nov. 13, 1986; "Miller's Dock: Landmark and a Battleground," *Camden County Tribune,* Aug. 4, 1989; Superintendent, CINS, to Steve Milton, Aug. 11, 1989, James E. Stein to Bob Baker, Aug. 3, 1989, CINS Central Files under "St. Marys Waterfront."

57. K. O. Morgan to Regional Director, SERO, Nov. 8, 1988, CINS Central Files, L1425.

58. James Stein to Bob Baker, Aug. 3, 1989.

59. "Miller's Dock: Landmark," Aug. 4, 1989.

60. "St. Marys Nixes Pavilion Giveaway," *Camden County Tribune,* June 15, 1990; "Rep. Thomas Addresses Current Issues," ibid., Aug. 18, 1989; Rolland Swain to Lindsay Thomas, Dec. 3, 1991, CINS Central Files, A3815.

61. "Miller's Dock Belongs on Register," *Camden County Tribune,* July 8, 1992; "Group Approves Plan to Replace Miller's Dock," ibid., Aug. 26, 1992; Institute for Environmental Negotiation, Univ. of Virginia, Oct. 1993, "Report on the St. Marys Waterfront Committee," CINS Central Files, A44, 4–6.

62. For summaries of the resource management issues, see CINS's Resource Management Plans for 1984, 1994, and 2000, CINS Library.

63. CINS, Aug. 28, 1981, "Southcut Fire Review," summary of an interagency meeting, CINS Central Files, Y26.

64. Ibid.

65. Ibid.; "Fire Blazes Again As Investigation Begins," *Southeast Georgian,* July 30, 1981.

66. CINS, Aug. 28, 1981.

67. "Mattingly Critical of Firefighting on Island," *Atlanta Journal,* July 22, 1981; "Mattingly Rips Report on Fire on Sea Island," *Savannah Morning News,* Sept. 9, 1981; "Embattled Boss of Cumberland Park Transferred," *Atlanta Journal and Constitution,* Aug. 1, 1981; "Superintendent Named for Cumberland Island," ibid., Sept. 12, 1981.

68. "Forestry Commission Willing to Negotiate with NPS," *Camden County Tribune,* Sept. 24, 1981; CINS, Oct. 30, 1981, "Cooperative Agreement for Fire Control on Cumberland Island and Little Cumberland Island between the U.S. Department of the Interior, Cumberland Island National Seashore, and State of Georgia, Georgia Forestry Commission," CINS Central Files, Y26; John E. Ehrenhard to Chief, Southeast Archaeological Center, Aug. 5, 1981, ibid.; H. B. Jordan and C. Ruckdeschel, Nov. 1982, "Observations on the Cumberland Island Fire of 16 July–24 Aug., 1981, and Subsequent Recovery, Aug. 1981–Aug. 1982: Fi-

nal Report Summary," ibid.; Kathryn Louise Davison, 1984, *Vegetation Response and Regrowth after Fire on Cumberland Island National Seashore, Georgia,* SERO Research/Resource Management Report SER-69; Sally Turner, 1985, *The Fire History of Cumberland Island National Seashore, 1900–1983,* Institute of Ecology, UGa, CPSU Technical Report no. 7.

69. Superintendent, CINS, to Director, Office of Environmental Review, NPS, July 9, 1980, CINS Central Files, L76; James H. Rathesberger to Commander E. R. Wilson, July 22, 1980, to E. R. Wilson, Oct. 24, 1980, E. R. Wilson to James H. Wilson, Jan. 23, 1981, ibid., D3219.

70. C. W. Ogle to Mack Mattingly, Mar. 8, 1985, SERO Cultural Resources Files under "Cumberland Island."

71. Ibid.; Crayton J. Lankford to Colonel Daniel W. Christman, Mar. 29, 1985, CINS Central Files, N2219; Richard C. Downing to U.S. Department of the Interior, NPS, Feb. 8, 1985, Robert Dolan to William Smith, Feb. 8, 1985, ibid., D3219; "An Act to Protect and Conserve Fish and Wildlife Resources, and for Other Purposes," 96 Stat. 1653, Oct. 18, 1982.

72. Michael J. Harris, July 1986, "Fish and Wildlife Coordination Act Report on Reroute of the Atlantic Intracoastal Waterway, Cumberland Sound, Georgia," U.S. Fish and Wildlife, Southeast Region, Atlanta, CINS Library.

73. Hans Neuhauser to Colonel Daniel W. Christman, Feb. 22, 1985, CINS Central Files, D3219; Charles W. McGrady to Commanding Officer, Naval Submarine Base, Kings Bay, Georgia, Apr. 23, 1985, ibid., A38; Sierra Club, Georgia Chapter, Oct. 24, 1986, "Conservationists Criticize Waterway Project," press release, ibid., D3219; David W. Carr Jr. to Kenneth O. Morgan Jr., Oct. 24, 1986, attached to a Southern Environmental Law Center report on relocating the waterway, ibid.; "Navy Scraps Plans to Reroute Intracoastal Waterway," *Florida Times-Union,* Mar. 3, 1987.

74. Stephen V. Cofer-Shabica, Darrell Molzan, and Joan Pope, July 1991, "Biological and Physical Aspects of Dredging, Kings Bay, Georgia, USA," in *Biological and Physical Aspects of Dredging Kings Bay, Georgia/Coastal Zone, '91 Conference, ASCE, Long Beach, California,* and Lindsay D. Nakashima, 1991, *Marsh, Mudflat, and Tidal Creek Assessment, Cumberland Island National Seashore,* Kings Bay Monitoring Program Report KBEMP-91/01, CINS Resource Management Library; "Cumberland Not Washing Away, Study Finds," *Florida Times-Union,* July 6, 1993.

75. Daniel J. Hippe, Feb. 2, 1999, "Cumberland Island National Seashore, Georgia, Project Plan for Level 1 Water-Quality Inventory and Monitoring," U.S. Geological Survey, CINS Resource Management Files under "Water & Resources Inventory," Abstract and 2–4; James B. Mack, Feb. 1994, *Field Investigation of Saltwater Intrusion, Cumberland Island, Georgia,* Georgia State Univ., Technical Report KBRPT 94/01, 92–96.

76. "Man and the Biosphere Program," 1999, UNESCO Internet site http://www.unesco.org/mab.

77. Ibid.; William Gregg, telephone interview with author, Feb. 25, 1997;

William Gregg to Carleton Ray, Aug. 6, 1982, NPS, Natural Resources Division Files, Washington, D.C., under "Man and the Biosphere."

78. Newton Sikes to Heather Smith, June 16, 1997, CINS Central Files, N16.

79. Guy R. McPherson, Aug. 1988, *Boundary Dynamics on Cumberland Island National Seashore*, Institute of Ecology, UGa, CPSU Technical Report no. 49; Davison 1984; Turner 1985; Susan P. Bratton and NPS Basic Vegetation Management Course, Oct. 1986, *Experimental Control of Tung Trees at Cumberland Island National Seashore*, ibid., no. 29; D. L. Stoneburner, Jan. 1981, *Summary of the Loggerhead Sea Turtle Research Conducted at Canaveral National Seashore, Cumberland Island National Seashore, and Cape Lookout National Seashore: A Final Report*, Institute of Ecology, UGa, Research/Resources Management Report no. 39; James I. Richardson, June 12, 1992, *Final Report: An Investigation of Survivorship, Mortality, and Recruitment of Adult Female Loggerhead Sea Turtles Nesting on Cumberland Island National Seashore, Georgia, 1987–1991*, Institute of Ecology, UGa, U.S. Fish and Wildlife Work Order no. 11; Larry J. Orsak, Susan P. Bratton, and Robert J. Warren, Feb. 1986, "Impact of Exotic Armadillo and Reintroduced Wild Turkey Populations on Natural Habitats of Cumberland Island National Seashore, Georgia," research proposal to NPS cooperating research program at Rutgers Univ., CINS Central Files, N1419; Mark E. Fene and David C. Ingram, 1986, "Distribution and Management Recommendations for Cumberland Island Turkeys," School of Forest Resources, UGa, ibid.; Barb Zoodsma and Susan P. Bratton, Mar. 27, 1986, *The West Indian Manatee (Trichechus manatus) in Cumberland Sound, Camden County, Georgia*, Institute of Ecology, UGa; Edward B. Harris, 1984, *Possible Extirpated Mammals and Reptiles of Cumberland Island, Georgia: A Survey of Five Species Being Considered for Reintroduction*, Institute of Ecology, UGa, CPSU Technical Report no. 9; Dean M. Simon, Monica G. Turner, Kit L. Davison, and Susan P. Bratton, 1984, *Habitat Utilization by Horses, Deer, and Rabbits on Cumberland Island National Seashore, Georgia*, ibid., no. 8; Charles R. Ford, 1987, *Spotlight Survey for White-Tailed Deer Population Trends on Cumberland Island National Seashore, Georgia*, ibid., no. 42; Robert J. Warren, Sept. 30, 1988, *Ecology of Feral Hogs on Cumberland Island National Seashore, Georgia*, School of Forest Resources, UGa, Cooperative Agreement CA-5000-48-005, Amendment no. 8; J. Ambrose, S. P. Bratton, K. Davison, J. Fitch, M. Goigel, F. Golley, F. Lemis, J. McMurtray, W. Querin, and D. Simon, 1983, *An Analysis of Feral Horse Population Structure on Cumberland Island*, Institute of Ecology, UGa, CPSU Technical Report no. 1; Monica Goigel Turner, May 1986, *Effects of Feral Horse Grazing, Clipping, Trampling, and a Late Winter Burn on a Salt Marsh, Cumberland Island National Seashore, Georgia*, ibid., no. 23; Susan P. Bratton, Oct. 1988, *Wood Stork Use of Fresh and Salt Water Habitats on Cumberland Island National Seashore*, ibid., no. 50; Susan P. Bratton, Mar. 1989, *Responses of Wading Birds to Natural and Unnatural Disturbances in Cumberland Sound*, ibid., no. 53.

80. Harris 1984.

81. SERO, 1988, "Finding of No Significant Impact on Environmental Assessment, Reintroduction of Bobcats, Cumberland Island National Seashore, Geor-

gia," CINS Central Files under "EA/FONSI Bobcats"; Robert J. Warren, Michael J. Conroy, William E. James, Leslie A. Baker, and Duane R. Diefenbach, 1990, "Reintroduction of Bobcats on Cumberland Island, Georgia: A Biopolitical Lesson," *Transactions of the 55th North American Wildlife and Natural Resources Conference, 1990,* 580–89; Robert J. Warren, n.d., "Turkey Mortality—With Special Reference to Cumberland Island and the Reintroduction of Bobcats," literature summary, CINS Central Files, L7617.

82. Warren, Sept. 30, 1988, 3–13; Hillrie Quin Jr. interviewed by Joyce Seward, Atlanta, Aug. 18, 1995, transcript in CINS Archives; CINS, Feb. 10, 1982, "Case Incident Report—Feral Hogs," Report 820002, CINS Law Enforcement Files; K. G. Kacer to RM & VP Staff, Aug. 28, 1984, CINS Central Files, N14.

83. Superintendent, CINS, to the Files, May 20, 1992, CINS Central Files, N2219; "Park Ends Hog Hunts with Dogs," *Florida Times-Union,* May 17, 1992; Rolland Swain to Beverly Babb, May 19, 1992, to Cheryl Washburn, May 19, 1992, Celenda H. Perry to Superintendent, CINS, June 6, 1992, CINS Central Files, N2219; "Park Service Reprimands Hog-Handlers," *Camden County Tribune,* July 8, 1992.

84. Rolland R. Swain to Celenda H. Perry, July 9, 1992, CINS Central Files, N2219; CINS, Aug. 1992, *Draft Environmental Assessment, Feral Hog Removal, Cumberland Island National Seashore,* ibid., under "Hog Removal"; CINS, Oct. 20, 2000, *Environmental Assessment of Management Alternatives, Feral Hog Population Control, Cumberland Island National Seashore,* ibid., N2219; John W. Crenshaw Jr., Ph.D., to Superintendent Rolland Swain, Sept. 29, 1992, ibid.; Thornton Morris to Carolyn Boyd Hatcher, Feb. 1, 1993, ibid., under "Hog Removal."

85. Susan P. Bratton, Mar. 5, 1983, *Feral Horse Census on Cumberland Island National Seashore,* Institute of Ecology, UGa, CINS Central Files, N2219; Robin B. Goodloe interviewed by Joyce Seward, Brunswick, Ga., July 25, 1995, transcript in CINS Archives; Robin B. Goodloe, 1991, "Census of the 1991 Feral Horse Population, Cumberland Island, Georgia," CINS Central Files, N2219.

86. Susan P. Bratton to Jay Gogue, Apr. 23, 1985, CINS Central Files, N1427; Goodloe, July 25, 1995; Robert J. Warren, Robin B. Goodloe, and Charles R. Ford, Apr. 1, 1986, "Immunosterilization and Genetic Management of Barrier Island Horse Populations," CINS Central Files, N2621.

87. Deputy Associate Regional Director, Natural Resources and Science, SERO, to Superintendents of Assateague Island, Cape Lookout, and Cumberland Island National Seashores, 1993 (no specific date), "Feral Horse Management Planning Project Update," CINS Central Files, N16; Goodloe, July 25, 1995; "Cumberland Horses Die from Encephalitis," *Camden County Tribune,* July 24, 1991; "Assateague Island Mares 'Shot' with Contraceptives," *Park Science,* 14, 3, unpaged; Bryan K. Burkingstock, Jason E. Long, and Linda M. Vallance, Nov. 22, 1994, "Management Plan for a Remnant Herd of Feral Horses on Cumberland Island," School of Forest Resources, UGa; CINS, Jan. 1995, "Feral Horse Management Planning Process," CINS Central Files, N16.

88. Lauren Lubin Zeichner, Mar. 1988, *Landscape Management Plan for Dungeness, Cumberland Island National Seashore, Georgia,* Institute of Ecology, UGa, CPSU Technical Report no. 44; Peggy Stanley Froeschauer, Aug. 1989, *The Interpretation and Management of an Agricultural Landscape—Stafford Plantation, Cumberland Island National Seashore, Georgia,* ibid., no. 59.

89. Robert M. Baker to Elizabeth Lyon, Jan. 12, 1984, SERO Cultural Resources Files under "Cumberland Island"; Elizabeth Lyon to Robert Baker, Oct. 26, 1982, CINS Central Files, H4217; Don L. Klima to Bob Baker, Dec. 20, 1982, ibid., H42.

90. Edwin C. Bearss to Chief of Registration Shull, Interagency Resources Division, Aug. 15, 1984, NPS, Park History Office Files, Washington, D.C., under "Cumberland Island"; Bennie C. Keel to Superintendent, CINS, Sept. 15, 1994, CINS Central Files, H2215.

91. CINS, 1981–95, "Superintendent's Annual Report" for 1980–94, CINS Central Files, A2621; James A. Giammo, July 28, 2000, "Line Item Construction Funding through FY 1998," NPS, Office of the Budget Files, Washington, D.C.

92. CINS, Jan. 6, 1988, "Degradation of Cultural Resources," briefing statement, CINS Central Files under "Briefing Statements."

93. John C. Garner Jr. to Superintendent, CINS, 1982 (no specific date), ibid., H42.

94. John E. Ehrenhard to Chief, Southeast Archaeological Center, Aug. 29, 1983, ibid., H2215; Regional Solicitor, SERO, to Regional Director, SERO, Sept. 2, 1983, ibid., H3015; Chief, Southeast Archaeological Center, to Regional Historian, SERO, Sept. 19, 1983, ibid., A9015; Len Brown to Bill Harris, Sept. 22, 1983, ibid., H42; K. O. Morgan to Charles W. McGrady, Oct. 28, 1985, ibid., A38.

95. K. O. Morgan to C. Jones Hooks, Nov. 15, 1983, ibid., H3015; Associate Regional Director, Operations, SERO, to Superintendent, CINS, Jan. 11, 1985, ibid., H30; Michael P. Doelger to John Murphy, Mar. 13, 1984, ibid., H3015; "Cumberland Mansion to Be Repaired," *Florida Times-Union,* Apr. 21, 1983; Bert Rhyne to Superintendent, CINS, June 12, 1985, CINS Central Files, H30.

96. "Cumberland Mansion," Apr. 21, 1983; CINS, Mar. 18, 1983, "Draft Minutes of Meeting Held on Cumberland March 11," CINS Central Files, A40.

97. CINS, Mar. 18, 1983, 8–16.

98. Ibid., 16–28.

99. SERO, Sept. 12, 1983, "Draft Minutes of the First Meeting of Cumberland Island–Plum Orchard Friends," SERO Planning Files under "Cumberland Island—Plum Orchard"; Cumberland Island Historic Foundation, Nov. 18, 1983, "Minutes of Meeting of Nov. 18, 1983," ibid.; CINS, Sept. 1984, *Draft Development Concept Plan and Environmental Assessment, Plum Orchard, Cumberland Island National Seashore, Georgia,* CINS Superintendent's Office; SERO, May 5, 1985, "Plan Approval/Finding of No Significant Impact, Plum Orchard Development Concept Plan/Environmental Assessment, Cumberland Island National Seashore, Georgia," SERO Planning Files under "Cumberland Island—Plum Orchard."

100. SERO, May 10, 1985.

101. Cumberland Island Historic Foundation, May 21, 1984, "Minutes of Meeting of May 21, 1984," SERO Planning Files under "Cumberland Island—Plum Orchard," 5; Margaret M. Graves, Oct. 14, 1995, "Plum Orchard," CINS Central Files, H3021.

102. CINS, Dec. 2000, *Draft Wilderness Management Plan, Appendix 1*, vol. 3, CINS Superintendent's Office, 65; CINS, Aug. 1984, "List of Cultural Landscape Features, Cumberland Island National Seashore," CINS Central Files, H3017.

103. Cumberland Island Historic Foundation, Nov. 9, 1988, "Minutes of Meeting of November 9, 1988," CINS Central Files, A3821; CINS, Oct. 25, 1989, "Staff Meeting Minutes," ibid., A22; Rolland R. Swain interviewed by author, CINS, June 11, 1996; Rodney M. Cook Jr. to Rolland R. Swain, Feb. 4, 1994, CINS Central Files, H3021.

104. Cook to Swain, Feb. 4, 1994; CINS, Jan. 1995, "Plum Orchard Issues Inventory, Draft Memorandum of Understanding Environmental Assessment, Cumberland Island National Seashore," CINS Central Files, A44; "Park Service Plans Hearing on Mansion," *Camden County Tribune,* May 3, 1995; "Park Officials Eye Plum Orchard Past," *Southeast Georgian,* Feb. 8, 1995.

8. Hope for the New Century

1. Former Superintendent, CINS, to Acting and New Superintendent, CINS, Sept. 1996 (no specific date), CINS Superintendent's Office Files.

2. CINS, Sept. 27, 1996, "Fire Management Plan for Cumberland Island National Seashore, 1996," ibid., under "Fire Management Plan" (this plan was never approved by SERO); Stephen Cofer-Shabica et al., Dec. 1997, "Kings Bay Environmental Monitoring Program: A Synthesis," prepared for the Department of the Navy, Technical Report KBEMP-96/01, vi–x, CINS Resource Management Library.

3. Jennifer Bjork, 2000, "Feral Hog Management," *Mullet Wrapper* (CINS employee newsletter), 12, 1, Dec. 3; CINS, Oct. 10, 2000, *Environmental Assessment of the Management Alternatives, Feral Hog Population Control, Cumberland Island National Seashore,* CINS Resource Management Library.

4. Feral Horse Project Coordinator, CINS, to Invited Participants, Jan. 18, 1995, CINS Central Files, N2219; CINS, Mar. 6, 1996, *Draft Environmental Assessment, Alternatives for Managing the Feral Horse Herd on Cumberland Island National Seashore,* ibid.

5. CINS, Apr. 16, 1996, "Public Meetings Scheduled—Topic: Horse Management on Cumberland Island," news release, ibid., K34; "Residents Speak Out for Horse Freedom on Cumberland," *Southeast Georgian,* May 8, 1996; Sonja Olsen Kinard to Rolland Swain, June 5, 1996, CINS Central Files, N2219; Betty Leiter and William Leiter to Roger Kennedy, July 7, 1996, Ruth M. Seppala to Roger G. Kennedy, July 26, 1996, ibid., A3815; Stacey Wasserman, June 9, 1996, "Horses Add Beauty," editorial in *Atlanta Journal and Constitution;* "Wild Horse Protection Program Sends Them to Slaughter," *Florida Times-Union,* Jan. 5, 1996.

6. "Park Service Should Do Its Job and Get Rid of the Horses," *Atlanta Journal and Constitution,* June 2, 1996.

7. "Cumberland Horse Plan Reined In?" *Florida Times-Union,* Aug. 16, 1996; "Horse Plan Haunts Kingston," ibid., Oct. 25, 1996.

8. "Horse Plan Haunts," Oct. 25, 1996.

9. "Artists' Retreat Criticized," *Florida Times-Union,* Mar. 14, 1995; "A Necessary Evil on Cumberland," editorial in *Atlanta Constitution,* May 22, 1995.

10. "Legislator Blasts Park Plan," *Southeast Georgian,* Feb. 15, 1995.

11. Don Barger to Rolland Swain, Mar. 2, 1995, CINS Central Files, H3021.

12. Brian A. Rosborough to GoGo Ferguson and Davis Sayre, Feb. 28, 1995, ibid.; "Artists' Retreat Criticized," Mar. 14, 1995.

13. CINS, Feb. 1996, "Finding of No Significant Impact on Environmental Assessment for Alternatives for Preservation of Plum Orchard Mansion in the Cumberland Island National Seashore between United States of America, Department of the Interior, National Park Service, Southeast Field Area, Cumberland Island National Seashore and The Plum Orchard Center for the Arts on Cumberland Island, Inc.," CINS Central Files, A64; Defenders of Wild Cumberland, Inc. v. United States; Bruce Babbit, Secretary of the Interior; Roger Kennedy, Director National Park Service; Robert Baker, Southeast Field Director, National Park Service; Rolland Swain, Superintendent, National Park Service, Cumberland Island National Seashore, 1996, U.S. District Court for Northern Georgia, case no. 1 96-CV-830-ODE; Robert M. Baker to Ms. Nancy Parrish and Ms. Janet (GoGo) Ferguson, June 14, 1996, CINS Central Files, A22.

14. John Mitchell interviewed by author, St. Marys, Ga., July 3, 2001.

15. Ibid.; Superintendent, CINS, to Chief, Museum Scientists, Harpers Ferry Center, NPS, Jan. 29, 1979, CINS Central Files, H2017; Chief, Division of Museum Services, to Regional Director, SERO (Attn: Bill Kay), Feb. 13, 1979, CINS Museum Files under "Accession No. 1 Plum Orchard."

16. Chief, Contracting and Property Management Division, to Superintendent, CINS, Mar. 6, 1979, Superintendent, CINS, to Chief, Contracting and Property Management Division, SERO, Feb. 23, 1979, CINS Museum Files under "Accession No. 1 Plum Orchard."

17. Mitchell 2001.

18. Ibid.; D. Warren-Taylor, Jan. 8, 1998, "Collection Condition Survey for Cumberland Island National Seashore Museum Collections," 2 vols., NPS contract no. 1443 PX 563097038, CINS Museum Files; author observation.

19. "Atlantan Plans to Build on Cumberland," *Atlanta Journal and Constitution,* Dec. 4, 1996; CINS, Mar. 19, 1998, "Superintendent's Annual Report for 1997," CINS Central Files, A2621; Denis Davis telephone interview with author, Aug. 9, 1997; Art Frederick interviewed by author, St. Marys, Ga., July 2, 2001.

20. "Superintendent's Report 1997"; "An Act to Establish a Land and Water Conservation Fund to Assist the States and Federal Agencies in Meeting Present and Future Outdoor Recreation Demands and Needs of the American People, and for Other Purposes," 78 Stat. 897, Sept. 3, 1964; Charles I. Zinser, 1995, *Out-*

door Recreation: United States National Parks, Forests, and Public Lands (New York: John Wiley), 38–39.

21. "Superintendent's Report 1997"; Nature Conservancy of Georgia, 1997, "Nature Conservancy Negotiates a 4-Year Purchase Plan to Protect 1,148 Acres on Cumberland Island," *Nature Conservancy of Georgia Newsletter,* Spring; "Kingston Urges Funding Delay in Cumberland Island Land Buy," *Florida Times-Union,* July 25, 1997; "Cleland Asks $6.4M for Cumberland Land Buy," ibid., July 11, 1997; "Legislators Divided over Cumberland Development," *Camden County Tribune and the Southeast Georgian,* July 25, 1997; Charlie Smith Jr. to Jack Kingston, Oct. 23, 1997, CINS Central Files, A3815.

22. Rolland R. Swain to Amber Reilly, Jan. 26, 1996, to Malcolm P. Smith, Feb. 20, 1996, Louisa Carl to Superintendent, CINS, 1996 (no specific date), Rolland R. Swain to Louisa Carl, Feb. 20, 1996, CINS Central Files, N1623; Rolland R. Swain to Dr. Charlotte Stephens, Feb. 21, 1996, ibid., K14.

23. Denis Davis to Eric D. Kimsey, July 15, 1997, Thornton Morris to Denis Davis, July 2, 1997, ibid., L1425.

24. "Cumberland Beach Driving Plan Reversed," *Florida Times-Union,* Mar. 18, 1998; Georgia Department of Natural Resources, May 27, 1998, "Public Scoping Meetings Announced," press release, CINS Superintendent's Office Files; "Public Meetings Address Driving on Beaches," *Camden County Tribune and the Southeast Georgian,* June 10, 1998; CINS, June 18, 1998, "Beach Driving Position Presented to Georgia Department of Natural Resources by National Park Service, Cumberland Island National Seashore," CINS Superintendent's Office Files.

25. "Beach Driving Revised," *Florida Times-Union,* Dec. 3, 1998; Hal F. Wright v. Shore Protection Committee, Coastal Resources Division, Georgia Department of Natural Resources, Final Decision, 1999, Office of State Administrative Hearings, State of Georgia, docket no. OSAH-DNR-SPC-9909337-020-WJB.

26. "A Cumberland Goodbye," *Florida Times-Union,* Aug. 31, 1996; Rolland Swain interviewed by author, CINS, July 23, 1996; "Davis to Head Cumberland NPS," *Southeast Georgian,* Nov. 20, 1996; Thornton Morris interviewed by author, Atlanta, Oct. 6, 1999.

27. CINS, June 18, 1998.

28. CINS, 2000, "Superintendent's Annual Report, 1999," CINS Central Files, A2621; Jennifer Bjork to Workshop Participants, May 26, 1998, ibid., A40; "Cumberland Island Wilderness Management Plan Forum," May 1–3, 1998, CINS, Ga., attended by the author; "Wilderness Plan Near?" *Florida Times-Union,* May 5, 1998.

29. GoGo Ferguson to All Participants in the Wilderness Management Plan Forum, May 5, 1998, includes "Transcript of Recommendations," CINS Central Files, N1623; Denis Davis to GoGo Ferguson, May 14, 1998, ibid., A3821.

30. Jack Kingston, June 23, 1998, "Kingston to Introduce Cumberland Preservation Act," news release, CINS Superintendent's Office Files; "Kingston Plan Divides Factions," *Florida Times-Union,* July 12, 1998; H.R. 4144, 105th Cong., 2d sess., June 25, 1998.

31. Denis Davis to Jack Kingston, July 17, 1998, CINS Central Files, W38; "Kingston Plan Divides," July 12, 1998; CINS, 2000, "Superintendent's Annual Report, 1999"; Robert Stanton to Max Cleland, Dec. 2, 1998, CINS Central Files, A98; "Park Service, Lawmaker at Odds over Seashore," *Atlanta Journal and Constitution*, Dec. 11, 1998; "Island Issues on Tap," *Florida Times-Union*, Dec. 13, 1998.

32. "Funds for Land Cleared," *Florida Times-Union*, Dec. 22, 1998; Max Cleland and Jack Kingston to Don Barry, John Berry, and Robert Stanton, Nov. 25, 1998, CINS Superintendent's Office Files.

33. CINS, 2000, "Superintendent's Annual Report, 1999"; CINS, Feb. 16, 1999, "Final Agreement," CINS Management Specialist's Office Files.

34. Andy Ferguson interviewed by author, St. Marys, Ga., July 2, 2001; CINS, 2000, "Superintendent's Annual Report, 1999"; Greyfield, Ltd., The Nature Conservancy, and The United States of America, June 30, 1999, "Substituted Agreement Option for Purchase of Real Property," CINS Central Files, L1425. See also individual "administrative determinations" for land tracts 02-208 through 02-210 in CINS Lands Files under "Greyfield."

35. Andy Ferguson, July 2, 2001.

36. CINS, 2000, "Superintendent's Annual Report, 1999."

37. National Parks and Conservation Association, Dec. 9, 1998, "Park Watchdog Group Honors Denis R. Davis," news release, CINS Superintendent's Office Files (see the announcement in the association's magazine, *National Parks*, 73, 3–4, Mar./Apr., A-1); Hal Wright to Don Barry, Jan. 11, 2000, CINS Central Files, A3821.

38. Andy Ferguson, July 2, 2001; Arthur Frederick interviewed by author, St. Marys, Ga., July 2, 2001.

39. SERO, Dec. 2000, *Draft Introduction to Planning Effort and Environmental Impact Statement for the Wilderness Management Plan, Long Range Interpretive Plan, Commercial Services Plan, Resource Management Plan (Cultural and Natural Resources)*, 3 vols. with appendixes, CINS Superintendent's Office.

40. According to the *Camden County Tribune and the Southeastern Georgian*, June 27, 2001, Andrew Carnegie III died on June 19, 2001; Andy Ferguson, July 2, 2001.

41. William E. Hammitt and Ingrid E. Schneider, 1996, "On-Site Recreation Conflict: An Investigation of Visitor Perception and Response, Cumberland Island National Seashore," Department of Parks, Recreation, and Tourism Management, Clemson Univ., CINS Resource Management Library; Margaret Littlejohn, Jan. 1999, *Cumberland Island National Seashore Visitor Study*, Visitor Services Project, Univ. of Idaho, CPSU Report no. 103; "Attracting More Tourists to Island Discussed at Park Service Forum," *Florida Times-Union*, July 27, 2000; "Citizens Ask Park Service for Visitation Study," *Camden County Tribune and the Southeast Georgian*, July 28, 2000.

42. Stewart Udall to Arthur Frederick, Mar. 21, 2001, SERO Planning Files under "Cumberland Island Plans."

43. George Sandberg to Thornton Morris, Feb. 25, 2001, ibid.

44. Arthur Frederick, July 2, 2001.

45. Richard Sussman telephone interview with author, July 13, 2001; SERO, Dec. 2000.

46. Bill Harlan to Denis Davis, June 22, 1997, CINS Central Files, L1425.

47. Gregory Paxton, July 28, 1998, "History Succumbs to the Wilderness," *Atlanta Journal and Constitution.*

48. William C. Warren III to Denis Davis, May 19, 1997, CINS Central Files, A3821.

INDEX

Acadia National Park, 77

Adamson, Douglas, 121–22

Advisory Board on National Parks, Historic Sites, Buildings, and Monuments, 81, 108, 176

Advisory Board on Wildlife Management, 167

Advisory Council on Historic Preservation, 178, 190, 191, 213, 218, 232, 261

African Americans, 2, 68, 257

Agriculture: Louisa Shaw period, 28–29; Lucy Carnegie period, 41–42; orchard, 39; trust years, 53–55

Ainsworth, A. A., 62

Aircraft and airports, 104, 205

Alberty, William, 68

Albright, Horace, 77, 171, 178

Allen, Arthur, 246

Allen, Chris, 248

Alligators, 16

Altamaha River, 23

Amelia Island, 12, 37, 169

American Cyanimid Company, 59

American Exploration and Mining Company, 61

American Smelting and Refining Company, 59

Andrew Mellon Foundation, 81, 101–2, 147

Angwin, Esther, 63

Animals, 53, 227, 240–43; extirpated, 175; species on Cumberland Island, 16–17, 175

Antiquities Act of 1906, 178

Appalachian Mountains, 13

Archaeological sites, 17–20, 22, 256; destruction of, 223, 234–35; on the mainland, 186; protection of, 135

Armadillos, 16, 176, 233

Aspinall, Wayne, 107

Assateague Island National Seashore, 1, 172, 229, 230

Atlanta, Ga., 68, 159, 210, 250

Atlanta Audubon Society, 199, 202

Atlanta Journal and Constitution, 93, 98, 104, 109, 120, 124, 140, 155, 198–99, 208, 220, 242

Atlantic coast, 78, 80

Atlantic coastal plain, 9–11

Avalon Foundation, 79, 101

Ayllon, Lucas Vasquez de, 21

Babb, Roger S., 207

Bachlott house, 216

Bainbridge, Jim, 140

Baker, Robert M., 199, 206, 208–9, 213, 220, 237, 244

Barbados, 30

Barger, Don, 244

Barrier islands, 9–11

Barrimacke, 23

Barry, Donald J., 253, 255

Bartram, William, 37

Baumgaertner, Martin, 132

Beach Creek, 12, 23, 63

Bearss, Edwin, 179, 191, 232

Belson, Jerry, 255

Benson, George W., 66

Bernardey, Peter (Pierre), 26, 180

Bible, Alan, 108–9

Bicycles, 207–8

Big South Fork National Recreation Area, 251

Bilbo archaeological phase, 16

Birds, 16

309

Bjork, Jennifer, 231, 240, 241

Black bears, 16, 175

Blue Ridge Mountains, 9

Boating, 145, 202–3, 212

Bobcats, 16, 242; reintroduction of, 175, 227–28

Bodie Island, 79

Bohlen, E. U. Curtis, 155, 157

Boll weevil, 31

Bratton, James, 115, 123, 132

Bratton, Susan, 227, 229–30

Brazell, Richard, 125

Brickhill, 18, 150; archaeological site at, 256; campground at, 196, 256; dock at, 146

Brickhill River, 116

Bridges, Sarah, 190

Brooke, Ernest, Jr., 101

Brower, David, 105

Brown, Joseph, 130, 147, 198, 199

Brown, Lenard, 181, 183, 190

Bruce, James, 127

Bruce, Walter, 139

Brunswick, Ga., 241

Brunswick Junior College, 87

Brunswick News, 61

Brunswick Pulp and Paper Company, 142, 145, 155, 157, 186

Bryan, Jonathan, 24

Bryant, John, 177

Bullard, Mary, 33, 47, 73, 75, 116, 135, 177, 180, 206

Bunkley, Robert, 67, 68

Bunkley, William R., 33, 66–67

Bunkley heirs, 128

Bunkley House, 66

Bunkley Road, 207

Burbank, M. T., 66, 68

Burial mounds, 18

Butler, Landon, 140

Butler, Nancy, 115, 234

Cabin Bluff, 99, 150

Caldwell, Horace, 88

Camden County, 53, 83, 87, 90, 109, 125, 145, 158, 210, 213, 228, 243; commission, 95, 98, 106–8, 170, 202–3, 210, 212; and the first seashore bill, 102–3; opposition to a national seashore, 94; superior court of, 73–74; support of seashore bill, 98–102, 152; and visitation on Cumberland Island, 143, 256

Camden County Historical Commission, 185, 191

Camden County Humane Society, 229

Camden County Recreational Authority, 94, 104

Camden County Tribune, 83, 218

Campgrounds, 158, 196, 210, 212, 213, 256

Candler, Howard, Jr., 84

Candler, Howard, Sr., 68

Candler, Sam, 94, 104

Candler family, 8, 36, 71, 93, 106, 117, 119, 120, 191, 201, 249, 254; and the National Park Service, 89, 148; retained rights, 130, 146; sale of High Point, 130–31

Cape Cod National Seashore, 80, 88, 154

Cape Hatteras National Seashore, 78–79, 80, 108, 154, 172, 222, 230

Cape Lookout National Seashore, 1, 230, 241

Carnegie, Andrew (I), 37, 43, 48, 79, 258

Carnegie, Andrew (II), 46–48, 50, 71–72

Carnegie, Andrew (III), 47, 89, 115, 256

Carnegie, Carter, 47–48, 55, 61, 73

Carnegie, Mrs. Carter, 44

Carnegie, Coleman, 46

Carnegie, Florence. *See* Perkins, Florence Carnegie

Carnegie, Frank, 46–47

Carnegie, George, 44, 47, 50, 258

Carnegie, Henry Carter, 47–48, 73, 89, 115, 123, 132

Carnegie, Lucy C. S. *See* Rice, Lucy Carnegie Sprague

Carnegie, Lucy Coleman, 36, 55, 63, 65, 66, 71, 75; children of, 43–48, 62; design of island trust, 49; division of island,

72–75; visit to Cumberland Island, 37–38

Carnegie, Margaret (daughter of Thomas I). *See* Ricketson, Margaret Carnegie

Carnegie, Margaret (mother of Thomas I), 37

Carnegie, Nancy. *See* Johnston, Nancy Carnegie Hever

Carnegie, Thomas (I), 37, 66, 71; purchase of Cumberland property, 38, 47, 75

Carnegie, Thomas (II, also called Morris), 44, 46–47, 51, 53, 55, 62, 72, 75, 115

Carnegie, Thomas (III), 44, 47, 65, 73

Carnegie, Thomas (IV), 47, 64, 89, 115, 123, 132

Carnegie, William (father of Thomas I), 37

Carnegie, William (son of Lucy), 43–44, 47, 48, 49, 64, 72

Carnegie Brothers and Company, 37

Carnegie-Cook Center for the Arts, 238, 243–44. *See also* Plum Orchard Center for the Arts

Carnegie estate, 64–65, 75; maintenance of, 55–58; supplying, 56

Carnegie family, 6, 33, 46–48, 63, 119, 245, 254; attempts to make money from Cumberland Island, 50–55; and the cattle plan, 53–54; disagreements during the trust years, 57, 61–62, 73; efforts to keep Cumberland Island as a family holding, 54, 72–73; family tree, 47; life on Cumberland Island, 43, 63–66; and the National Park Service, 81, 86, 95–96, 148, 261; sale of Cumberland Island property, 89, 113–32

Carnegie Office Building, 49

Carnegie Trust (Lucy), 63, 65, 112, 206; design of, 49; life and economy during, 49–63; maintenance of Cumberland estate during, 55–58

Carolinian-South Atlantic Biosphere Reserve, 226

Carter, Jimmy, 120–22, 127

Cattle, 53, 172–74, commercial production of, 53–54; destruction by, 54–55, 171; population on Cumberland Island, 54

Causeway (to the mainland), 63, 83, 98, 106; prohibition of, 108

Cemeteries: Carnegie, 65, 73, 116; Greene-Miller, 39, 179

Century Theater Building, 143

Charleston, S.C., 22

Chayote, 55

The Chimneys: the Carnegie House called, 44, 56, 116; remains of slave quarters, 31–32, 44, 187, 189

Chincoteague National Wildlife Refuge, 229

Christmas Creek, 12

Citrus fruit, 55

Civilian Conservation Corps, 63, 78

Civil War, 31, 66, 171

Cleland, Max, 249, 253–54

Clement, Charles, 106

Cleveland, Ohio, 59

Clubb, Rebecca, 33

Clubb family, 66

Coastal Bank Building, 218, 247

Coastal Barrier Resources Act of 1982, 224

Coastal Georgia Audubon Society, 200, 201

Coastal surveys. *See* Seashore surveys

Colorado River, 251

Concessions, 158

The Conservation Foundation, 152–54, 158

Cooper, Cynthia, 117, 121

Copp (née Rockefeller), Nancy, 47, 118, 184

Coral Gables, Fla., 62

Coram, Robert, 128, 198–99, 242

The Cottage, 44–45, 56; replacement house, 73

Cotton, 29; gin, 25

Crater Lake National Park, 202

Crawford, Henry A., 118, 126

Crawford, John, 161

Creek Indians, 22

Crusoe, Donald, 187

Cumberland Island: Carnegie division of, 73–75; climate, 13; description, 1, 12; ecology of, 12–17; Fraser's plan for, 89–90; geology of, 9–12, 59–60, 167–69; Greene-Lynch division of, 26; income from, during the trust years, 49–50; location, 1, 9; mapping, 72–73; marshes, 15–16; north end, 66–71; population of, 31; as a possible national park, 80; as a possible spaceport, 63; as a possible wildlife refuge, 86, 107; proposed as a park, 81; suggestions for economic use, 50–55; topography, 12; during the trust period, 49–66; vegetation, 13- 16

Cumberland Island Advisory Committee, 103, 109

Cumberland Island Agreement, 254–55

Cumberland Island Authority, 81, 85

Cumberland Island Club, 67–68

Cumberland Island Company, 73, 85, 86–87

The Cumberland Island Conservancy, 90

Cumberland Island Conservation Association, 93, 102

Cumberland Island Historic Foundation, 236–38, 243

Cumberland Island Holding Company, 89–90, 123

Cumberland Island Hotel, 67

Cumberland Island Master Plan, 99, 150, 154, 171, 177, 193, 198

Cumberland Island National Seashore: bills to create, 86, 105–10; budgets, 139–40, 185, 196, 269; comanagement by residents, 141; economic benefits of, 99; enabling act, 8, 109; funding, for historic preservation, 185, 233, 235, 253–54; —, for land acquisition, 119, 253–54; general management plan for, 149–64, 196–200, 208–13; hearings, 155, 157, 159–62, 200–202, 241–42, 257; inholdings, 111; land acquisition, 111–32, 248–49; land donations to, 123–24; museum holdings, 245–48, personnel, 267; planning, 148–64, 196–200, 208–13; pressure to open, 143; proposed development by National Park Service, 86–87, 99; removal of abandoned vehicles, 140–41; retained rights, 88, 132–36, 146–47; visitor carrying capacity, 152–55, 157–59; visitor numbers on, 108, 198–99, 209, 236, 237, 243, 254; wilderness management, 157, 204–8, 251–59; wilderness proposals, 159–61, 200–204. *See also* National Park Service; Resources

Cumberland Island pocket gopher, 176

Cumberland Island Preservation Act, 253

Cumberland Island Preservation Society, 251, 252

Cumberland Island Properties, Inc., 115, 123

Cumberland Island Study Committee, 83, 85

Cumberland Island Wharf, 66, 150

Cumberland Island wilderness act, 200–207

Cumberland River, 251

Cumberland Sound, 12, 115, 169–71, 222–25

Dairying, 41

Darcy Lumber Company, 52

Davis, Denis, 251, 253, 255

Davis, George T., 57

Davis, Robert, 90, 108, 120, 122, 125–26; island development by, 88, 100, 117–18; land sales, 117–18

Davis, William G. M., 33, 38

Davis Land Company, 117, 126

Davisville, 118, 119, 124–26, 129, 134

"The debatable land," 20

Deer, 16, 33, 53, 175–76

Defenders of Wild Cumberland, 244, 250, 252

Dennis, Frank E., 53

Denver Service Center, 139, 180, 189, 210

Deptford archaeological phase, 17–18, 186

Deptford Tabby House, 179

Des Moines, Iowa, 77

De Soto, Hernando, 21

D'Ewart, Wesley, 83

Dickey, Alvin, 129

DiLorenzo, Peter, 128

Docks, 66, 134, 212, 256; private use of, 145–46

Dolan, Robert, 224

Dredging, 63, 169, 222–25; spoils from, 63, 169–70, 224

Driving. *See* Vehicle use

Drum Point Island, 131, 169, 223, 224, 260

Drury, Carl, 107

Drury, Newton, 172

The Duck House, 210, 212, 213, 232–33

Duck House Road, 115, 135, 148

Duck houses, 45

Dunes: destruction of and encroachment by, 167, 171, 174- 75, 179, 227; formation of, 12

Dunfermline, Scotland, 37

Dungeness, 12, 40–41, 63, 137, 143, 150, 158, 181, 188, 203; carriage house, 141, 181; dock, 18, 41, 73, 123, 134, 142, 144, 146, 164, 170; estate, 40–42, 45, 231; garden, 40; hunting lodge, 23; map of, 42; plantation, 28, 34, 41

Dungeness Historic District, 198, 231, 232

Dungeness mansion of Catherine Greene Miller, 26–28, 35, 38

Dungeness mansion of Lucy Carnegie, 38–39, 44, 56, 64; abandonment of, 50; demolition of, 57; destruction by fire, 57–58; stabilization by the Park Service, 181–82, 184

Dungeness Recreation House. *See* Recreation House

Earthwatch, 244

Eastern Wilderness Act, 162

Eaton, William F., 33

Edmunds, Al, 80, 81

Edwards, Max, 87, 98

Ehrenhard, John, 17, 33, 179, 188, 189, 222, 234–35

Eisenhower administration, 79, 85

Ellis, Vincent, 91, 96

Ely, Margaret Gertrude, 43

English: forts, 26; on Cumberland Island, 22–24; during the War of 1812, 29

Environmental Protection Agency, 140

Erosion, coastal, 167–69, 222–25, 240

Everett, Richard A., 101

Everglades National Park, 219

Everhardt, Gary, 122–23, 143, 188

Everhart, Bill, 79, 81, 183

Executive Order 11593, 178, 189

Fader Creek, 127, 129

Fauna of the National Parks, 172

Faust, Richard, 186

Federal Register, 208

Ferguson, Janet "GoGo," 47, 236, 238, 244, 252–53

Ferguson, Lucy Ricketson, 44, 47, 61, 75, 91, 94, 107, 117, 118, 127, 161, 203; and cattle, 53–54, 75, 147, 172–74; concerns about land retention, 96, 106, 154; lands of, 73, 130; and pigs, 53, 75, 171–74; relations with the National Park Service, 88, 96–97, 154–55, 171–74, 180

Ferguson, Oliver, 47, 172

Ferguson, Robert D., 49, 54, 57–58, 72

Ferguson, Robert W., 48, 49, 83, 94; and the cattle plan, 53–54, 83, 93

Ferguson, William, 200–201, 203

Ferguson family, 84, 89, 96, 106, 120, 206, 212

Fernandina Beach, Fla., 38, 53, 57, 145, 155, 158, 218, 241

Ferry service, 90, 98, 99, 106, 143, 145, 150, 176, 209

Figs, 55

Fire Island National Seashore, 88

Fires, 219–22

First African Baptist Church, 2, 68, 248

First National Bank of Brunswick (Ga.), trustee, 49, 59, 61, 63, 73

Fishing, 64, 106

Flexer, Judge W. D., 75

Florida Aquifer, 225

Florida panther, 16

Florida Times-Union, 124, 174, 225, 242

Ford Foundation, 100

Forests on Cumberland Island, 16–17, 33; acreage of, 51–52

Fort Frederica, 22; National Monument, 88, 109, 139

Fort Prince William, 23, 24, 150

Fort St. Andrew, 22, 24, 68, 150

Foster, Franklin, 107, 108, 120, 122, 123–24, 234

Foster, Lucy Carnegie Sprague, 46–47, 66, 116–17, 122, 123- 24, 131, 187

Foster Beach House, 256

Foster family, 132, 187, 256

Fraser, Charles, 62, 99, 100, 101, 116, 117, 125, 132, 140, 157, 196, 219; and the Carnegie family, 90–95, 113; as a concessioner, 95, 103; development of roads and airport, 104–5; financial status, 93, 105; and Hilton Head, 89–90; legislative efforts, 94–95; and the National Park Service, 91; opposition to development plans of, 91–95; proposed development on Cumberland Island, 90–91; publicity campaign, 93; purchase of Cumberland land, 89; retained right, 102, 122–23; sale of Cumberland property, 105, 113, 115

French Huguenots, 21

Froeschauer, Peggy S., 33

Gale, Judith, 226

Gambrell, David, 105

Gannon, Michael, 22

Garbage collection, 56, 140

Garner, John, 177, 181, 183, 191

General Authorities Act, 152

Georgia, state of, 78, 90, 112, 258; control of beach and marshes on Cumberland Island, 104, 131–32; general assembly, 83; gift to Nathaniel Greene, 25; interest in Cumberland Island, 81–83, 94–95, 103–4; legislature, 81–83, 94–95, 103–4, 203; settlement of, 24; Supreme Court, 62

Georgia Coastal Area Planning and Development Commission (Committee), 93, 214

Georgia Coastal Islands and Marshlands Planning Commission, 104

Georgia Coastal Island Study Committee, 103

The Georgia Conservancy, 106, 107, 109, 155, 199, 200, 209, 210, 224; and Charles Fraser, 91, 94, 148; and wilderness, 202

Georgia Department of Natural Resources, 108, 131, 162, 176, 190, 201, 212, 224, 227; and beach driving, 250–51

Georgia Forestry Commission (Department), 52, 219–20, 222

Georgia Marine Institute, 90

Georgia Marshland and Island Foundation, 104

Georgia Ocean Science Center, 103

Georgia-Pacific Corporation, 116

Georgia Planning Department, 93

Georgia Sierran, 91

Georgia State Game and Fish Department, 94, 145

Georgia State Historic Preservation Office, 188–91, 213, 218, 231–32

Georgia State Recreation Council, 93

Georgia State Tourism Division, 93

Georgia Teachers Association, 67

Georgia Trust for Historic Preservation, 190–91

Gillette, Martin, 118
Ginn, Ronald "Bo," 147–48, 183, 199, 200, 201
Giobbi, Elinor, 118
Glacier Bay National Park, 77
Glacier National Park, 255
Glen Canyon National Recreation Area, 251
Glidden Company, 59, 61–62
Glynn County, Ga., 87, 99
Godley, James E., 99, 107–8, 118, 126–27
Golf, 44, 64, 89
Goodloe, Robin, 230–31
Goodsell, Richard, 123, 134
Gordillo, Francisco, 21
Gossypium barbadense, 30
The Grange, 41, 73, 115, 256
Graves, Joseph, Jr., 86–87, 163
Graves, Lucy, 47, 115–16, 132, 256
Gray, Edwin, Jr, 56
Great Depression, 78
Great Lakes coast, 78, 80
Great Smoky Mountains National Park, 79
Greene (later Miller), Catherine, 25–30
Greene, General Nathaniel, 18, 25
Greene-Lynch division of Cumberland Island, 26
Gregg, William, 226
Greyfield: corporation, 249; estate, 45, 54, 56, 75, 130; mansion, 44, 73
Greyfield Inn, 96, 98, 146, 154, 205–8, 251, 258
Griffen, Governor Samuel M., 82, 83
Guale Indians, 18
Gulf of Mexico coast, 78, 80
Guse, Neal, 205–6

Hadley, Lawrence C., 120–21
Haleakala National Park, 77
Half Moon Bluff, 71, 137, 150, 159, 163, 191
Hall, S. J., 51–52

Hamilton, Jiles, 125
Hanie, Bob, 104
Hannaford, George, 126–27
Hannaford, M. W., 126
Harlan, Bill, 258
Harpers Ferry, W.Va., 246
Harrell, Riley, Jr., 118
Harriet's Bluff, 83; Road, 150
Harris, A. M., 59, 61, 62, 86
Harris, Edward B., 227
Harris, William, 209–10, 212–13, 222
Harrison, Kenneth, 108
Harrison, Robert, 94, 117, 119
Hartzog, George, 91, 96, 101–2, 104–5, 108, 177, 178, 206, 257
Haskins, Clarence, 103
Hatteras Island, 78, 79
Hauser, Charles, 187, 232, 234–35
Hawaii Volcanoes National Park, 77
Hawkins, Thomas D., 33
Hawkins Creek, 146
Hearst Castle, 235
Henderson, David A., 180, 183, 185
Henderson, J. Grover, 119, 124–25, 129, 134, 190
Hercules Powder Company, 52
Hernandez, N. S., 57
Herrington, Charles A., 189
Hever, James, 47–48
Hibbard, Wallace, 212–13, 222
Hickory Hill, 196
High Point, 102, 117, 137, 150, 153, 188; acquisition, by Candlers, 68; —, by the National Park Service, 130–31; and the Greyfield land, 249, 253; hotel at, 66; map of, 71; and wilderness, 159
High Point Cumberland Island Company, 68–69
High Point–Half Moon Bluff Historic District, 71, 162–63, 191, 198, 203, 213
High Point Road, 68
Hillestad, Hilburn O., and study team, 165, 171, 175

Hilton Head, S.C., 89

Historic furnishings, 245

Historic preservation, 178–79, 252–55, 257–58, 260–61; complexes, 45, 166, 181; districts, 188–90; of museum objects, 245–48; in St. Marys, 217–18, 255; of structures, 3, 179, 234

Historic resources management, 177–93, 231–38

Historic Sites Act of 1935, 178

Hogendof, Morgan, 68

Horses, 1, 33, 54, 172, 212; control of, 229–31, 240–42; destruction by, 171, 174, 229; diseases of, 229; mortality of, 229; population, 229; sale of, 53; stables for, 150

Horton, Virginia Olsen, 127

Hotels, 66–68, 150, 191

Humphrey Gold Company, 59

Hunting: camps, 212; deer, 212; with dogs, 228–29, 240; for income, 53; by island residents, 146; in national parks, 106; recreational, 64; for wildlife population control, 228–29, 240

Hurricanes, 16, 28, 171

Hutchison, Ira, 112, 163

Ickes, Harold, 78

Ike, Albert, 152–53

Ilmenite, 59

Indigo, 55

Ingle, Robert, 199, 209

Interpretation of Cumberland Island, 139, 145, 181

Intracoastal Waterway, 63, 223–25, 240

Izaak Walton League, 162

Jacksonville, Fla., 51, 59, 203

James D. Lacey Company, 51–52

Jarvis, T. Destry, 176

Jekyll Island, 12, 37, 81, 88, 103, 105, 107, 109, 121

Jenkins, Ben, 115, 146, 256

Jennings, Judy, 242

Jetties, 169, 240

Johns, Betty, 129

Johnson, Herbert, 94

Johnston, Coleman, 47–48, 59, 115–16, 135, 186–87

Johnston, Marius, 47–48

Johnston, Marius, Jr., 47–48, 81, 115–16, 132

Johnston, Nancy Carnegie Hever, 45, 47–48, 115

Johnston, Thomas, Jr., 115–16

Johnston, Thomas M. C., 47–48, 75, 81, 115, 187, 235

Johnston family, 75, 135, 184, 247; and the National Park Service, 81, 83, 87, 89, 115–16

Jones, Alfred W. "Bill," 88, 100, 101, 109

Joyce, Dwight, 59

Judd, Hank, 182

Katmai National Park, 77

Kelly, J. Pat, 56–57

Kennedy, John F., Jr., 248

Kerr, G. Robert, 212–13

Kings Bay Army Terminal, 91, 99, 131, 155, 169–70

Kings Bay Naval Base, 170, 214, 222–23, 225

Kingsland, Ga., 241, 242, 250

Kingston, Jack, 242–43, 249, 253–55

Kirkland, Zack, 174

Koester, Patricia, 198

Kriz, William, 119, 124

Labor on Cumberland Island, 56–57

Lafayette National Park. *See* Acadia National Park

Lake Whitney, 102, 210, 212, 213

Land: acquisition, by Carnegies, 38, 63; —, by the National Park Foundation, 113–17; —, by the National Park Service, 111–32, 248–49, 253, 255–56; appraisals, 54, 84, 112, 122, 125; for coastal recreation, 78–80; condemnation, by National Park Service, 87,

105, 127–30, 214; —, by U.S. Army, 63; division, by Carnegies, 73–75; —, by Greene and Lynch families, 26–27; donated to the seashore, 123–24; map of, 27, 74, 114; offers for, 62–63; for parks, 79; prices, 72, 117–21, 126

Land and Water Conservation Fund, 119, 249

Landowners: on Cumberland Island, 98, 100; on Little Cumberland Island, 95, 98

Landscapes: of cattle raising, 54; historic, 231; plantation era, 25;

Larkin, Leo, 52

Larsen, Lewis, 182–83, 188–89

Laughlin, Margaret, 47–48, 117, 121

Law, George, 125–26

Least terns, 176

Lee, General Henry "Light-Horse Harry," 29

Lee, Robert E., 29

Lee, Ronald F., 81

Lee, Thomas, 119

Leopold, A. Starker, 167

Leopold Report, 167, 172

Ley, Frederick, Jr., 183

Lippincott's Monthly Magazine, 38

Little Cumberland Island, 9, 12, 26, 85, 89, 95, 105, 163, 177

Little Cumberland Island Association, 85, 99–100

Littlefield plantation, 30

Live-oaking, 17, 24

Lopez, Father Baltasar, 21

Lynch, Thomas, 24, 25, 26

Lyon, Elizabeth, 191, 232

MacDonnell Building, 143

MacIntosh Creek, 116

Mackay, Hugh, 22–23

Macon Company, 67

Maddox, Governor Lester, 103

Main Road, 32, 56, 93, 116, 148, 243, 253, 254; as a historic site, 189, 232; resi-

dents' rights to drive, 73, 134–35, 205–6; visitor transportation on, 143–45, 203, 258; wilderness status, 159, 163, 200, 201, 205–7, 213, 252–53

Maine, state of, 77

Mainland visitor center and embarkation points, 143, 155- 56, 158, 164, 198, 209, 210, 213–16

Mammoth Cave National Park, 79

Man and the Biosphere Program, 226–27

Manatees, 176

Mankin, William, 212, 228–29

Marshes, 163; Georgia control of, 104, 131–32

Masland, Frank, 173–74

Masters, Dr. Hugh B., 93, 94

Mather, Stephen, 76, 77

Mattingly, Mack, 201, 220

McCollum, R. R., 118

McCrary, Paul, 122, 173, 174, 176, 210, 219–20, 222, 246, 247

McDowell, Putnam B., 86–87, 91–92, 96–97, 107, 154, 173

McFadden, Nancy, 107

McIntosh, George, 26

McKee, Heloise, 118

McKee, Louis, 118–19, 125; and Carol Ruckdeschel, 128–29

McKey, Robert M., 60, 84, 112

McLaren, F. W., 52

McLauchlan, Cindy, 236

McNamara, Verna, 200

McPhee, John, 105

Megafauna, 20

Mellon, Andrew, 79

Mellon, Paul, 81, 102

Mellon Foundation. *See* Andrew Mellon Foundation

Meltzer, Milton, 38

Merrou, Quash, 68

Merrow, George, 131

Miami Herald, 172

Miller, Laurence A., 67, 68, 119, 126

Miller, Mary, 119, 127, 161

Miller, Phineas, 25–26

Miller's Dock, 216–18

Miller subdivision, 119, 126, 127

Mining on Cumberland Island, 58–62, 73, 83–84; map of, 60

Mission, 66, 79, 150, 187

Missionaries, 21

Missions, Spanish, 21–22

Mitchell, Primus, 68, 70

Monks, Robert, 121–22, 132

Morgan, Kenneth O., 204, 205, 217, 218, 235

Morris, Leander, 38

Morris, Thornton, 96, 102, 104, 154, 173, 209, 229, 236, 257; and Charles Fraser, 93–95; retained rights of, 116; and the seashore bill, 97–98, 102; and South Cut Road, 147–48; and wilderness, 161, 200–201, 203

Morton, Rogers, 120–21

Mott, William Penn, 235–36

Mulberry Grove plantation, 25

Museum collection, 245–48

National Aeronautics and Space Administration (NASA), 63

National Association for African American Historic Preservation, 257

National Audubon Society, 107

National Council of State Historic Preservation Officers, 232

National Environmental Policy Act (NEPA), 149, 152, 224, 227, 231, 236–37, 241, 244

National Historic Preservation Act of 1966, 178–79, 194, 244; section 106 of, 178, 190

National Lead Company, 58–59

National Park Foundation, 101, 105–6, 113–17, 123, 131, 132, 146, 157, 247, 257

National parks, 77, 111; criteria for identifying, 80; purpose of, 77

National Parks and Conservation Association, 107, 176–77, 227, 244, 255

National Park Service, 3, 36, 54, 63, 71, 75; act establishing, 3, 76, 77; and Camden County, 98–102; and Charles Fraser, 91, 94; and coastal recreation, 76–81; condemnation of lands by, 87, 127–30; development plans, 99; fire management, 219–22; hearings, 155–56, 159–62, 241–42, 257; historic resource management, 177–94, 218, 231–38, 243–48; land acquisition, 87, 111–13, 119–32, 214, 248–49, 254–55; meeting with Cumberland Island Company, 86–87; natural resource management, 167–77, 219–31, 239–43, 260; publicity, 87; and seashore surveys, 78–81; secrecy, 96; studies of Cumberland Island, 81; visitor management, 77

policies: early, 77, 278 (n. 2); fire management, 220; historic resources, 177–79, 190–91, 253–55; land acquisition, 111–12; management of the system, 152, 196, 265–66, 271- 72 (n. 2); for national seashores, 149; natural resource management, 167; personnel, 267–68; on retained rights and estates, 97

National Parks for the Future, 152

National Parks Omnibus Bill of 1978, 127

National Park Trust Fund Board, 101

National Register of Historic Places, 5, 134, 178, 179, 182, 195, 213, 246, 252; nominations for, 187–92, 232; in St. Marys, 216

National significance, as a criterion for parks, 80

National Trust for Historic Preservation, 191

Native Americans, 5, 13, 17–22, 258

The Nature Conservancy, 100–101, 109, 249, 253, 254

Naval stores, 51

Neuberger, Senator Richard, 85–86

Neuhauser, Hans, 109, 152, 155, 159, 163, 201, 212, 213, 224

New York City, 62

New Yorker magazine, 105

Nightingale, Phineas Miller, 26, 33; sale of Cumberland land, 30

Nightingale: Avenue, 158; Beach, 212

Nixon, Richard, 8, 105, 109, 178

North Carolina, state of, 78

North Cut Road, 219, 260

North Georgia Mountains Authority, 94

Nunn, Sam, 120–21, 148, 201

Oakland plantation, 29

Ober, Frederick A., 38

Ocracoke Island, 79

Odom, John, 81–82

Oglethorpe, James, 22

OGR subdivision, 118–19, 121, 126

Okeefenokee Swamp, 21

Old Clubb Road, 68

Old Dominion Foundation, 79, 101

Old House Creek, 12, 86, 115, 146

Old House Field, 64

Olive production, 29, 33, 40–41, 55

Olmsted, Frederick Law, Sr., 37

Olsen, O. H. "Bubba," 127

Olsen family, 112, 127

Orange Production, 29, 55

Oregon Dunes National Seashore, 86

Oriental House, 66

Osborne, Henry, 26

Osvald, Stephen, 170

Our Vanishing Shoreline, 80–81

Outer Banks, N.C., 226

Owen, Norman, 244

Pacific coast, 78

Padre Island, Tex., 80

Page, William, 41, 62

Palmetto, 17, 219

Paper mills, 53

Park, Parkway, and Recreational Area Study Act, 78

Parrish, Nancy, 244

Patterson, Miller, Crawford, and Arensberg, Inc., 49

Paxton, Gregory, 258

Peabody and Stearns, architects, 39, 44

Peachtree Sportsmen's Club, 161–62

Peeples, J. B., 53, 57, 118–19, 127, 173

Peoples First National Bank (Pittsburgh), trustee, 49, 52

Perkins, Coleman, 47–48, 73, 75, 91, 116

Perkins, Florence Carnegie, 36, 47–48, 57, 61–62; death of, 73, 85

Perkins, Frederick, Jr., 47–48

Perkins Beach Road. *See* South Cut Road

Perkins family, 84, 89, 93, 147–148

Phillips, S. Larry, 123, 134

Piehl, Tom, 129

Pigs, 171–74, 228–29; destruction by, 171, 228, 240; hunting and trapping of, 53, 174, 228–29, 240; population of, 53, 174; sale of, 53

Pine plantations, 51–52

Pittsburgh, Pa., 37, 49

Planning: at Cumberland Island, 99, 137, 139, 148–64, 196- 200, 208–13; hearings, 155, 157, 159–62; at St. Marys, 214–19; for wilderness management, 208, 239, 251–59

Plantation period, 24–33; landscape of, 25

Pleistocene Silver Bluff formation, 12

Plum Orchard, 56, 102, 115, 137, 150, 158, 179, 181, 200, 203, 223; Historic District, 190, 198, 231, 232

Plum Orchard Center for the Arts, 244, 251. *See also* Carnegie-Cook Center for the Arts

Plum Orchard mansion, 73, 88, 96, 115, 124, 130, 180–85, 247; attempts to lease or use, 183–84, 235–38, 243–45, 254, 256; construction of, 44–45; maintenance of, 205, 233, 235, 252, 254; visitors to, 145, 212, 252, 256, 258

Poachers on Cumberland Island, 57–58

Point Peter, 130, 155, 158, 162, 164, 198, 213–15, 236

Point Reyes National Seashore, 208
Pollution, 53
Ponce de Leon, Juan, 21
Price, Jackson E., 86

QRC Corporation, 184

Raccoon Keys, 131, 169, 224, 260
Raccoons, 175–76
Rathesberger, James, 222
Rayburn, Finis, 125
Rayfield Archaeological District, 232
Read, Robert, 127
Readdick, Wilbur, 118–19, 127
Reagan, Ronald, 202, 203, 249
Recreation House, 218, 253, 256; con-
    struction of, 39–40; damage to, 56–57,
    184–85; preservation of, 180, 182, 185;
    restoration plans for, 150, 190
Recreation in America, 76–79
Recreation on Cumberland Island: of
    Carnegies, 64–66; planning for, 150–51
Red wolf, 176
Reed, Nathaniel, 108–9, 173–74
Report to the 1956 Session of the Georgia
    Assembly, 83
Reptiles on Cumberland Island, 16, 172
Research on Cumberland Island: ar-
    chaeological and historical, 179–80;
    scientific, 165–69, 176–77, 225, 227–31,
    239–40, 301 (n. 79)
Resources: management of, 165–94, 219–
    38, 239–48; prioritization on Cum-
    berland Island, 177–78, 183
Retained rights and estates, 88, 97–98,
    102, 113, 124–27, 129, 130, 132–36, 146–
    47, 204–6, 256, 261–62, 296 (n. 26);
    construction and modification of
    structures, 134, 187; cutoff date on
    Cumberland Island, 106–8, 124; to
    lease homes, 134; National Park Ser-
    vice policy on, 97; and small land-
    holders, 124–27; See also Vehicle use
Revolutionary War, 25

Ribault, Jean, 20
Rice, Lucy Carnegie Sprague, 46, 47, 56,
    61, 73, 75, 89, 116, 121, 122
Richards, Margaret, 115
Richards, Tom, 100
Richardson, Ingram H., 100
Richardson, James, 106, 152–53, 176
Ricketson, Margaret Carnegie, 44, 46–
    48, 116
Ricketson, Oliver G. (I), 44
Ricketson, Oliver G. (II), 47, 72–74
Ricketson, Oliver (III), 47, 73, 75, 117–19
Rinck, Anthony, 170
Roads, 31, 134–35, 201, 207, 219, 256
Roberts, Bertram C., 126, 140, 158, 172–
    74, 183
Roberts and Eichler, Inc., 214, 216
Rockefeller, Andrew, 97–98, 248
Rockefeller, James Stillman (I), 46
Rockefeller, James Stillman (II, "Pebble"),
    47–48, 129–30
Rockefeller, Nancy Carnegie, 46, 54, 57,
    64–65, 73, 92, 107; and mining, 59, 61–
    62; and the National Park Service, 84,
    89; retained estate of, 129
Rockefeller family, 75, 106, 117, 120, 131
Rogers, William, 125
Roosevelt, Franklin D., 78
Rose, Alexander, 24, 25
Rose (née Rockefeller), Georgia, 107, 248
Ruckdeschel, Carol, 132, 134, 176, 190,
    191, 207, 228–29; and Louis McKee,
    128–29; and seashore plans, 161, 212-
    13; and wilderness, 161
Ruckelshaus, William, 154

S. C. Loveland Company, 170–71
St. Augustine, Fla., 21
St. Catherine's Island, 88
St. Johns River, 21
St. Marys, town of, 130, 143, 155, 158, 198,
    210, 228, 247; as National Park Service
    headquarters, 213–19
St. Marys Historical Commission, 218

St. Marys Inlet, 12

St. Marys River, 22, 169

St. Simons Island, 22, 88

Sandberg, George, 96, 113, 115–16, 147, 177, 206, 257

Sand dunes. *See* Dunes

San Francisco, Calif., 61

San Pedro de Mocama (mission), 21

San Pedro y San Pablo de Puritiba (mission), 21

San Simeon, Calif., 235

Santee Delta–Cape Romain, S.C., 226

Sapelo Island, 78, 226; Foundation, 109

Satilla River, 12

Savannah, Ga., 250

Savannah Morning News, 95, 214

Save America's Vital Environment (SAVE), 107

Sax, Joseph, 111

Schwartz, Gertrude, 115, 146, 256

Sea Camp, 21, 30, 102, 123, 129, 143, 158, 164, 196, 248; campground, 196, 210; visitor center, 115;

Sea Island Company (and Resort), 88, 100, 101

Sea island cotton, 30–31, 41, 55

Sea Islands of Georgia, 9, 20, 37, 87

Sea level change, 10–11, 17

Sea Pines Plantation (Company), 89–90, 157

Sea Winds Development Corporation, 216–18

Seashore surveys, 78–81

Serendipity, 44

The Settlement, 68, 71, 119, 128, 134, 163, 190, 191, 207, 252, 253, 256, 258

Settlement on Cumberland Island; African Americans, 68; English, 22–24; Native Americans, 5; pirates, 23; plantation period, 25–26; Spanish, 21–22

Shaw, Louisa Greene, 28–29

Shenandoah National Park, 79

Sherman, David, 185, 188–89

Sierra Club, 91, 93–94, 106, 107, 109, 148, 199, 202, 205, 209, 212, 224, 229, 242, 244, 253, 258

Sieur de Monts National Monument. *See* Acadia National Park

Sikes, Newton, 227

Slaves, 30–33; cabins of, 179; descendants of, 68, 119

Sloss, H. H., 63, 81

Smethhurst, Lucy, 155

Smith, Charlie, Jr., 243

Snakes, 16

Social life on Cumberland Island: during the Greene-Miller period, 29; during the Lucy Carnegie period, 63–66

Soil, on Cumberland Island, 51

South Cut: Fire, 219–22; Road, 116, 147–48, 200, 207, 256, 257

Southeast Archaeological Center, 179, 187, 246

Southeast Georgian, 243

Southeast Regional Office of the National Park Service, 119, 123, 132, 174, 188; and land acquisition, 139

Spanish, 21–23

Spillman, Donald M., 187

Sprague (née Ricketson), Margaret, 47, 73, 116

Sprague, Phineas, 47, 122, 129

Stafford, Elizabeth, 31, 33

Stafford, Robert, 26, 30

Stafford, Robert, Jr., 30, 31, 33, 180

Stafford, Thomas, 26, 30

Stafford, 210; Beach House, 122; estate, 56, 102, 137, 150, 158; field, 33, 54, 56; golf course, 44; mansion, original, 31, 39, 43; —, Carnegie replacement, 43, 45, 73, 75, 116, 134, 234; plantation, 38, 41, 68, 231

Stafford Chimneys. *See* The Chimneys

Stafford Historic District, 232, 234

Stafford Island, 12

Stanley, John H., 56, 87

Stanton, Robert, 253

State parks, 77, 80

Stevens, Stoddard, 101–2

Stoneburner, D. L., 174

Storey, Frederick G., 214, 236

Storey, Jean Lucas, 214

Stuckey, Williamson "Bill," 92, 95, 96, 98, 108; and bill to create Cumberland Island National Seashore, 101–3, 105–7

A Study of the Park and Recreation Problem of the United States, 78

Subdivisions, 68–69, 88, 114, 117–19; proposed by Fraser, 90

Submarines, 222–25

Sussman, Richard, 256

Sutton, Jack, 202–3, 210, 212

Swain, Rolland, 208, 229, 239, 243, 244

Sweet potatoes, 55

Swem, Theodore, 103

Tabby House, 28, 39, 179, 180, 181

Table Point, 30, 56, 186, 232; Fire, 220

Table Point Company, 75, 116

Tacatacuru, 18, 23

Tallahassee, Fla., 246

Talmadge, Herman, 105, 148

Tamarack, 240

Tanner, Joe D., 108, 201

Taro, 55

Technical Information Center. *See* Denver Service Center

Termite damage, 56–57

Terrapin Point, 150

Thaw, Evelyn Nesbit, 44

Thaw, Harry, 44

Thaw, Margaret, 44–45

Thomas, Judge Douglas F., 61, 73

Thomas, Lindsay, 218

Thompson, Ben, 86–87

Thompson, David, 122, 145, 169, 179

Timber industry and logging, 24, 75; S. J. Hall plan, 51-52

Timucuan Indians, 18, 20–21

Titanium, 58–62; map of, 60

Tobacco, 55

Tomkins, John, 33

Torres, Louis, 23, 180, 181, 188, 190

Tourism: at Cumberland Island, 143–45, 261; at the north end, 66–68; postbellum, 37–38; in St. Marys, 214, 216. *See also* Visitors

Trails, 254, 256

Transportation on Cumberland, 143–45, 150, 158, 162–64, 196, 198, 203, 212, 257

Treaty of Paris, 23

Trimmings, Charlie, 68

Trimmings house, 134

Troy, H. Philip, 122

Trust for Public Lands, 130

Tung trees, 55, 227, 239

Turtles, 16, 64, 176–77, 239, 250

Tuten, J. Russell, 87–88, 98

Udall, Stewart, 87, 100, 206, 257; and the Cumberland Island bill, 97; and the Leopold Report, 167; visit to Cumberland Island, 87–88, 98

Union Carbide Company, 59

United Nations Educational, Scientific, and Cultural Organization (UNESCO), 226

U.S. Air Force, 63

U.S. Army, 169–70

U.S. Army Corps of Engineers, 63, 131, 169–70, 224, 252

U.S. Bureau of Land Management, 242

U.S. Bureau of the Budget, 85

U.S. Congress, 78, 113, 116, 157, 169, 178, 213, 249, 252, 254; appropriations to the seashore, 131; and the Cumberland Island National Seashore Act, 102-11; parks established by, 79; reports to, 78, 109, 199–200. *See also* U.S. House of Representatives; U.S. Senate

U.S. Department of Defense, 224

U.S. Department of the Interior, 109, 222, 255

U.S. Fish and Wildlife Service, 171, 176, 222, 224

U.S. Forest Service, 202, 219–20
U.S. General Accounting Office, 249
U.S. General Services Administration, 143
U.S. House of Representatives, 86; bills, 102, 105–10, 200; Committee on Interior and Insular Affairs, 107, 201–2; hearings of, 107–8, 200–201; Subcommittee on National Parks and Recreation, 107–8; Subcommittee on Public Lands and National Parks, 200. *See also* U.S. Congress
U.S. Navy, 62, 131, 170, 222–26, 233, 240
U.S. Senate: bills, 107–10, 201–3; Committee on Energy and Natural Resources, 202; Committee on Interior and Insular Affairs, 107–8; hearings of, 108, 202; Subcommittee on Public Lands and Reserved Waters, 202. *See also* U.S. Congress
University of Georgia, 176, 205, 227, 229; Bureau of Business and Economic Research, 99; College of Business Administration, 106; Institute of Community and Area Development, 152–54; Institute of Ecology, 219
Utley, Robert, 177, 181–82

Van Cleve, Robert, 118, 127
Vegetation communities, 13–16, 51–52, 220, 227; on dunes, 171; impact of livestock on, 54, 171
Vehicle use: on the beach, 135, 176–77, 249–51, 256; on Cumberland roads, 134–35, 147–48; in wilderness, 205–7, 257, 258
Virginia, state of, 29
Visitors: carrying capacity, 152–55, 157–59; limits, 157- 59, 164, 198–99, 209, 214, 236, 237, 243, 254, 261
Voight, William, Jr., 108, 162

Wallop, Malcolm, 202
Warren, Robert J., 227, 229–30
Warren, William C., III, 92, 94, 258–59
Water resources, 11, 225–26
Watt, James, 200, 220
Watts Engineering Company, 72–73
Webster, Hal, 103
Weems, Sam, 99, 172, 178
Whalen, William, 147, 163, 177
Wharton, Carol, *See* Ruckdeschel, Carol
White, Stanford, 44–45
Whitney, Eli, 25, 29
Wilderness Act of 1964, 131, 157, 161, 204, 207, 253, 258
Wilderness on Cumberland Island, 149, 261–62; feasibility study, 109; management of, 157, 204–8, 251–59; planning for, 159–61, 208, 239, 254–59; proposals for, 158–61, 198, 251–52
The Wilderness Society, 130, 148, 199, 202, 212
Wilderness Southeast, 161
Wilderness Watch, 255
Wildlife Management in the National Parks. See Leopold Report
Wilson, Woodrow, 77
Wirth, Conrad, 81, 84
Woodbine, Ga., 103, 155
World War II, 79, 172
Wright, Hal, 244, 250, 252, 255
Wright, Margaret, 47, 52, 61, 75, 84, 88, 103, 115–16, 132, 247

Yachts, 63
Yankee Hill, 196
Yarn, Jane, 106, 235
Yellowstone National Park, 167, 222
Yosemite National Park, 158
Youth Conservation Corps, 140, 233